METHODS IN MOLECULAR BIOLOGY

Series Editor
John M. Walker
School of Life and Medical Sciences
University of Hertfordshire
Hatfield, Hertfordshire, UK

For further volumes:
http://www.springer.com/series/7651

For over 35 years, biological scientists have come to rely on the research protocols and methodologies in the critically acclaimed *Methods in Molecular Biology* series. The series was the first to introduce the step-by-step protocols approach that has become the standard in all biomedical protocol publishing. Each protocol is provided in readily-reproducible step-by-step fashion, opening with an introductory overview, a list of the materials and reagents needed to complete the experiment, and followed by a detailed procedure that is supported with a helpful notes section offering tips and tricks of the trade as well as troubleshooting advice. These hallmark features were introduced by series editor Dr. John Walker and constitute the key ingredient in each and every volume of the *Methods in Molecular Biology* series. Tested and trusted, comprehensive and reliable, all protocols from the series are indexed in PubMed.

CryoEM

Methods and Protocols

Edited by

Tamir Gonen

Howard Hughes Medical Institute Departments of Biological Chemistry and Physiology, University of California, Los Angeles, Los Angeles, CA, USA

Brent L. Nannenga

Chemical Engineering, School for Engineering of Matter, Transport and Energy; Center for Applied Structural Discovery Biodesign Institute, Arizona State University, Tempe, AZ, USA

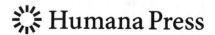 Humana Press

Editors
Tamir Gonen
Howard Hughes Medical Institute
Departments of Biological Chemistry
and Physiology
University of California, Los Angeles
Los Angeles, CA, USA

Brent L. Nannenga
Chemical Engineering, School for Engineering of Matter,
Transport and Energy; Center for Applied Structural Discovery
Biodesign Institute
Arizona State University
Tempe, AZ, USA

ISSN 1064-3745   ISSN 1940-6029   (electronic)
ISBN 978-1-0716-0968-2   ISBN 978-1-0716-0966-8   (eBook)
https://doi.org/10.1007/978-1-0716-0966-8

This Humana imprint is published by the registered company Springer Science+Business Media, LLC, part of Springer
Nature.
The registered company address is: 1 New York Plaza, New York, NY 10004, U.S.A.

Preface

The past decade has seen the interest in the field of electron cryomicroscopy (cryo-EM) greatly expand. This has been driven by new advancements in the design of the transmission electron cryomicroscope (cryo-TEM) and the development of high-speed direct electron detectors together with powerful data processing software. These advancements, along with many others, have brought cryo-EM to the forefront of structural biology. Various modalities of cryo-EM exist including tomography, single particle analysis, 2D electron crystallography, and microcrystal electron diffraction (MicroED) of 3D crystals. This volume brings together commentary and detailed methods on cutting-edge cryo-EM techniques from leading practitioners. The volume is organized into four parts each of which focuses on a particular methodology within the overall field of cryo-EM. While each chapter is placed within a single method, several chapters will also have valuable insights applicable across all cryo-EM modalities.

In Part I of this volume, the emphasis is on electron cryotomography, where multiple images from a sample are collected at different angles and a 3D volume is reconstructed from these series of images. Tomography is able to provide unprecedented views on the inner workings of cells and tissues. Chapters within Part I focus on methods for sample preparation, and overviews on how data are collected and processed to produce final 3D reconstructions. Single particle analysis (SPA) is the focus of Part II of this book with several chapters devoted to this cryo-EM technique. SPA makes use of images of purified biomolecules suspended in vitreous ice to produce high-resolution 3D reconstructions of the sample. Methods related to single particle sample screening and preparation, microscope hardware, and data processing and refinement, among others, are found in this chapter.

Parts III and IV focus on the crystal-based cryo-EM methods of electron crystallography of 2D crystals and microcrystal electron diffraction (MicroED) using 3D crystals, respectively. The chapters in Part III highlight the methodology for studying membrane protein 2D crystals and new algorithms for processing the data. In Part IV, chapters on the recently developed MicroED cryo-EM method for the study of 3D crystals are presented. In these chapters, procedures for crystal screening, automated data collection, structure solution, as well as a general overview for how to best collect data to minimize the effects of radiation damage in MicroED, are presented.

The broad scope of this volume on cryo-EM methodology should serve as an excellent resource to current users of cryo-EM and can serve as the foundation for new comers to this growing field in structural biology.

Los Angeles, CA, USA   *Tamir Gonen*
Tempe, AZ, USA   *Brent L. Nannenga*

Contents

Contributors

SRIRAM AIYER • *The Salk Institute for Biological Studies, La Jolla, CA, USA*

PHILP R. BALDWIN • *The Salk Institute for Biological Studies, La Jolla, CA, USA*

GUANHONG BU • *Chemical Engineering, School for Engineering of Matter, Transport and Energy, Arizona State University, Tempe, AZ, USA; Biodesign Center for Applied Structural Discovery, Biodesign Institute, Arizona State University, Tempe, AZ, USA*

GUILLERMO CALERO • *Department of Structural Biology, University of Pittsburgh School of Medicine, Pittsburgh, PA, USA*

STEPHEN D. CARTER • *Division of Biology and Biological Engineering, California Institute of Technology, Pasadena, CA, USA*

KA-YI CHAN • *School of Molecular Sciences, Arizona State University, Tempe, AZ, USA; The Biodesign Institute, Arizona State University, Tempe, AZ, USA*

PO-LIN CHIU • *School of Molecular Sciences, Arizona State University, Tempe, AZ, USA; The Biodesign Institute, Arizona State University, Tempe, AZ, USA*

GEORGES CHREIFI • *Division of Biology and Biological Engineering, California Institute of Technology, Pasadena, CA, USA*

TRISTAN CRAGNOLINI • *Institute of Structural and Molecular Biology, Department of Biological Sciences, Birkbeck College, University of London, London, UK*

KAREN M. DAVIES • *Department of Molecular and Cell Biology, University of California, Berkeley, CA, USA; Molecular Biophysics and Integrative Bioimaging Division, Lawrence Berkeley National Laboratory, Berkeley, CA, USA*

M. JASON DE LA CRUZ • *Structural Biology Program, Sloan Kettering Institute, Memorial Sloan Kettering Cancer Center, New York, NY, USA*

CHLOE DU TRUONG • *School of Molecular Sciences, Arizona State University, Tempe, AZ, USA; The Biodesign Institute, Arizona State University, Tempe, AZ, USA*

DEBNATH GHOSAL • *Division of Biology and Biological Engineering, California Institute of Technology, Pasadena, CA, USA; Department of Biochemistry and Molecular Biology, Bio21 Molecular Science and Biotechnology Institute, The University of Melbourne, Parkville, VIC, Australia*

SHANE GONEN • *Department of Molecular Biology and Biochemistry, University of California Irvine, Irvine, CA, USA*

JOHAN HATTNE • *Howard Hughes Medical Institute, Department of Biological Chemistry, David Geffen School of Medicine, University of California, Los Angeles, CA, USA*

MARK A. HERZIK JR • *Department of Chemistry and Biochemistry, University of California, San Diego, La Jolla, CA, USA*

RYAN K. HYLTON • *Department of Biochemistry and Molecular Biology, The Penn State College of Medicine, Hershey, PA, USA*

GRANT J. JENSEN • *Division of Biology and Biological Engineering, California Institute of Technology, Pasadena, CA, USA; Howard Hughes Medical Institute, California Institute of Technology, Pasadena, CA, USA*

MATTHEW C. JOHNSON • *Structural Biology, Genentech, San Francisco, CA, USA*

MOHAMMED KAPLAN • *Division of Biology and Biological Engineering, California Institute of Technology, Pasadena, CA, USA*

VINSON LAM • *University of California, San Diego, La Jolla, CA, USA*

GUOWU LIN • *Department of Structural Biology, University of Pittsburgh School of Medicine, Pittsburgh, PA, USA*

DMITRY LYUMKIS • *The Salk Institute for Biological Studies, La Jolla, CA, USA*

LAUREN ANN METSKAS • *Division of Biology and Biological Engineering, California Institute of Technology, Pasadena, CA, USA; Howard Hughes Medical Institute, California Institute of Technology, Pasadena, CA, USA*

TAKANORI NAKANE • *MRC Laboratory of Molecular Biology, Cambridge Biomedical Campus, Cambridge, UK*

BRENT L. NANNENGA • *Chemical Engineering, School for Engineering of Matter, Transport and Energy, Arizona State University, Tempe, AZ, USA; Biodesign Center for Applied Structural Discovery, Biodesign Institute, Arizona State University, Tempe, AZ, USA*

KASAHUN NESELU • *School of Biological Sciences, Georgia Institute of Technology, Atlanta, GA, USA*

WILLIAM J. NICOLAS • *Division of Biology and Biological Engineering, California Institute of Technology, Pasadena, CA, USA; Howard Hughes Medical Institute, California Institute of Technology, Pasadena, CA, USA*

YU-PING POH • *The Biodesign Institute, Arizona State University, Tempe, AZ, USA*

RICARDO RIGHETTO • *Center for Cellular Imaging and NanoAnalytics, Biozentrum, University of Basel, Basel, Switzerland*

JOSE A. RODRIGUEZ • *Department of Chemistry and Biochemistry, UCLA-DOE Institute for Genomics and Proteomics, STROBE, NSF Science and Technology Center, University of California, Los Angeles (UCLA), Los Angeles, CA, USA*

AMBARNEIL SAHA • *Department of Chemistry and Biochemistry, UCLA-DOE Institute for Genomics and Proteomics, STROBE, NSF Science and Technology Center, University of California, Los Angeles (UCLA), Los Angeles, CA, USA*

MICHAEL R. SAWAYA • *Departments of Biological Chemistry, Chemistry & Biochemistry, and Molecular Biology Institute, Howard Hughes Medical Institute, UCLA-DOE Institute, Los Angeles, CA, USA*

SJORS H. W. SCHERES • *MRC Laboratory of Molecular Biology, Cambridge Biomedical Campus, Cambridge, UK*

INGEBORG SCHMIDT-KREY • *School of Biological Sciences, Georgia Institute of Technology, Atlanta, GA, USA; School of Chemistry and Biochemistry, Georgia Institute of Technology, Atlanta, GA, USA*

VICTORIA H. SEADER • *Department of Biochemistry and Molecular Biology, The Penn State College of Medicine, Hershey, PA, USA*

DANIEL SERWAS • *Department of Molecular and Cell Biology, University of California, Berkeley, CA, USA; Molecular Biophysics and Integrative Bioimaging Division, Lawrence Berkeley National Laboratory, Berkeley, CA, USA*

HENNING STAHLBERG • *Center for Cellular Imaging and NanoAnalytics, Biozentrum, University of Basel, Basel, Switzerland*

AARON SWEENEY • *Institute of Structural and Molecular Biology, Department of Biological Sciences, Birkbeck College, University of London, London, UK*

MATTHEW T. SWULIUS • *Department of Biochemistry and Molecular Biology, The Penn State College of Medicine, Hershey, PA, USA*

MAYA TOPF • *Institute of Structural and Molecular Biology, Department of Biological Sciences, Birkbeck College, University of London, London, UK*

YUSUF M. UDDIN • *School of Biological Sciences, Georgia Institute of Technology, Atlanta, GA, USA*

SANDRA VERGARA • *Department of Structural Biology, University of Pittsburgh School of Medicine, Pittsburgh, PA, USA*

ELIZABETH VILLA • *University of California, San Diego, La Jolla, CA, USA*

SIMON WEISS • *Department of Structural Biology, University of Pittsburgh School of Medicine, Pittsburgh, PA, USA*

CHIH-TE ZEE • *Department of Chemistry and Biochemistry, UCLA-DOE Institute for Genomics and Proteomics, STROBE, NSF Science and Technology Center, University of California, Los Angeles (UCLA), Los Angeles, CA, USA*

CHENG ZHANG • *The Salk Institute for Biological Studies, La Jolla, CA, USA*

WEI ZHAO • *Division of Biology and Biological Engineering, California Institute of Technology, Pasadena, CA, USA; Howard Hughes Medical Institute, California Institute of Technology, Pasadena, CA, USA*

Part I

Electron Cryotomography

Chapter 1

Getting Started with In Situ Cryo-Electron Tomography

Daniel Serwas and Karen M. Davies

Abstract

Cryo-electron tomography (cryo-ET) is an extremely powerful tool which is used to image cellular features in their close-to-native environment at a resolution where both protein structure and membrane morphology can be revealed. Compared to conventional electron microscopy methods for biology, cryo-ET does not include the use of potentially artifact generating agents for sample fixation or visualization. Despite its obvious advantages, cryo-ET has not been widely adopted by cell biologists. This might originate from the overwhelming and constantly growing number of complex ways to record and process data as well as the numerous methods available for sample preparation. In this chapter, we will take one step back and guide the reader through the essential steps of sample preparation using mammalian cells, as well as the basic steps involved in data recording and processing. The described protocol will allow the reader to obtain data that can be used for morphological analysis and precise measurements of biological structures in their cellular environment. Furthermore, this data can be used for more elaborate structural analysis by applying further image processing steps like subtomogram averaging, which is required to determine the structure of proteins.

Key words Cell biology, Cellular morphology, Cryo-electron tomography, Mammalian cells, Segmentation, Structural cell biology, Tilt series, Vitrification

1 Introduction

Electron microscopy (EM) was a key driver in the establishment of the cell biology field in the mid twentieth century as it provided (and still provides) novel ultrastructural insights into cellular organization [1–3]. However, over the years, EM lost its popularity potentially due to its labor-intensive sample preparation steps, including fixation, dehydration, and heavy metal treatments, which are prone to artifact generation [4–7]. The development of cryo-EM has overcome these limitations as the sample preparation process allows for imaging of rapidly frozen (vitrified) but still hydrated biological specimens, e.g. mammalian cells, in close to native conditions [8–12]. These samples can be imaged by cryo-electron tomography (cryo-ET), which is a specialized EM technique that can generate 3D volumes of unique cellular samples

Tamir Gonen and Brent L. Nannenga (eds.), *CryoEM: Methods and Protocols*, Methods in Molecular Biology, vol. 2215,
https://doi.org/10.1007/978-1-0716-0966-8_1, © Springer Science+Business Media, LLC, part of Springer Nature 2021

[11, 13]. Although cryo-ET provides impressive detail of cellular features, this method is not widely used by cell biologists mainly due to the perceived complexity of the technique.

The basic workflow of a cryo-ET experiment involves the vitrification of samples, followed by data acquisition and processing [14]. For the preservation of thin samples, e.g. purified proteins or a monolayer of cells, vitrification can be achieved by plunging the specimen surrounded by a thin layer of liquid into a cryogen (e.g. liquid ethane). Excess liquid is blotted away using a filter paper prior to plunging [8–12]. Devices for plunge freezing range from custom-made manual plungers to commercial automated instruments such as the Vitrobot (FEI, now Thermo Fisher), GP2 (Leica), or Cryo-Plunge 3 (Gatan). For thicker samples, e.g. small multicellular organisms like *C. elegans* or tissue biopsies, high-pressure freezing can be used to achieve vitrification [15, 16]. The vitrified samples need to be kept below −140 °C to prevent devitrification, which damages samples through the formation of ice crystals. To prevent devitrification, samples are stored in a liquid nitrogen dewar and are transferred under liquid nitrogen conditions to the cryo-stage of a cryo-cooled transmission electron microscope for imaging [14].

To perform cryo-ET, 2D projection images of the vitrified sample are recorded at different tilt angles. This collection of images is called a tilt series and normally covers a tilt range from about −60° to +60° with an increment of typically 2°–3° per image [17]. The single images are usually recorded at negative defocus (~4–8 μm) [18, 19]. This allows images of sufficient contrast to be generated but at the cost of resolution. If resolution is paramount, a phase plate can be used which enhances contrast of close to focus images [20, 21]. However, using a phase plate currently adds complexity to data collection and processing.

Biological samples experience radiation damage when exposed to an electron beam. Radiation damage causes the breakage of chemical bonds and in severe cases, the complete destruction ("boiling") of a sample [22, 23]. To limit radiation damage, the total electron dose that is applied to the sample needs to be limited and controlled over the whole tilt series [14]. Tilt series are collected using a process called low dose where focusing is performed off axis, and the area of interest is exposed at high magnifications only when the image is being recorded, thus reducing the required total electron dose [24, 25]. However, even under the best imaging conditions, radiation damage will accumulate over the tilt series range.

Sample imaging is also limited by poor electron penetration depth, which directly affects obtainable resolution [13]. As a guide, sample thickness should not exceed 600 nm for descent quality tomograms when a 300 kV cryo-transmission electron microscope (cryo-TEM) with energy filter is used. As the sample is tilted, its thickness increases and thus higher electron doses are required to obtain images of adequate contrast [13, 26]. Furthermore, high tilt images suffer more from effects of drift, doming, and defocus gradients which reduce image quality and obtainable resolution [27–31]. Thus, to obtain the best quality tilt series, the low tilt images should be collected first when the radiation damage is minimal.

Two different tilt-schemes have been developed to maximize the quality and resolution of information contained in a tilt series. The two tilt-schemes are called bidirectional and dose symmetric. For the bidirectional tilt-scheme, image collection starts at a low angle (e.g. 0° or −20°) and the tilt angle is gradually increased to the maximum tilt angle (e.g. +60°). The stage is then reset to the start tilt angle and images are recorded in the other direction towards the negative maximum tilt angle (e.g. −60°). For the dose symmetric scheme, tilt series recording starts at a low tilt angle and alternates between an increasing negative and positive tilt angle [32].

As the total electron dose applied to a tilt series needs to be limited to prevent excessive radiation damage, individual images have a low signal-to-noise ratio resulting in low contrast images. The contrast and data quality of individual images can be increased during data collection by using an energy filter which screens out inelastically scattered electrons, and a camera system which directly detects electrons [33–37]. Furthermore, phase plates can also be used to increase contrast [20].

After tilt series recording, a 3D volume (tomogram) is computationally reconstructed. Two different algorithms are routinely used: "weighted-back projection" and "simulative iterative reconstruction technique (SIRT)" [38, 39]. The later generates higher contrast tomograms, which are beneficial for morphology studies. The first has a proven record of maintaining high-resolution information and is the most popular algorithm especially if the final goal is determining the protein structure.

The contrast of the final reconstructed tomogram can be further enhanced by the use of different imaging filters. These filters can be applied to either the 2D projection images of the tilt series or the final 3D volume. Cellular features, like cytoskeletal elements and organelles, can be highlighted (segmented) in the tomogram to generate a 3D model. This helps visualize complex 3D arrangements of proteins and membranes and allows for precise measurements [14]. Additionally, structures of proteins can be obtained by applying a process called subtomogram averaging [40]. Typical resolutions of the protein structures obtained by subtomogram averaging are in the range of 10–40 Å [18, 41, 42]. However, it is possible to obtain protein structures at significantly higher resolution [29, 43].

In this book chapter, we aim to guide the reader through the principle steps required to generate and analyze a cryo-electron tomogram of mammalian cells. To achieve this, we provide a step-by-step protocol for the initial sample preparation and suggest a general method for data collection, tomographic reconstruction, and analyses. We hope that this chapter will encourage more people to pursue cryo-ET.

2   Materials

2.1   Growing Cells on Grids

2.1.1   Grid Preparation

1. Gold EM grids, 200 mesh with holey carbon support films (*see* **Note 1**).
2. Tweezers, Type 5 (*see* **Note 2**).
3. Two 10 cm Ø Pyrex® petri dishes.
4. Acetone (≥99.5%).
5. Fixed speed rocker with platform to accommodate a 10 cm Ø petri dish.
6. Filter paper (Whatman No. 1, 90 mm Ø).
7. EM Grid storage box (*see* **Note 3**).
8. Glow discharging device.
9. 70% ethanol solution in H_2O: For 1000 μl, mix 700 μl anhydrous ethanol with dH_2O.

Carbon Coating (Optional)

1. Muscovite Mica sheets, V-1 Quality (*see* **Note 4**).
2. Carbon rods (*see* **Note 5**).
3. Carbon evaporator.
4. Smith Grid Coating Trough.

Extracellular Matrix Coating (Optional)

1. Sterile solutions of extracellular matrix components (e.g. poly-L-lysine).

2.1.2   Cell Seeding

1. Biosafety cabinet.
2. 6-Well cell culture plate (sterile).
3. Cell culture media: For SK-MEL-2 cells, use Gibco™ DMEM/F12 supplemented with 10% fetal bovine serum (premium grade, VWR Seradigm Life Science) and 1% Gibco™ Penicillin/Streptomycin (*see* **Note 6**).
4. 10 cm Ø plastic cell culture dish with SK-MEL-2 cells (ATCC® HTB-68™) grown to 75–90% confluency (*see* **Note 7**).
5. Cell culture incubator set to 37 °C and 5% CO_2.
6. Gibco™ Dulbecco's Phosphate Buffered Saline (PBS) without Calcium and Magnesium (sterile).

7. 0.05% Trypsin-EDTA solution (Gibco™).

8. Sterile plastic pipettes (10 and 5 µl).

9. Pipette controller ('Pipetboy').

10. Conical falcon tube (15 µl).

11. Centrifuge (for pelleting cells at ~180 × *g*; should be able to accommodate 15 µl conical falcon tubes).

12. Inverted light microscope equipped with a 4× and 10× Objective to monitor cell growth control.

2.2 Sample Vitrification

1. 10 nm Gold Fiducial Markers (*see* **Note 8**).

2. Two microcentrifuge tubes (1.5 µl).

3. Micropipettes (1000, 200 and 20 µl) and pipette tips.

4. Benchtop centrifuge which handles microcentrifuge tubes.

5. Teflon sheet (*see* **Note 9**).

6. Scissors.

7. Filter paper (Whatman No. 1, 55 mm Ø).

8. Heat block set to 37 °C (should be able to accommodate 6-well plate).

9. Heat block set to 37 °C (should be able to accommodate 1.5 µl microcentrifuge tubes).

10. Automatic plunge freezing device (*see* **Note 10**).

11. Liquid nitrogen.

12. Ethane gas (99.999% pure).

13. Conical falcon tube with string (50 µl, for storage of cryo-grid boxes with vitrified samples, *see* **Note 11**).

14. Liquid nitrogen dewars (4 L, smaller ones for sample transport in 50 µl conical falcon tube).

15. Cryo-EM grid boxes (*see* **Note 12**).

16. Gripper tool for cryo-grid boxes (*see* **Note 12**).

17. Liquid nitrogen sample storage dewar.

2.3 Data Collection

1. 300 kV cryo-TEM (e.g. Krios G3i, Thermo Fisher Scientific, previously FEI) equipped with energy filter (e.g. Bioquantum, Gatan) and direct electron detecting device (e.g. K2). Optional equipment: Phase plate (e.g. Volta Phase Plate).

2. Latest version of SerialEM software installed on computer controlling microscope camera ([44], http://bio3d.colorado. edu/SerialEM/).

2.4 Data Processing

1. Workstation(s) (Linux, Windows or Mac operating systems can be used) for data processing. Minimal configuration: Desktop computer with modern Intel or AMD processor, modern

Nvidia-based video card (4 GB or higher; AMD cards can be used for Windows or Mac) and 16 GB RAM *See* also: https://bio3d.colorado.edu/imod/, **Note 13**).

2. Installation of latest stable IMOD version ([45], https://bio3d.colorado.edu/imod/).

---

3 Methods

3.1 Growing Cells on Grids

All work involving mammalian cells is performed in a biosafety cabinet unless otherwise stated. Grids should always be handled with fine forceps.

3.1.1 Grid Preparation

1. When handling grids, make sure to touch only the outer metal part of the grid with the tweezers to prevent damage to the supporting film (Fig. 1a).

2. In a fume hood, place the required amount of holey carbon grids in a 10 cm Ø Pyrex® petri dish containing acetone and gently agitate on a rocker for 15–30 min at RT to remove residual plastic backing (*see* **Notes 14** and **15**).

3. Dry grids on a filter paper and store in grid box.

4. If required, apply a thin carbon film (~8 nm) to the cleaned EM grids (*see* **Notes 15–18**).

5. Plasma clean grids in a glow discharging device to make the surface hydrophilic.

6. Sterilize grids by placing grids in a 10 cm Ø Pyrex® petri dish with 70% ethanol in H_2O and gently agitate on a rocker for 15–30 min at RT.

7. Move to a biosafety cabinet.

8. If required: coat grids with extracellular matrix to improve cell adhesion. Remove grid from ethanol and drain excess ethanol by touching grid edge to surface of ethanol. (Do not completely dry grid). Place grid into 6-well plate containing ~1.5 µl of extracellular matrix protein solution. Ensure carbon side faces up (*see* **Note 19**). Incubate at 37 °C in cell culture incubator for 1–2 h.

9. After coating or if no coating is required, pick up grids and drain off excess liquid by touching the edge of grid to the surface of the solution.

10. Place EM grid in a 6-well plate where each well contains 2 µl of cell culture media. Place only one grid per well with carbon side up (*see* **Note 20**, Fig. 1b).

11. Incubate overnight at 37 °C and 5% CO_2 in a cell culture incubator (*see* **Note 21**).

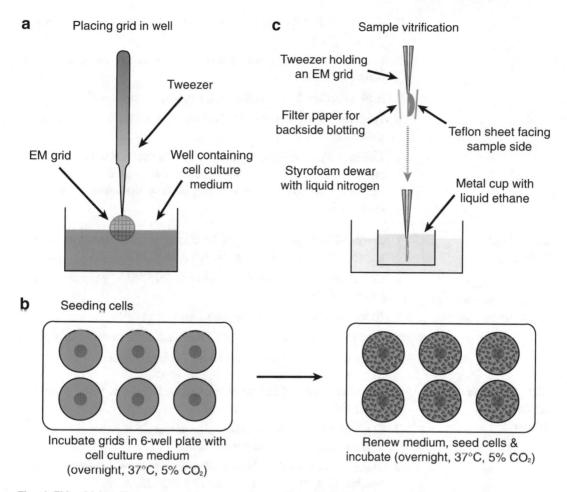

Fig. 1 EM grid handling and sample preparation outline. (**a**) Schematic showing how to hold a grid and submerge it in cell culture media. EM grid should be held with the outer metal ring only and inserted into the media orientated 90° to liquid surface. Once submerged, the grid can be tilted and laid flat on the well floor with carbon side up. (**b**) Workflow of EM grid seeding. *Left:* EM grids are incubated overnight, carbon side up in 6-well plates containing 2 ml of cell culture medium. *Right:* After incubation, cell culture medium is renewed, and cells added. Cells are allowed to settle on grids overnight before plunge freezing. (**c**) Seeded EM grids are removed from cell culture plate and placed in freezing device with cells facing Teflon sheet. After backside blotting, grids are plunged into liquid ethane and transferred to storage device under liquid nitrogen

3.1.2 Cell Seeding

1. Take 10 cm plastic cell culture dish where cells are grown to 75–90% confluency.

2. Wash cells with PBS without calcium and magnesium.

3. Replace PBS with 0.05% Trypsin-EDTA solution (3 µl) and incubate for 2 min at 37 °C to detach cells.

4. Inactive Trypsin-EDTA by adding 7 µl of cell culture media and transfer the entire volume to a 15 µl falcon tube.

5. Harvest cells by centrifugation (2 min, RT, ~180 × g).

6. Remove supernatant and resuspend in 9 µl cell culture media.

7. Remove the cell culture media in the 6-well plate that contain the EM grids.

8. Place 0.5 µl of the resuspended cells into each well of the 6-well plate (*see* **Note 22**).

9. Add a further 1.5 µl cell culture media to each well.

10. Incubate overnight at 37 °C and 5% CO_2 in a cell culture incubator.

11. Using a light microscope, check whether cells are well dispersed on the EM grid and a sufficient number have spread optimally on the grid producing thin cytoplasmic regions (*see* **Note 23**).

3.2 Sample Vitrification

3.2.1 Preparation of Concentrated Gold Fiducial Marker Solution

1. Place 600 µl of 10 nm gold fiducial marker solution in two microcentrifuge tubes (*see* **Notes 8** and **24**).

2. Centrifuge microcentrifuge tubes at $20,000 \times g$ for 30 min to pellet gold beads.

3. Remove 400 µl of supernatant from each tube.

4. Resuspend the pellets in the remaining liquid and combine.

5. Store at 4 °C and warm up to 37 °C before use.

3.2.2 Filter Paper and Teflon Sheet Preparation

1. Take one piece of Teflon sheet and one piece of filter paper (*see* **Note 25**).

2. Cut Teflon sheet to the same size as the filter paper and make a 2 cm Ø hole in the center of both.

3. Place the cut Teflon sheet and filter paper on separate pads in the blotting device as indicated by manufacture.

3.2.3 Plunge Freezing

1. Turn heat block on and set to 37 °C.

2. Switch on automatic plunge freezing device and set temperature to 37 °C and humidity to 90–100%.

3. Leave both devices to reach required temperature and humidity for 15–30 min.

4. Precool the Styrofoam cryo-dewar of the automatic plunge freezing device by filing with liquid nitrogen. Ensure liquid nitrogen also enters the metal pot.

5. Once the liquid nitrogen has evaporated from inside the metal pot, top-up the liquid nitrogen surrounding the metal pot holder (*see* **Note 26**).

6. Wearing a face shield or safety glasses and insulated gloves, prepare the liquid ethane by focusing a jet of ethane gas against the inside surface of the cryo-cooled metal pot of the Styrofoam cryo-dewar (*see* **Notes 27** and **28**).

7. Place 6-well plate containing EM grid on heat block.

8. Pick up an EM grid from the 6-well plate using tweezers.

9. Transfer grid to the tweezers used by the plunge freezing device.

10. Using a micropipette and filter paper, dropwise add and remove a total of 20 μl concentrated 10 nm gold fiducial marker solution to the EM grid. Do not allow the grid to dry during this process.

11. Place a final volume of 3 μl concentrated 10 nm gold fiducial marker solution to the grid after last blot (*see* **Note 29**).

12. Place tweezers in plunge freezing device. The side of the grid with adhered cells should face the Teflon sheet (*see* **Note 30**, Fig. 1c).

13. Set blotting condition (*see* **Note 31**).

14. Initiate plunge freezing routine.

15. After plunge freezing, transfer grids into cryo-grid boxes and store them in liquid nitrogen (*see* **Note 32**). Ensure grid does not warm up during transfer by always using pre-cooled tweezers and keeping the grid in liquid nitrogen or liquid nitrogen atmosphere.

16. Record blotting conditions and location of grid.

17. Repeat **steps 8–16** for each grid.

18. Store grids in cryo-grid boxes under liquid nitrogen conditions until imaging (*see* **Note 33**).

3.3 Data Collection

To image cells, it is recommended to use a 300 kV cryo-TEM with energy filter and direct electron detecting device. A phase plate can also be useful. In this section, we focus on describing the principle steps for recording tilt series of cells rather than microscope alignment. We presume data collection will occur at a user facility where a specialist is available to align the microscope and assist the user. The data collection strategy presented here is for use with the data collection software, SerialEM (*see* **Note 34**). The basic workflow involves obtaining a low magnification overview of the entire grid (whole grid montage) to identify regions of interest, then a medium magnification montage (polygon map) is recorded of those regions to locate points of interest, and then tilt series at the points of interest are recorded.

3.3.1 Sample Loading and Collection of Whole Grid Montage

1. Ensure microscope is aligned and grids are loaded.

2. Insert one grid into microscope column.

3. Set magnification to 81× (~0.2 μm pixel size).

4. Find eucentric height: Place easily detectable feature at center of screen. Tilt to +20° and bring feature back to center by

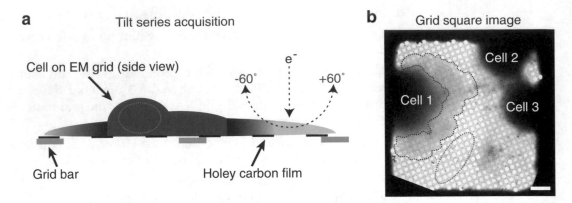

Fig. 2 Tilt series acquisition. (**a**) Schematic of cell growing on holey carbon grid (side view). Brighter areas indicate regions suitable for tilt series acquisition. Tilt series should be recorded over a hole and typically covers a tilt range of −60° to +60°. (**b**) Cryo-EM image of a grid square with three cells. Cell labels indicate location of nucleus. Red-dashed outline indicates area suitable for tilt series collection. Blue-dashed circle indicates carbon film holes devoid of vitrified ice or cells. Scale bar: 10 μm

changing z-height. Tilt to 0° and recenter screen using x/y controls. Repeat until change in position of detectable feature is minimal.

5. Collect montage of whole grid (software path: Navigator → Montaging & Grids → Setup Full Montage). Ensure image overlap is set to 20%.

3.3.2 Collection of Polygon Maps

1. Switch magnification and setup low dose (*see* **Note 35**). Suggested mode magnifications are: Record/focus/preview 0.3 nm/pixel, View ~3 nm/pixel.

2. Align whole grid montage to view mode (*see* **Note 35**).

3. On montage map, locate squares containing good ice and thin cell regions with no obvious damage to carbon support film (Fig. 2).

4. For each square, collect a polygon map covering the thin cell region. Ensure image overlap is 20–25%.

3.3.3 Identifying Points of Interest and Tilt Series Collection

1. Based on project goals, zoom into each polygon map and identify appropriate cell features which are located above holes in carbon support film. Mark these areas using the "add points" function in the navigator window (*see* **Note 36**).

2. Deselect or order points so that no tilt series is collect over a pre-exposed area, e.g. take careful note of focus area location. The edge of the focus beam should be at least 1 μm from edge of the record image.

3. Set up automatic tilt series collection. The parameters suggested in Table 1 are optimized for generating data that can be used for both morphological and structural studies.

Table 1
Suggested imaging parameters

Parameter	Suggested setting
Autofocus	Perform at least every 6° (tilt series parameters)
Autofocus area	>3 μm from record area, along tilt axis, away from thick sample area
Defocus (-Volta phase plate)	−4 to −6 μm
Defocus (+Volta phase plate)	−0.5 to −2 μm
K2 frame rate	0.25 s/frame
K2 mode	Super resolution
Pixel size	2.5 to 6 Å
Slit width (energy filter)	25 to 30 eV
Tilt angle increment	2°
Tilt-scheme	Bidirectional, starting at 20°
Tilt series range	±60°
Total dose	100 to 120 e⁻/Å²

4. Ensure that dose-fractionation is enabled for record images and the frames are saved to a specific directory with tilt angle and navigator position appended to the file name.

5. For each tilt series collected, you should obtain an filename.mrc, filename.mrc.mdoc, and filename.log file as well as the individual dose-fractionated images which are saved to a separate directory.

3.4 Tomogram Generation

Tomogram generation is performed using Etomo in IMOD. A tutorial for reconstructing tomograms with Etomo is available on the IMOD website (https://bio3d.colorado.edu/imod/doc/etomoTutorial.html). In this section, we will mainly focus on the modification to the tutorial for the processing of the dataset obtained in Subheading 3.3. An example tomogram is shown in Fig. 3.

1. Move each tilt series filename.mrc together with respective filename.mrc.mdoc, and filename.log files into individual directories/folders on your processing workstation.

2. For morphological studies, filename.mrc tilt series file from SerialEM should be sufficient. For subtomogram averaging, the dose-fractionated frames should be aligned with Motion-Cor2 and assembled into a filename.mrc tilt series stack. This can be done by using the "newstack" function in IMOD.

3. Using the command line, move into the directory containing the tilt series for reconstruction.

4. Start Etomo by typing "etomo" in the command line.

5. When the "Front Page" window opens, select "Build Tomogram".

6. On the "Setup Tomogram" window, select the name of the filename.mrc file in the dataset name bar and choose "cryo-Sample.adoc" as the "system template".

7. Press "Scan Header", which should automatically fill in "Pixel size (nm)" and "Image rotation" (degrees). If not, fill in manually (*see* **Note 37**).

8. Add the value for "Fiducial diameter (nm)" (e.g. 10).

9. Enable "Parallel Processing" when working on multi-workstation clusters and "Graphics card processing", if available.

10. For bidirectional tilt series, ensure "Series was bidirectional from" is activated under Axis A and the correct start angle is displayed (e.g. 20.0).

11. Press "View Raw Image Stack" to open the tilt series.

12. Manually inspect the quality of all individual images in the tilt series stack. Take note of images which are unfocused, blurry, or too dark to detect gold particles. Add the numbers of these images to the "Exclude views" section under Axis A and select "remove excluded views" in the section above Axis A.

13. Press "Create Com Scripts". This will generate the "align.com" file inside the working directory and open the window for the tomogram generation workflow. Close this window.

14. At the command line, open the "align.com" file in a text editor, (e.g. "vi align.com"), go to line 52 and change the "XTiltOption" from 0 to 4. Save changes keeping original file name.

15. Restart Etomo by typing "etomo filename.edf" in the command line.

16. Follow the tomogram generation workflow as outlined in the online tutorial (https://bio3d.colorado.edu/imod/doc/etomoTutorial.html). Ignore the steps for dual-tilt reconstruction and under "Fiducal Model gen." choose "Make seed model manually".

17. For tomogram generation, you can choose either the back projection or SIRT algorithm (~15 iteration should be sufficient). SIRT will give a higher contrast tomogram than back projection but with the potential loss of high-resolution information. Thus, for subtomogram averaging, back projections are usually used. The final tomogram will be saved as a filename.rec file in your working directory.

18. Cryo-tomograms normally suffer from low signal-to-noise ratio and thus further processing to enhance contrast is required. There are several procedures for this, and the best option needs to be determined empirically for each sample. We commonly use the smooth filter option in the IMOD clip program or the "Nonlinear Anisotropic Diffusion" filter.

3.5 Segmentation Model Generation

Segmentation models are a great way to extract essential information from tomograms of the crowded cell interior and an excellent way to visualize its complex three-dimensionality. Segmentation model generation can be performed with the IMOD software package and a detailed tutorial can be found on the IMOD website (https://bio3d.colorado.edu/imod/doc/3dmodguide.html). In brief, models in IMOD are structured in three levels:

1. "Object" (e.g. all microtubules in a tomogram).

2. "Contours" (e.g. an individual microtubule).

3. "Points" (points that are placed to trace an individual microtubule; *see* below). In other words, an "Object" is made of "Contours" that consist of "Points". Furthermore, segmentation models can also be used to pursue subtomogram averaging, for example, using the IMOD software PEET (http://bio3d.colorado.edu/PEET/). Here, we will briefly explain the generation of segmentation models on three different cellular features. These are:

   (a) Filaments, e.g. microtubules.

   (b) Membrane vesicles.

   (c) Small globular densities, e.g. ribosomes (Fig. 3b, c).

3.5.1 Generate Segmentation Model

1. Open tomogram in 3dmod.

2. In the "Information window" choose "Model" under mode.

3. Make a new object or use the current one if not in use already.

4. Example 1: Segmenting filaments.

   (a) Set "Object type" to "Open".

   (b) Open the "Slicer window" by using the hot key "\".

   (c) Alter rotation angles and "View axis position" so that the filament of interest appears as straight as possible.

   (d) Start at one end of filament and place points along the center of the filament axis until the other end of the filament is reach. Adjust tomogram z-slice as necessary.

   (e) Make sure that all points selected in (**step d**) are contained in the same contour and appear in a linear order.

   (f) To trace a second filament, generate a new contour by using the hot key "n" and repeat the **steps c–e**.

(g) To generate tube, open the "Objects" window and choose meshing.

(h) Select tube and set "Diam" to a pixel value that corresponds to the filament diameter.

(i) Press "mesh one", this will generate a tube that represents the traced filament. If "Cap" is activated, this will generate a capped tube.

5. Example 2: Segmenting of vesicles or other membranous organelles.

(a) Segmentation of vesicles and other membranous compartments is performed using the special drawing tools plugin (*see* **Note 38**).

(b) Generate a new object by selecting in the information window edit > object > new.

(c) Set "Object type" to "Closed".

(d) Open drawing tools toolbar by clicking Special > Drawing tools from the menu bar in the information window.

(e) Use the "Sculpt" function and trace the membrane of the organelle across *z*-slices in the "ZaP Window". It is not necessary to do this on all slices, every second or third should be sufficient.

(f) To generate a mesh for the vesicle, open the "Objects" window and choose meshing. If membrane tracing was not performed on every slice, activate "Skip" option in the meshing dialog window and then press "mesh one".

6. Example 3: Segmenting of small globular features, e.g. ribosomes.

(a) Generate a new object by clicking on edit > object > new in the information window.

(b) Select "Object type" to "Scattered".

(c) In the "ZaP Window" or "Slicer", place a point in the center of the feature of interest.

(d) Under "Points" in the "Object"' window, set "Sphere Size" to a value that corresponds to the size of the feature of interest and set drawing style to "Fill".

(e) To adjust the smoothness of the sphere, modify the "Object Quality" and "Global Quality" values.

3.6 Analysis of Segmentation Model

IMOD has a series of command line programs for the analyses of segmented data. A complete list with precise description can be found at: https://bio3d.colorado.edu/imod/doc/program_list ing.html. Below is a list of programs we find particularly useful.

1. Use the program "imodinfo" to extract quantitative information about the segmented objects, e.g. filament length, surface areas, and volume.

2. Use the program "imod-dist" to extract information on distances between features.

3. Use "boxstartend" to extract subtomograms around the modeled features.

4. Use "model2point" to extract coordinates in ascii (text format).

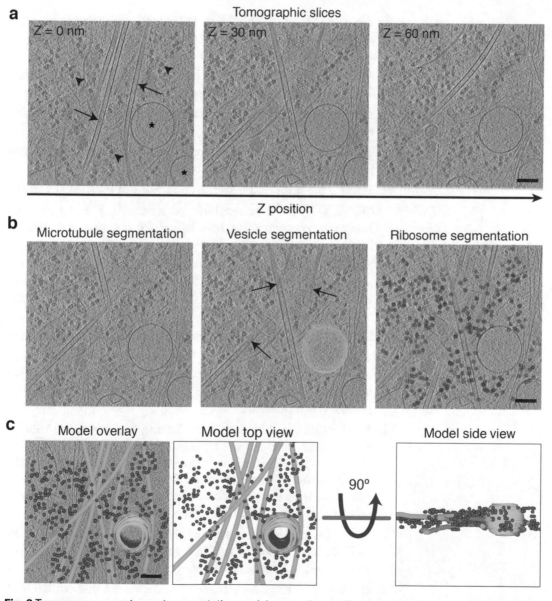

Fig. 3 Tomogram processing and segmentation model generation. (**a**) Tomographic slices of an SK-MEL-2 cell at $Z = 0$, 30, and 60 nm. Tilt series was recorded on a Jeol 3100 equipped with a K2 direct electron detector with pixel size 3.98 Å. The tilt series was binned by a factor of 2, and the reconstructed tomogram was filtered using the "clip smooth" function in IMOD. Cellular features including microtubules (arrows), ribosomes (arrowheads), and vesicles (*) can be observed. (**b**) Segmentation of MT (green), vesicle (cyan), and ribosomes (pink). Protofilaments are clearly visible in center image (arrows). (**c**) Meshed segmentation model of the features in (**b**). *Left*: tomographic slice with overlaid model. *Center/right*: model only. Scale bars: 100 nm

4 Notes

1. We use Quantifoil® gold 200 mesh EM grids with 2 μm holes and 1 μm spacing (R2/1). Smaller holes sizes can be used if higher magnifications than described in this chapter are used for imaging. We prefer Quantifoil over C-flats as the carbon film is stronger.

2. We use Dumont, N5-PO, Dumoxel®. Alternatively, anti-capillary tweezers can be used that prevent the uptake of liquid and thus might enable easier release of the grid.

3. Any grid box similar to the ones from Electron Microscopy Sciences (EMS), cat #71155 can be used.

4. Mica can be purchased from any supplier of EM consumables, e.g. EMS, cat # 71855-05.

5. Thickness and grade of carbon rods depend on evaporator type. Check manufacturer instructions.

6. Prepare medium under sterile conditions in biosafety cabinet.

7. SK-MEL-2 cells were used for work described in this chapter. However, protocols can be adapted for the use of other cell lines.

8. We used 10 nm BSA Gold Tracer (EMS, cat # 25486). It is important to have the gold colloidal conjugated to a protein or PEG coating. Uncoated gold colloidals aggregate when added to solutions containing salt.

9. We use Gold Seal Specialty Papers PTFE Sheets from Amazon.

10. We use a Vitrobot Mark IV (Thermo Fisher Scientific, previously FEI). Alternatively, manual plunge freezers or other automatic devices such as the Leica GP2 or the Gatan Cryo-plunge 3 can be used.

11. Small holes are drilled into the lid and top of conical tube to allow liquid nitrogen gas to escape and to attach string. The purpose of the string is to retrieve the conical tube from the liquid nitrogen sample storage dewar. Small metal screws or bolts can be added to the conical tube to ensure the tube is submerged in the liquid nitrogen.

12. Cryo-EM grid boxes can be purchased from standard suppliers of EM consumables (e.g. EMS, cat #71166-40). To open the grid boxes, use the gripper tool (EMS, cat #71166-SP).

13. It is important to have fast access to data and sufficient RAM to hold a tomogram in memory (32 or 64 GB are good, 128 GB is excellent). GPUs enhance the speed of tomogram reconstructions.

14. Quantifoil® carbon support film is made with a plastic backing. This should have been removed before grids are delivered but some batches still contain the plastic which deforms during tilt series collection and destroys the tomogram. Soaking grids in acetone helps to remove the plastic. Chloroform and acetone can be used instead of acetone. Acetone is very useful for particularly troublesome batches of Quantifoil®.

15. When placing grids into a solution, hold them perpendicular to the surface of the solution, push them under until the grid is fully submerged and then lay the grid down on a filter paper or flat surface with carbon side facing up.

16. Some cell lines may not spread on holey carbon. If this is the case, adding a thin layer of carbon will encourage them to spread. Furthermore, certain cellular events (e.g. cell adhesion formation) can only occur where the cell is attached to the surface and thus a thin layer of carbon is required in order to be able to image the event over a hole in the holey carbon support film.

17. To add thin carbon to holey carbon EM grids, first evaporate a thin layer of carbon onto a mica sheet using a carbon evaporator device. After a delay of at least 24 h to allow the carbon molecules to fuse, cut the mica to the size of an EM grid, float the carbon film onto the surface of a water drop by inserting the mica into the water drop at a 45° angle, carbon side up. Then push an EM grid into the water drop from the side, near the bottom of the drop, and bring it up and out through the carbon film. Multiple grids can be coated at once using a device like a "Smith Grid Coating Trough". Allow the coated grids to dry overnight before use.

18. To help remove the carbon from the mica, gently breathe on the carbon surface. The moisture from the breath helps dislodge the carbon from the mica surface.

19. Quantifoil® indicates in which direction the carbon-coated grid surface faces inside the grid box. Visually, the carbon-coated side appears less shiny compared to the uncoated side.

20. To prevent removal of thin carbon film, ensure the grid enters the cell culture media at an angle of 90° to the liquid surface.

21. If grids were coated with extracellular matrix components, this incubation step is not necessary.

22. If cells density appears too high or too low and cells do not spread well as a result, adjust the volume of cell suspension accordingly.

23. It requires some experience to judge cell thickness by light microscopy. Therefore, it might be helpful to take images on

the light microscope and compare with cell appearance by cryo-TEM.

24. The gold fiducial markers are required for later tilt series alignment as they act as fiducial markers. Alternatively, "Fiducialless alignment" (without gold tracer) can be performed in IMOD. However, we highly recommend the use of gold tracer as fiducial marker.

25. For a one-sided blotting device, e.g. the Leica GP2, the Teflon sheet is not required.

26. Always ensure liquid nitrogen in Styrofoam dewar is at an adequate level. This should be just above the metal bar that holds the grid boxes.

27. Splashes from liquid ethane can cause cryo-burns instantaneously. Proper personal protection equipment should be worn including eye protection, lab coat, and gloves. We like using thin cotton gloves with plastic finger tips (electrician gloves) covered with nitrile gloves rather than the thick blue cryo-protection gloves. This maintains flexibility and dexterity while protecting against splashes.

28. When liquifying ethane, aim the nozzle against a cryogen-cooled metal surface. As the ethane starts to liquify, keep the nozzle just below the surface of the liquid ethane. When removing the nozzle from the liquid ethane, turn off gas to prevent splashing as the nozzle is removed. Take extra care when placing the nozzle down as it often contains liquid ethane.

29. If "Fiducialless alignment" is preferred, wash grid with cell culture media or PBS (with calcium and magnesium) and place 3 μl of the respective liquid on grid after last blot.

30. For single-sided blotting devices, ensure that cells face away from pad with filter paper. If cells face blotting paper, they can rupture or be pulled off the grid.

31. Blotting conditions need to be determined empirically for each sample and each device. Suggested starting conditions for the Vitrobot Mark IV are: 2.5–4 s with blot force 10 and 0–1 s drain time after blotting before plunge freezing.

32. To prevent sample contamination by breathing on sample or liquid nitrogen container, consider wearing a mask or face shield.

33. For work on Titan Krios, grids need to be clipped in autogrid holders using c-clips and base rings. These consumables can be obtained from Thermo Fisher Scientific.

34. Helpful tutorials for general usage of SerialEM can be found here: https://www.youtube.com/user/BL3DEMC.

35. Information on how to set up and use grid montaging and low dose imaging can be found at https://bio3d.colorado.edu/SerialEM/hlp/html/about_low_dose.htm, https://www.youtube.com/watch?v=N6A_BVR13Gc

36. If feature of interest is not detectable by this approach, you may consider using a correlative light and electron microscopy approach. Here, the feature of interest can be identified by fluorescence light microscopy first (*see* [46]).

37. To check the image stack file has the correct pixel size, type "header filename.st" at the command line. If the pixel size is wrong (as can happen if MotionCor2 is used), type "alterheader filename.st" and use the "del" option to change the pixel value.

38. Instructions on how to use the drawing tool plugin for 3dmod can be found at https://www.youtube.com/watch?v=BsNSVLIQ-cE.

Acknowledgments

This work was supported in part by the Office of Science of the US Department of Energy DE-AC02-O5CH11231 (K.M.D.), the Human Frontier Science Program fellowship LT000234/2018-L (D.S.) and UCB Start-up funds (K.M.D.).

References

1. Porter KR, Claude A, Fullam EF (1945) A study of tissue culture cells by Electron microscopy. J Exp Med 81:233–246. https://doi.org/10.1084/jem.81.3.233

2. Palade GE (1952) The fine structure of mitochondria. Anat Rec 114:427–451. https://doi.org/10.1002/ar.1091140304

3. Dalton AJ, Felix MD (1954) Cytologic and cytochemical characteristics of the Golgi substance of epithelial cells of the epididymis–in situ, in homogenates and after isolation. Am J Anat 94:171–207. https://doi.org/10.1002/aja.1000940202

4. Geuze HJ (1999) A future for electron microscopy in cell biology? Trends Cell Biol 9:92–93. https://doi.org/10.1016/S0962-8924(98)01493-7

5. Heuser J (2002) Whatever happened to the 'microtrabecular concept'? Biol Cell 94:561–596. https://doi.org/10.1016/S0248-4900(02)00013-8

6. Small JV (1981) Organization of actin in the leading edge of cultured cells: influence of osmium tetroxide and dehydration on the ultrastructure of actin meshworks. J Cell Biol 91:695–705

7. Maupin-Szamier P, Pollard TD (1978) Actin filament destruction by osmium tetroxide. J Cell Biol 77:837–852

8. Dubochet J, McDowall AW, Menge B et al (1983) Electron microscopy of frozen-hydrated bacteria. J Bacteriol 155:381–390

9. Dubochet J, McDowall AW (1981) Vitrification of pure water for Electron microscopy. J Microsc 124:3–4. https://doi.org/10.1111/j.1365-2818.1981.tb02483.x

10. Dubochet J, Adrian M, Chang J-J et al (1988) Cryo-electron microscopy of vitrified specimens. Q Rev Biophys 21:129–228. https://doi.org/10.1017/S0033583500004297

11. Medalia O, Weber I, Frangakis AS et al (2002) Macromolecular architecture in eukaryotic cells visualized by Cryoelectron tomography. Science 298:1209–1213. https://doi.org/10.1126/science.1076184

12. Adrian M, Dubochet J, Lepault J, McDowall AW (1984) Cryo-electron microscopy of viruses. Nature 308:32–36. https://doi.org/10.1038/308032a0

13. Grimm R, Singh H, Rachel R et al (1998) Electron tomography of ice-embedded prokaryotic cells. Biophys J 74:1031–1042. https://doi.org/10.1016/S0006-3495(98)74028-7

14. Lučić V, Rigort A, Baumeister W (2013) Cryo-electron tomography: the challenge of doing structural biology in situ. J Cell Biol 202:407–419. https://doi.org/10.1083/jcb.201304193

15. Serwas D, Su TY, Roessler M et al (2017) Centrioles initiate cilia assembly but are dispensable for maturation and maintenance in *C. elegans*. J Cell Biol 216:1659–1671. https://doi.org/10.1083/jcb.201610070

16. Moor H (1987) In: Steinbrecht RA, Zierold K (eds) Theory and practice of high pressure freezing BT-cryotechniques in biological electron microscopy. Springer, Berlin, Heidelberg, pp 175–191

17. Frank J (2006) Electron tomography-methods for three-dimensional visualization of structures in the cell. Springer-Verlag, New York

18. Lin J, Nicastro D (2018) Asymmetric distribution and spatial switching of dynein activity generates ciliary motility. Science 360: eaar1968. https://doi.org/10.1126/science.aar1968

19. Guichard P, Chretien D, Marco S, Tassin AM (2010) Procentriole assembly revealed by cryo-electron tomography. EMBO J 29:1565–1572. https://doi.org/10.1038/emboj.2010.45

20. Danev R, Buijsse B, Khoshouei M et al (2014) Volta potential phase plate for in-focus phase contrast transmission electron microscopy. Proc Natl Acad Sci U S A 111:15635–15640. https://doi.org/10.1073/pnas.1418377111

21. von Loeffelholz O, Papai G, Danev R et al (2018) Volta phase plate data collection facilitates image processing and cryo-EM structure determination. J Struct Biol 202:191–199. https://doi.org/10.1016/j.jsb.2018.01.003

22. Glaeser RM (2008) Retrospective: radiation damage and its associated "information limitations". J Struct Biol 163:271–276. https://doi.org/10.1016/j.jsb.2008.06.001

23. Talmon Y (1987) In: Steinbrecht RA, Zierold K (eds) Electron beam radiation damage to organic and biological cryospecimens · BT-cryotechniques in biological electron microscopy. Springer, Berlin, Heidelberg, pp 64–84

24. Dierksen K, Typke D, Hegerl R et al (1992) Towards automatic electron tomography. Ultramicroscopy 40:71–87. https://doi.org/10.1016/0304-3991(92)90235-C

25. Koster AJ, Chen H, Sedat JW, Agard DA (1992) Automated microscopy for electron tomography. Ultramicroscopy 46:207–227. https://doi.org/10.1016/0304-3991(92)90016-D

26. Baumeister W, Grimm R, Walz J (1999) Electron tomography of molecules and cells. Trends Cell Biol 9:81–85

27. Xiong Q, Morphew MK, Schwartz CL et al (2009) CTF determination and correction for low dose tomographic tilt series. J Struct Biol 168:378–387. https://doi.org/10.1016/j.jsb.2009.08.016

28. Fernández JJ, Li S, Crowther RA (2006) CTF determination and correction in electron cryo-tomography. Ultramicroscopy 106:587–596. https://doi.org/10.1016/j.ultramic.2006.02.004

29. Turoňová B, Schur FKM, Wan W, Briggs JAG (2017) Efficient 3D-CTF correction for cryo-electron tomography using NovaCTF improves subtomogram averaging resolution to 3.4Å. J Struct Biol 199:187–195. https://doi.org/10.1016/j.jsb.2017.07.007

30. Henderson R, Baldwin JM, Ceska TA et al (1990) Model for the structure of bacteriorhodopsin based on high-resolution electron cryo-microscopy. J Mol Biol 213:899–929. https://doi.org/10.1016/S0022-2836(05)80271-2

31. Fernandez J-J, Li S, Bharat TAM, Agard DA (2018) Cryo-tomography tilt-series alignment with consideration of the beam-induced sample motion. J Struct Biol 202:200–209. https://doi.org/10.1016/j.jsb.2018.02.001

32. Hagen WJH, Wan W, Briggs JAG (2017) Implementation of a cryo-electron tomography tilt-scheme optimized for high resolution subtomogram averaging. J Struct Biol 197:191–198. https://doi.org/10.1016/j.jsb.2016.06.007

33. Grimm R, Koster AJ, Ziese U et al (1996) Zero-loss energy filtering under low-dose conditions using a post-column energy filter. J Microsc 183:60–68. https://doi.org/10.1046/j.1365-2818.1996.77441.x

34. Langmore JP, Smith MF (1992) Quantitative energy-filtered electron microscopy of biological molecules in ice. Ultramicroscopy 46:349–373. https://doi.org/10.1016/0304-3991(92)90024-E

35. Schröder RR, Hofmann W, Ménétret J-F (1990) Zero-loss energy filtering as improved imaging mode in cryoelectronmicroscopy of

frozen-hydrated specimens. J Struct Biol 105:28–34. https://doi.org/10.1016/1047-8477(90)90095-T

36. McMullan G, Chen S, Henderson R, Faruqi AR (2009) Detective quantum efficiency of electron area detectors in electron microscopy. Ultramicroscopy 109:1126–1143. https://doi.org/10.1016/j.ultramic.2009.04.002

37. Milazzo A-C, Cheng A, Moeller A et al (2011) Initial evaluation of a direct detection device detector for single particle cryo-electron microscopy. J Struct Biol 176:404–408. https://doi.org/10.1016/j.jsb.2011.09.002

38. Gilbert P (1972) Iterative methods for the three-dimensional reconstruction of an object from projections. J Theor Biol 36:105–117. https://doi.org/10.1016/0022-5193(72)90180-4

39. Radermacher M (2006) In: Frank J (ed) Weighted Back-projection methods BT-electron tomography: methods for three-dimensional visualization of structures in the cell. Springer, New York, NY, pp 245–273

40. Wan W, Briggs JAG (2016) Chapter thirteen-cryo-electron tomography and subtomogram averaging. In: Crowther RA (ed) Methods enzymol. Academic Press, New York, pp 329–367

41. Mahamid J, Pfeffer S, Schaffer M et al (2016) Visualizing the molecular sociology at the HeLa cell nuclear periphery. Science 351:969–972. https://doi.org/10.1126/science.aad8857

42. Mühleip AW, Dewar CE, Schnaufer A et al (2017) In situ structure of trypanosomal ATP synthase dimer reveals a unique arrangement of catalytic subunits. Proc Natl Acad Sci U S A 114:992–997. https://doi.org/10.1073/pnas.1612386114

43. Schur FKM, Hagen WJH, de Marco A, Briggs JAG (2013) Determination of protein structure at 8.5Å resolution using cryo-electron tomography and sub-tomogram averaging. J Struct Biol 184:394–400. https://doi.org/10.1016/j.jsb.2013.10.015

44. Mastronarde DN (2005) Automated electron microscope tomography using robust prediction of specimen movements. J Struct Biol 152:36–51. https://doi.org/10.1016/j.jsb.2005.07.007

45. Kremer JR, Mastronarde DN, McIntosh JR (1996) Computer visualization of three-dimensional image data using IMOD. J Struct Biol 116:71–76. https://doi.org/10.1006/jsbi.1996.0013

46. Hampton CM, Strauss JD, Ke Z et al (2017) Correlated fluorescence microscopy and cryo-electron tomography of virus-infected or transfected mammalian cells. Nat Protoc 12:150–167. https://doi.org/10.1038/nprot.2016.168. http://www.nature.com/nprot/journal/v12/n1/abs/nprot.2016.168.html#supplementary-information

Chapter 2

Cryo-Electron Tomography and Automatic Segmentation of Cultured Hippocampal Neurons

Ryan K. Hylton, Victoria H. Seader, and Matthew T. Swulius

Abstract

Cryo-electron tomography is fast becoming a preferred method for studying intracellular environments at the molecular scale. Increases in data collection throughput means that large numbers of tomograms can be generated at rates too fast for humans to easily explore quantitatively. Currently, there is a large effort to make data collection and segmentation tools more automated. Here, we describe a workflow for preparing cultured neurons on electron microscopy grids, batch tomographic data collection, reconstruction and automatic segmentation using freely and commercially available software.

Key words Cryo, Cryo-EM, Cryo-ET, ECT, CET, Tomography, Neuron, Segmentation, Neural Network, IMOD, EMAN2

1 Introduction

While all cells are fascinating, the neuron is of specific importance to the diversity of animal behavior and psychological experience. It is not surprising, then, that they take on very complex morphologies themselves and have much to reveal about the structural mechanisms underlying many dynamic cellular processes. Fortunately, cryo-electron tomography (cryo-ET) is a useful method for imaging the fine biological architecture of neurons, where the direct imaging of cryo-preserved cytoplasm allows for unprecedented three-dimensional clarity at the molecular scale [1, 2].

One of cryo-ETs major limitations is sample thickness. There is a practical cut-off at approximately 1 μm, but to achieve the highest contrast and resolution, the sample should be no more than a few 100 nm thick. Due to this, mammalian cells are typically inaccessible without the use of cryo-sectioning or cryo-focused ion milling [3, 4]. Luckily, the long branches of neurons grown in culture are ideal for cryo-ET (<500 nm-thick) and provide very high quality cryotomograms reliably.

Tamir Gonen and Brent L. Nannenga (eds.), *CryoEM: Methods and Protocols*, Methods in Molecular Biology, vol. 2215, https://doi.org/10.1007/978-1-0716-0966-8_2, © Springer Science+Business Media, LLC, part of Springer Nature 2021

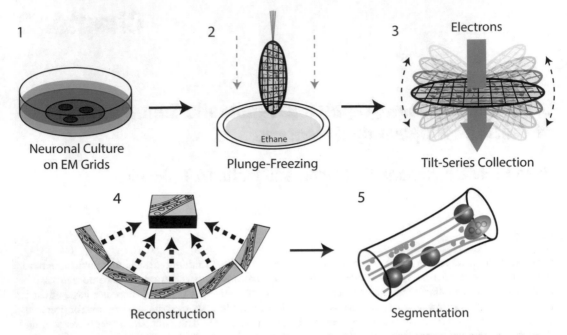

Fig. 1 Workflow for cryo-ET of cultured neurons. (*1*) Cells are cultured on top of gold EM grids in a glass bottom culture dish. (*2*) Grids are plunge frozen in liquid ethane (vitrification). (*3*) Vitrified sample is transferred into a cryo-TEM and tilt-series are collected. (*4*) Tilt-series are used to generate tomographic volumes. (*5*) Tomograms are segmented to generate a 3D model for easy visualization and analysis

In the last decade or so, tomographic data collection has become semi-automated and a variety of software packages have been developed for collecting image tilt-series in large batches. This increased throughput has made it possible to routinely collect tens of tomograms per day and upwards of 100 under ideal conditions. Such a large amount of data is difficult to annotate and quantify, so methods for automated tomogram segmentation have become a necessity [5, 6]. In response, the software package EMAN2 was updated in 2017 to include convolutional neural networks, which can be trained to segment and annotate tomographic volumes [7].

Here, we detail and discuss a workflow for investigating the ultrastructure of cultured hippocampal neurons using cryo-ET. The workflow (Fig. 1) starts with culturing neurons on an electron microscope (EM) grid. The grid is then plunge frozen in liquid ethane, and the vitrified sample is transferred into a cryo-transmission EM (cryo-TEM). Many thin regions of the cell are imaged across a range of angles by incrementally tilting the cryo-stage. These images are computationally aligned and back-projected to generate a tomographic volume. Finally, the tomographic volume is segmented for visualization and analysis. While this protocol addresses the use of neurons and a specific set of software, many of the same principles will apply across a range of cellular targets, regardless of the software used to collect and process the data. Though the methods described are not exhaustive in terms of all the possible options and microscope configurations, much of this methodology can serve as a resource for others venturing into the study of 3D cellular architecture in general.

2 Materials

2.1 Grid Preparation

1. Sterilized, fine-tipped tweezers (Dumont No. 5).
2. 200 mesh gold QUANTIFOIL R 2/2 holey carbon grids. *See* **Note 1**.
3. Glow Discharge System.
4. Glass slide and glass petri dish. Manufacturer and size of the slide and dish do not matter, as long as the dish is large enough to hold the slide.
5. 70% ethanol.

2.2 PDL Coating

1. Poly-D-Lysine solution.
2. 0.2 µm PES vacuum filter.

2.3 Neuronal Culture

1. Hippocampal Neurons. We use a Complete Kit for E18 Sprague Dawley rat hippocampus (BrainBits). *See* **Note 2**.
2. 35 mm glass bottom cell culture dishes. *See* **Note 3**.
3. Culture Media: We use NbActiv4 media (BrainBits) with 1% Pen-Strep, but standard neurobasal media and other variants have been used successfully (*see* **Note 2**).
4. Trypan Blue Solution, 0.4%.

2.4 Fiducial Gold

1. Lyophilized BSA.
2. 10 nm colloidal gold nanoparticles.

2.5 Vitrification

1. Whatman Grade 1 Qualitative Filter Paper; 55 mm circles.
2. Plunge Freezer. We use the Thermo Fisher Scientific Vitrobot Mark IV, but any should be able to work, even completely manual setups without strict environmental controls.
3. Ethane Canister.
4. Liquid Nitrogen.
5. Liquid Nitrogen dewars for transport and storage.
6. CryoEM grid storage boxes.
7. Long (>6 in.) pair of blunt tweezers.

2.6 Cryo-Electron Tomography

1. Microscope: 300 kV cryo-TEM.
2. Data Collection Software: Batch tomography collection software, such as Tomography (Thermo Fisher Scientific), SerialEM (Boulder, CO), Leginon (NRAMM), or UCSFTomo (UCSF, CA).

2.7 Image Processing Software

1. Reconstruction: IMOD 4.9.12 (UC Boulder, CO).
2. Segmentation and Visualization: EMAN 2.31 (Baylor College of Medicine, TX) and Amira (Thermo Fisher Scientific).

3 Methods

3.1 Grid Preparation

1. UV and ethanol sterilize a biosafety cabinet.
2. Ethanol sterilize a glass slide and a 100 mm petri dish. Place the slide inside the petri dish (*see* **Note 4**).
3. In the biosafety cabinet, use autoclaved forceps to remove the grids from the storage box and set them carbon-side up on the glass slide.
4. Examine grids' carbon supports under a stereo microscope to ensure that the majority of the surface is intact.
5. Glow discharge grids to produce a hydrophilic surface (*see* **Note 5**).
6. Carefully place the glass top on the petri dish before transfering the grids into the biosafety cabinet. UV sterilize forceps and grids for 5–10 min.

3.2 Poly-D-Lysine Coating

1. Dilute Poly-D-lysine (PDL) to 100 μg/mL in Milli-Q H_2O. Filter sterilize PDL in biosafety cabinet through a 0.2 μm PES vacuum filter.
2. Add enough PDL solution to completely cover the bottom of the dish. 2 mL is sufficient in a 35 mm glass bottom dish.
3. Using forceps, submerge the grids carbon-side up in the PDL solution and place on the glass bottom of the dish (*see* **Note 6**).
4. Place dish containing grids into a 5% CO^2, 37 °C incubator overnight.
5. The next morning, wash grids three times with 2 mL of autoclave-sterilized Milli-Q H_2O, being careful not to disturb the grids. The wash step is very important; excess PDL in the growth media can inhibit neurite outgrowth.
6. Leave grids in water or growth media and place back in the incubator until neurons are to be plated.

3.3 Neuronal Culture

1. Obtain E18 rat hippocampal neurons. We use a kit from Brain-Bits, but any protocol for obtaining primary neurons should suffice. The kit can be stored at 4 °C for up to 3–4 days, but there may be a drop-off in the total number of living cells acquired from the tissue.

2. If using the kit, follow the protocol outlined below (or follow the kit protocol):

    (a) Make the cell dissociation solution by dissolving the provided 6 mg of Papain in 3 mL Hibernate E without Calcium (HE-Ca) in a 30 °C water bath for 10 min.

    (b) Using a Pasteur pipette, place hippocampi in the cell dissociation solution and incubate in a 30 °C water bath for 10 min.

    (c) Use the Pasteur pipette to return the hippocampi to its original vial/media.

    (d) Triturate tissue for ~1 min to dissociate cells into a single cell suspension (*see* **Note 7**).

    (e) Let heavy debris settle to the bottom of the tube for at least 1 min, before carefully transferring the cell suspension into a sterile 15 mL conical (leaving behind any debris).

    (f) Spin single cell suspension at $256 \times g$ for 2 min (*see* **Note 7**).

    (g) Discard supernatant and resuspend the pellet in 1 mL of growth media.

3. Count cells in a hemocytometer using trypan blue exclusion.

4. Dilute cell in growth media to desired plating density, assuming 2 mL of media per 35 mm dish.

5. Aspirate water or media from 35 mm dishes and plate cells at 20-40 k cells/cm^2 over the grid-containing plates. Plating 192 k, 288 k, or 384 k cells/dish in 2 mL of media will result in 20 k, 30 k, and 40 k cells/cm^2, respectively.

6. If grids move during plating, separate them by gently pushing them away from one another with forceps.

7. Check cell distribution on a light microscope to ensure that cells are monodispersed.

8. Feed cells every 4 days by exchanging half of the media (1 mL) with fresh media, out to the number of days in vitro desired (*see* Fig. 2).

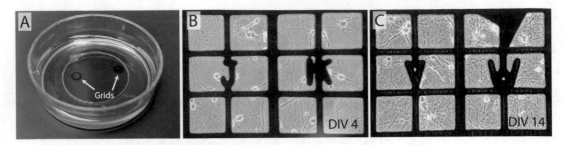

Fig. 2 Neuronal Culture on gold holey carbon grids. (**a**) 35 mm glass bottom petri dish plated with dissociated primary hippocampal neurons. PDL-coated grids were placed at the bottom of the dish prior to plating with cells. (**b**, **c**) Hippocampal neurons growing on gold 200-mesh Quantifoil R2/2 finder grids at DIV 4 and 14, respectively, viewed by a light microscope. By DIV 14, a dense network of neuronal branches can be seen

3.4 Preparation of BSA-Coated Gold Fiducials

1. Prepare 5% (weight/volume) BSA stock solution in water.

2. Mix 4 volumes of gold bead solution with 1 volume of 5% BSA and vortex. Final BSA concentration is 1%. To make 8 mL, add 6.4 mL gold solution to 1.6 mL 5% BSA.

3. Put 1 mL of mixture into 8 × 1.5 mL microfuge tubes and centrifuge at maximum speed for 30 min at RT in a benchtop centrifuge.

4. Carefully pipette off as much supernatant as possible (the pellet can be disturbed easily; it is not critical to completely remove the supernatant).

5. Rinse each pellet with 1 mL of water and vortex. Centrifuge at max speed for 30 min at RT.

6. Pipette off as much of the supernatant as possible without disturbing the pellet and resuspend the pellet in any remaining supernatant.

7. Combine all resuspended gold into one tube. If solution is not a mostly opaque, dark red, centrifuge at max speed for 30 min at RT. Remove approximately half of the supernatant and resuspend.

8. Store at 4 °C.

3.5 Vitrification

Although any plunge freezer can be used, this protocol describes the use of Thermo Fisher Scientific's Vitrobot Mark IV.

1. Set environmental chamber to maintain 70% humidity at room temperature (20–22 °C).

2. Set "blot time," "blot force," and "blot total" to zero to ensure that the blotting pads do not blot the grid during the process. This allows for manual blotting of the grid from behind (*see* **Note 8**), using a long set of forceps and a small rectangle of Whatman filter paper (Fig. 3a).

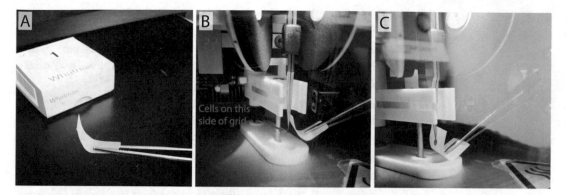

Fig. 3 Single-sided manual blotting in the Vitrobot. (**a**) Cut a small rectangle of filter paper and bend toward the center, lengthwise. (**b**) Use a set of long sturdy forceps to reach through the open door on the side of the Vitrobot chamber and make sure the paper is aligned with the back-side of the grid. (**c**) Press the blotting paper flush against the back of the grid and watch the water wick away. Counting 2 s is a good place to start with neurons (or other adherent cells)

3. In the options section of the screen, select the following: "Use foot pedal", "Humidifier off during process", and "Skip grid transfers."

4. Cool the copper insert, and Styrofoam cup with liquid nitrogen. Once cool, fill the inner chamber to the top with liquid ethane by putting a 1000 μL pipette tip in the ethane hose, pressing the tip into the corner of the ethane cup, and slowly filling with ethane until there is a convex liquid surface at the rim of the cup. Quickly remove the pipette tip and allow time for the ethane to cool properly (*see* **Note 9**).

5. Label cryo grid boxes and submerge in liquid nitrogen. On the aluminum platform in the Styrofoam cup, there are cutouts where the grid boxes will sit in place.

6. Transport culture dishes containing grids to the plunge freezer and place on a 37 °C heating block (or bench top incubator, if available).

7. Dilute BSA-coated gold fiducials 1:4 in culture media from dish and maintain at 37 °C on the heating block.

8. Carefully pick up grid from the bottom of dish with tweezers, keeping the cell-side up. Use clip on tweezers to lock them in the clamped position.

9. Set tweezers down carefully (cell-side up), and pipette 3 μL of fiducial gold/media mixture on to the grid from the edge, being careful not to touch and bend the grid with the pipette tip.

10. Attach the tweezers to the plunging apparatus with the cells facing away from the side that will be blotted. For example: if blotting from the right, cells should be facing the left side of the blotting chamber.

11. Using the foot pedal, cycle through the following steps: Raise sample into the blotting chamber, raise the ethane cup to the bottom of the blotting chamber, and lower the tweezers into the sample addition/hand blotting position (typically used for adding sample to the grid).

12. Insert long tweezers holding the folded blotting paper, from the side opposite the cells. Gently place the blotting paper flush against the back of the grid, being careful not to bend the grid.

13. Blot for ~2 s, making sure the media has blotted into the paper (*see* **Note 10** and Fig. 3b, c).

14. Using the foot pedal, plunge the grid into the liquid ethane.

15. Transfer grid from liquid ethane to the liquid nitrogen quickly, but smoothly. Place grid inside of a grid box for storage under liquid nitrogen until ready for imaging.

3.6 Imaging

This section describes the use of Thermo Fisher's "Tomography" software for the batch collection of tilt-series on a Titan Krios cryo-TEM equipped with a Gatan Imaging Filter and K2 direct electron detector. There are many other configurations of microscope hardware and software that can achieve similar results, and many of the same principles discussed here (especially in the detailed notes) will apply to any tomography data collection session. This protocol assumes the following: (a) The sample has been loaded into the Thermo Fisher "Autogrid" assembly and transferred under LN into the Krios, (b) the sample has been loaded on the stage, (c) the microscope is properly aligned (including the GIF), and (d) the "Tomography" software is properly calibrated for batch data collection. For a detailed video protocol of all these steps, please consult the curriculum online at Thermo Fisher Scientific's CryoEM University [8], which was a joint effort with Caltech professor Grant Jensen. *See* **Note 11**.

3.6.1 Set Imaging Modes

In order to collect a batch of multiple tilt-series using Thermo Fisher's "Tomography" software package, the user is required to define eight different optical modes in the "Preparation" tab. Once established, the setting can be saved and recalled in future sessions. See Table 1 for an example of these settings. The magnification for "Tracking," "Exposure," "Focus," and "Zero Loss" should be the same value and are typically the only modes that are adjusted between imaging sessions. These optical setting should be established over a broken hole in the carbon.

To set the exposure mode for imaging with the K2 in counted mode, find a region of representative ice and center the stage over a cellular feature similar to ones you will collect data from during your session (in this case, a neurite). Change to the exposure mode and use Gatan's Digital micrograph software to start the camera's

Table 1
Example of optical mode definitions in Thermo Fisher's "Tomography" software package

Imaging mode	Magnification	Pixel size (ang/pixel)	Binning	Exposure (s)	Linear/ counted	Slit width (eV)	Defocus (μm)	Spot size
Atlas	220	508	2×	1	Counted	No slit	−20	8
Overview/ positioning	720	155	2×	1	Counted	No slit	−20	8
Search/ template	4600	24	2×	1	Counted	40	−20	8
Eucentric height	7200	16	2×	1	Counted	40	−20	8
Tracking	26,000	4.3	2×	2	Counted	10	−20	8
Exposure	26,000	4.3	1×	Variable	Counted	10	−8	Variable
Focus	26,000	4.3	2×	2	Counted	10	0	8
Drift	26,000	4.3	2×	1	Counted	No slit	−10	8
Zero loss	26,000	4.3	2×	1	Linear	No slit	0	3

These settings can be saved and recalled in future sessions

live view. Adjust the beam intensity and spot size until you have the desired illumination area and a dose measurement between 8 and 10 electrons/pixel/s. Save the current beam setting by selecting "get" on the preparation tab. This is the optimal range for the K2 to detect single electron events. Now move back over the broken hole in the carbon and read the electrons/pixel/sec without the sample in the beam path. This number represents the electron dose your sample is actually experiencing. Use the following formula to calculate the exposure time with the desired beam setting:

$$\text{Exposure time (s)} = \left[\text{Dose}_{total} \left(e/\text{Å}^2 \right) * \left(\text{Pixel size (Å)} \right)^2 \right]$$
$$/ \left[\text{Dose}_{measured} \left(e/pix/s \right) * \left(\text{Total images} \right) \right]$$

where

Dose$_{total}$ is the desired total cumulative electron dose, typically 100–200 e/Å^2.

Pixel size is determined by the collection magnification.

Dose$_{measured}$ is the number of electrons measured with no sample.

Total images is the number of images collected during the tilt-series, which depends on the tilt range and increment. For example: there are 121 images in a tilt-series from −60 to 60 in one-degree increments.

3.6.2 Collect Atlas	Switch to the "Atlas" tab and collect a low-magnification atlas of the entire grid, starting from the center and spiraling out. This will take about 10 min and the atlas should be used for all long-range navigation of the grid. Right-click on any location on the atlas and select "Move Stage Here". The red cross-hair that indicates current stage location will move to the new location on the atlas. Navigate to and store the coordinates for a large hole in the carbon support. Setting the dose will require imaging through a vacuum with nothing in the beam path.
3.6.3 Set Eucentric Height	Move to a grid square near the center of the grid, or near the center of the grid-region where data will be collected. Then, in the "Search/Template" mode, locate and center a visible feature. Turn on the stage wobbler and adjust the stage Z-axis on the control panels until lateral translations in the object are minimized as the stage wobbles.
3.6.4 Grid Square Exploration	Using the atlas, identify grid squares to further examine if they contain neurons (*see* Fig. 4a, b). In the "Preparation" tab, select the "Overview/Positioning" imaging mode from the drop-down menu. Load the saved settings by clicking the "Set" button. Navigate to the center of each grid square of interest and collect an image. This setting should produce an image of the entire grid square, giving a broad overview of what the cells on the grid surface look like. Save the stage coordinates for easy navigation back to these locations at the next higher magnification.
3.6.5 Search for Target Areas	In the "Preparation" tab, select the "Search/Template" imaging mode from the drop-down menu and click "Set". This magnification should be set to display the largest area of the grid possible, where features of the cell are still identifiable as target areas for data collection. For a neuronal branch, ~10 μm of neurite length, mitochondria, and other subcellular organelles can be used as target guides (*see* Fig. 4c). Insert the objective aperture, if needed. Use the stage coordinates saved during grid square exploration to revisit specific grid squares of interest. Examine them in live view using the K2. Move around the grid square using minimal exposure time and dose to locate cellular regions of interest. Center these areas quickly and save the stage coordinates (*see* **Note 12**). Repeat until each target has been identified and locations saved.
3.6.6 Select Specific Targets for Data Collection	In the "Preparation" tab, select the "Exposure" imaging mode from the drop-down menu and click "Set". This imaging mode should be set to the data collection parameters. Recall the first stage position saved as a target coordinate and switch to live view. Briefly center the target and stop the live view to prevent further electron dosing. If necessary, acquire an image for a final assessment of the target region with better signal-to-noise. Centering targets at the

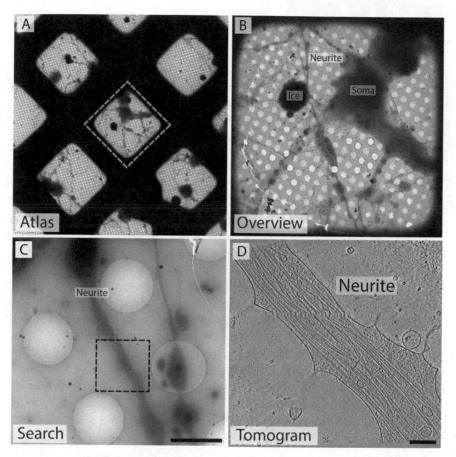

Fig. 4 Grid Exploration and Imaging. (**a**) The grid atlas should allow the user to assess which grid squares contain neurons and are worth inspecting closer. The orange dashed box represents the area in "B". (**b**) The "Overview/Positioning" mode should allow the user to determine whether the grid square contains intact vitrified neurons, as well as assess the feasibility of tomographic data collection. (**c**) The search mode should provide enough morphology to determine which areas are optimal targets for tilt-series collection. (**d**) 8-nm thick slice through a tomographic reconstruction of the boxed region in "C". Microtubules and other molecular assemblies are easily resolved

exposure setting is only used to optimize the exact positioning of the target. The time-step for the live mode view should be set to a minimum that allows for just enough signal to center the object as quickly as possible.

3.6.7 Saving Tilt-Series Positions

In the "Tomography" tab, select "Apply", then click "Acquire" in the "Prepare Position" section of the "Tomography" tab. The first time this tab is used, a prompt will appear to create a new session and determine where the files will be stored during data collection. Set the specimen type to "slab-like" and check the "Low Dose" box. To collect more than one tilt-series, check the "Batch" box. Wait for the image of the target to appear. There should be three colored circles over the image. The yellow circle represents the size of the beam in "Exposure" mode and the square at its center represents the detector, the green circle represents the size of the beam in "Tracking" mode and the blue circle represents the size of the beam in "Focus" mode. Select the green circle and move it along the tilt-axis (the only direction it will move) to a region that does not overlap the region of interest or another target area. Move the blue circle to overlap with the green one. Ensure that the ice is not too thick in this region. These are the regions where the tracking and autofocus images will be collected, so it is important that there is good contrast for the features to track easily. In the "batch" section of the interface, give the file a name and set the defocus (*see* **Note 13**). Press the "Add Position" button to add this location to the list of tilt-series. Move to the next saved target and repeat the process until all the desired targets have been added. Once you have all of the tilt-series positions added, go to the "View Positions" section of the "Tomography" tab. Press the "Refine All" button on the menu bar. This procedure will have the scope revisit each position, perform an auto eucentric height and focus, and collect reference images for several imaging modes that it will use during data collection to find and center the target again. It takes ~5 min per target position. Positions that cannot be refined can be deleted from the list by right clicking on them and selecting "Delete". Before deleting, it can be useful to right click and on the position in the list and "refine" again. Observe where the procedure fails and decide if there is something that can be done to fix the issue.

3.6.8 Start Data Collection

In the "Tilt-Series" section of the "Tomography" tab, press the "Parameters" button and fill out the fields. Table 2 (left side) contains an example of values appropriate for data collection of neuronal branches in culture. Once the parameters are specified, press the "Corrections" button and fill out the fields. Table 2 (right side) contains the values appropriate for data collection of neuronal branches in culture. Press the "Acquire" button. The dataset will collect automatically, but it is recommended to observe the first tilt-series begin with no errors.

3.7 Preprocessing and Reconstruction

This section covers the basics of how to reconstruct tomograms using the "Etomo" program within the IMOD image processing suite [9] (https://bio3d.colorado.edu/imod/). While there are multiple packages for doing tomographic reconstruction, IMOD has long been many labs' program of choice due to the fact that: (a) It is well maintained and easy to install/update, (b) the documentation is extensive and the developers are fast to respond, and (c) it is capable of reconstructing tomograms individually and in batches. In addition to this outline, it is recommended that the official IMOD guide be consulted when necessary at http://bio3d. colorado.edu/imod/doc/tomoguide.html. Unless otherwise specified, the default values are used.

Table 2
Example values used for "Parameter" and "Corrections" sections of the tilt-series collection page

Parameters		Corrections	
Start angle (degrees)	0	Check focus before acquisition	Yes
Max. negative angle	−65	Auto focus periodicities	
Max. positive angle	65	High tilt	5
Stage relaxation time (s)	1	Low tilt	5
Tilt scheme	Linear	Switch angle	80
Low tilt step	1	Track after acquisition	Yes
High tilt switch	80	Track before acquisition	No
High tilt step	1	Tracking before periodicities:	
Adjust exposure time	No	High tilt	100
		Low tilt	1
		Switch angle	80
		Holder prediction	XY and focus
		Auto zero loss	Check adjust
		Periodicity	200

1. Create a new directory for each tomogram and navigate to the directory in the command terminal. IMOD requires this because of the extensive number of files generated during reconstruction.

2. If data is collected on a 4 k × 4 k detector (or larger), the reconstruction process is faster with a two-fold bin. This is optional, but effective, unless the data is specifically intended for high-resolution averaging. Bin the data two-fold in the x and y axis using the following command:

```
binvol -x 2 -y 2 -z 1 <input_file_name>
<output_file_name>
```

3. Inspect the tilt-series by opening it with the "3dmod" program within IMOD to ensure that the target was successfully tracked and that there are no high-angle images that need to be removed from the reconstruction. This is typically due to a grid bar entering view at high-angle or some large surface contamination. If the tilt-series appears to be of very low contrast, proceed with the protocol to see if the issue is resolved during the "Pre-Processing" step. Use the following command to open the tilt-series stack:

```
3dmod <filename>
```

4. If images on either high-tilt end of the series need to be removed use the "trimvol" command to generate a trimmed stack:

```
trimvol - z <# of starting image>,<# of end image>
<input file name> <output file name>
```

For example: if you need to remove images 1–10 of a 120-image stack, use the command:

```
trimvol -z 11,120 <input file name> <output file name>
```

5. Begin the reconstruction with the command "etomo", then select the "Build Tomogram" option in the pop-up window.

6. A new dialogue box will open. Select the tilt-series file in the "dataset name" field.

7. Set the "System template" to "cryoSample.adoc".

8. Select "Scan Header" to have the program extract metadata from the file header.

9. Fill in the fiducial diameter in nanometers (*see* **Note 14**).

10. Select "Create Com Scripts". A new dialogue box will open showing the reconstruction workflow.

3.7.1 Pre-processing

The preprocessing step finds individual pixels that are very far from the mean value and replaces them with the local average. In nearly all cases, the default values work perfectly, but they can be changed if too many or too few pixels are being targeted for removal. Select "Create Fixed Stack". Once done, select "View Fixed Stack". If the fixed stack is adequate, close the 3dmod window and select "Use Fixed Stack". Select "Done".

3.7.2 Coarse Alignment

Select "Calculate Cross-Correlation". Then, select "Generate Coarse Aligned Stack". View the aligned stack in 3dmod to ensure the alignment is satisfactory. Close the 3dmod window and select "Done".

3.7.3 Fiducial Model Generation

Under "Seed Model" tab, select "Make Seed and Track" and "Generate Seed Model Automatically". Select between 10 and 20 fiducial points to track. If there are less than ten fiducials within the field of view, use the number of fiducials visible (*see* **Note 15**). Click "Generate Seed Model". Open the seed model and ensure that fiducials were selected correctly and evenly across the field of view. Under the "Track Beads" tab, select "Track Seed Model". The project log window will display how many missing points are left after the automatic tracking.

If there are missing points: Select "Track with Fiducial Model as Seed". This will use the existing fiducial model as a seed and often allows more of the beads to be tracked. Repeat this tracking with the fiducial model as the seed until the number of missing points no longer decreases. If there are still missing points, select "Fix Fiducial Model".

On the bead fixer window, select "Autocenter" and input the diameter of your fiducial. Select "Go to Next Gap". Use the page up/page down keys to scroll between images and follow one bead through the tilt-series. Middle-click on the center of the bead to add model points to the images where fiducial tracking failed. Press "S" on the keyboard to save the modified model file (*see* **Note 16**). Select "Track with Fiducial Model as Seed" again and select "Done".

3.7.4 Fine Alignment

Select "Compute Alignment" and check the residual error mean and standard deviation in the log window. A mean near 0.5 is ideal, but not routine for thicker samples; typically, the mean fluctuates between 0.5 and 1.5. For details on improving the residual error, consult the IMOD users guide at the URL listed above. Select "Done".

3.7.5 Tomogram Positioning

For neuronal branches, use a sample tomogram thickness of ~800. Select "Create Whole Tomogram" with a binning of three and then select "Create Boundary Model." Rotate the tomogram 90° under the menu "Edit → Image → Flip/Rotate". Define the upper and lower boundaries of the cell by drawing a horizontal line above and below the region of interest near one of its edges. Lines are drawn by creating two model points with the middle mouse button, and they should be automatically connected. Do the same at the center and far end of the object. Save the model, and select "Create Final Alignment".

3.7.6 Final Aligned Stack

Determine which binning/pixel size allows for the best contrast without losing the desired detail. To increase contrast and make future segmentation calculations faster, bin the tilt-series another two-fold to create a final aligned stack that is 1 k × 1 k. To set the binning to two-fold, change the "Aligned Image Stack Binning" parameter to 2. Select "Create Full Aligned Stack" and then view the stack to make sure the alignment is satisfactory. Select "Done".

3.7.7 Tomogram Generation

If using parallel processing, ensure that "Parallel Processing" is checked, and then select the # of CPUs to use. If using a CUDA-enabled GPU, select "GPU Processing". If a boundary model was generated in the "Tomogram Positioning" section, the "Tomogram Thickness in Z" parameter will already be set. If a boundary model was not generated, set it to a relatively large number of pixels compared to how thick the sample is. For neuronal branches that are only a few 100 nm thick, setting the parameter to 800 pixels should be enough. Select "Generate Tomogram" and view once finished. If the whole volume of your target is not contained within the tomogram, reset the tomogram thickness to a higher value and repeat until the entire object is captured in the reconstruction. Select "Done".

3.7.8 Post-processing

If there is too much empty space after reconstruction, use the "Volume Trimming" section to set the X, Y, and Z ranges for trimming the volume. If the tomograms will be used for automated segmentation, uncheck the "convert to bytes" box. EMAN2 (used in the "Segmentation" section below) typically displays the non-converted images better. Check the "Rotate Around X Axis" box in the "Reorientation" section. The rotated volume will open displaying the x-y orientation automatically within 3dmod and EMAN2. This reorientation will not overwrite the original file (which will be denoted by a "_full" suffix), but it does double the amount of data produced in the reconstruction process. Select "Trim Volume" and view the volume once finished (*see* Figs. 4d and 6a). Select "Done". Choose the "Clean Up" tab. Delete any intermediate/excess files (*see* **Note 17**). Select "Done".

3.8 Automated Segmentation with EMAN2

This section outlines a basic workflow for using EMAN2.31's convolutional neural networks for automatically segmenting both membranes and microtubules from cryotomograms of neurons [7] although the segmentation of other cellular features would follow the same basic protocol (*see* **Note 18**). The training process is supervised and each cellular component will need a separate, specifically trained network for identifying/segmenting this component. That being said, using the default preprocessing settings (a bandpass filter) to remove high and low frequency noise from the image, worked well in these examples. Exploring different frequency ranges within your image may prove fruitful for increased fidelity (*see* **Note 18**). In this protocol, all options are left as default, unless otherwise noted.

1. Open the EMAN2 project manager using the command "e2projectmanager.py". The command terminal window will provide updates to the progress of each step described below.

2. Select the "Tomo" workflow mode.

3. Select "Raw Data", then "Import Tomograms". Browse and select the desired tomogram file. Select "Launch". In the "Segmentation" tab, select "Preprocess tomograms". Browse to the imported tomogram. It should have a .hdf file extension. Select "Launch".

4. Select the "Box training references" tab. Browse and find the imported, pre-processed tomogram, as indicated by the "_pre-proc.hdf" file extension. Select "Launch". This action will open the particle list, options, and the main windows.

5. In the main window, middle-clicking the tomogram will bring up a GUI where contrast, brightness, and magnification can be adjusted. Typically, the auto contrast and a magnification of 1.0 works well. On the right side of the tomogram display window is a slider for scrolling through tomogram slices. After boxing particles for training, their location within the tomogram will be displayed here.

6. Begin boxing "positive particles", which contain the feature (s) of interest (FOIs), by left-clicking FOIs in the "Main Window". Use the slider to scroll through the tomogram and pick particles from different Z-heights. The recommended minimum is 10 but more is better to a certain extent. We typically choose between 20 and 50 particles. The most important principle is to pick particles that include the variety of shapes, sizes, angles, and cross-sections that fully represent the FOIs. This is critical for segmenting the features fully in all dimensions because the software segments the FOIs within individual two-dimensional slices of the tomographic data.

7. In the "Options" window, select the set "00" and rename it "ptcls_good". Be sure the check box next to the "ptcls_good" set is ticked, and select "save" to output the particle stack file. The filename suffix that displays should be "ptcls_good".

8. Close "Main Window" and select the "Segment training references" tab. Browse for the "good" particles file and click "Launch". The purpose of this step is to use the pen tool to "paint" over the features in each boxed particle. This step is critical because the hand-segmented data is used during neural network training to define which features it will consider for segmentation. We find it is easier if you zoom in and typically place the magnification at ~2.0 or higher for this step. The pen size should be chosen based on the features being segmented and on the fineness of the segmentation needed. Smaller pen sizes give you finer control but take longer to paint large areas of the tomogram (*see* **Note 19**). Paint by clicking the left mouse button and dragging the cursor across the feature. Use the arrows on the keyboard to move between particles. The program autosaves for this step, so simply exit when done.

9. Again select "Box training references", and click "Launch". In the "Options" window, create a new set and name it "ptcls_bad". Now that more than one particle set has been created, newly boxed particles will be added to whichever set is highlighted in the "options" window. Highlight the

"ptcls_bad" set, and then using the left mouse button, box a set of "negative particles", which should consist of features in the tomogram that represent things that are not your FOIs. The recommended minimum is 100 but more is better, if possible. Pick as many different aspects of the tomogram as possible (other cellular components, carbon hole boundaries, gold fiducials, cytoplasm, extracellular space, etc.) that are not the FOIs. In the main window, the boxes that belong to this particle set should now be represented as a different color from the "ptcls_good" set. Be sure the check box next to the "ptcls_bad" set is the only one ticked and select "save" to output the particle stack file. The filename suffix that displays should be "ptcls_bad".

10. Close the "Main Window" and select the "Build training set" tab. This step will combine the positive and negative particle files as well as the hand-painted segmentation/mask for each positive particle into one larger file. Under "particles_raw", browse for the good/positive particles file. For "particles_label", find the segmented/painted particles file. In "boxes_negative" locate the bad/negative particles file. Select "Launch".

11. Select "Train the neural network". Under "trainset" find the trainset file (*see* **Note 20**). Name the output file in the "nettag" box. Select "Launch". The resulting training can be viewed by opening the e2display browser window using this command: e2display.py. From there, find the trainout_file and open it using the "Show Stack" option. An example result is shown in Fig. 5, where the left-most column shows the region of interest in the tomogram. The middle column shows the hand-drawn segmentation while the right-hand column is the AI training result. The training is considered successful if the middle and right-hand columns closely match one another.

12. In this step, the trained neural network will be used to perform the final feature segmentation of the tomogram. Select the "Apply the neural network" tab. Select the imported, pre-processed tomogram in the "tomograms" box. Find the trained neural network from the previous step under "nnet". Name the output file and click "Launch". The output file will show up in the "segmentations" subdirectory within your project's directory. The resulting volumes will be the same dimensions as the input tomogram, and bright pixels represent regions that have been segmented by the neural network (*see* Fig. 6).

3.9 Visualization of Segmentation Results

Visualization and analysis of the resulting segmentation volumes can be performed in your favorite volume rendering software, but here we describe a relatively simple workflow for visualization using the commercially available software Amira. UCSF Chimera is a

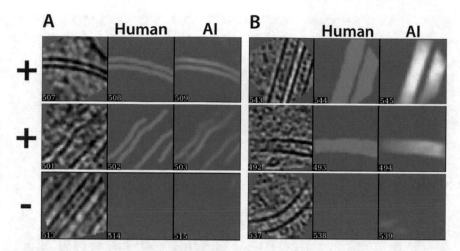

Fig. 5 Training the neural network. A properly trained neural network for segmenting membranes (**a**) and microtubules (**b**). Column one shows two representative input images containing true (+) and false (−) objects. (**a**) The true input contains membranes (each individual line is a single bilayer), and the false input contains microtubules. (**b**) The reverse is true. It is clear that the AI is distinguishing between membranes and the similar linear densities observed in microtubules

freely available visualization package that can display segmented tomograms as well. In this workflow, Amira is used also to manually classify different membrane types within the resultant segmentation (*see* Fig. 6d).

1. To start, convert both the membrane and microtubule segmentation output volumes (.hdf) to .mrc files using the e2proc3d conversion command: e2proc3d.py <input_file. hdf> <output_file.mrc>.

2. Open the converted microtubule segmentation volume (.mrc) in Amira, and browse to the "Segmentation Editor" tab. Under the Selection window, activate the "Threshold" tool, and adjust the mask until voxels representing the microtubules are highlighted sufficiently.

3. In the "Materials" window, right-click and add a material called "Microtubules". Add these voxels to the new material by pressing "Select Masked Voxels".

   For membranes, complete the two previous steps in the same manner as above. In order to classify membranes for display/modeling purposes, the 3D membrane segmentation can be subdivided into individual membrane types (mitochondria, plasma membrane, etc.) using the following steps: After thresholding all membranes as above, add another material and label it as the compartment you are about to segment (e.g., plasma membrane). This time, under the "Selection" tab, click "Pick & Move". In the right-hand window, click on all plasma membrane segments. Again "Select Masked Voxels". From the

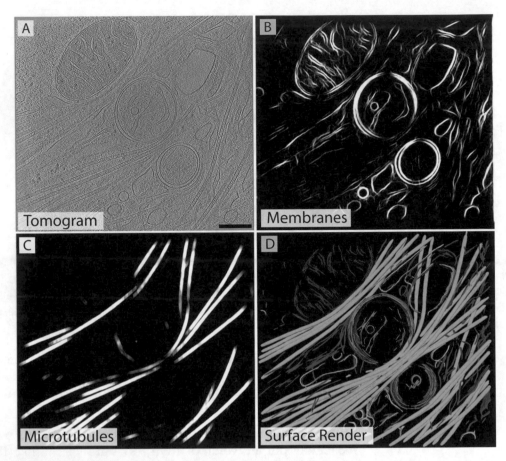

Fig. 6 Segmentation and visualization of a whole tomogram. (**a**) 20-nm thick tomographic slice through neuronal cytoplasm. Microtubules, a mitochondrion, and multiple membrane-bound organelles and vesicles can be seen. Scale bar represents 100 nm. (**b**) 20-nm thick slice through AI segmented membranes. (**c**) 20-nm thick slice through AI segmented microtubules. (**d**) Three-dimensional surface rendering of segmented volumes using Amira

"Materials" tab, select "Add". Repeat this for each membrane compartment (endoplasmic reticulum, vesicles, mitochondria, etc.).

4. To visualize each segmentation, open the "Project View" window. Generate a surface by right clicking the "labels" data, selecting "Generate Surface", and clicking "create". Under the "Properties" tab, adjust the smoothing and select "Apply".

4   Notes

1. Using holey carbon grids serves two purposes: (a) imaging through the holes (when possible) provides images free from carbon background and (b) the holes allow more water from

around the cells to wick away while blotting from the back side of the grid. A 200 mesh grid is the best compromise for tomography in terms of the tradeoff between carbon stability and imageable area.

2. Any established neuronal culture protocol can be used, with the major difference being that EM grids are added to the bottom of the dish. NbActiv4 has worked well in this workflow, but a variety of growth medias have been used successfully in other studies [10–12].

3. Whatever dish you use, it is important to test how easily you can lift the grid from the bottom before using them. 35 mm dishes are a good comprise in terms of minimizing the number of cells needed to plate at the appropriate density, and having enough room to easily get the forceps underneath the grid when removing for plunge freezing. Six-well plates can be used, but the approach with the forceps is from a higher angle, making it more difficult to get under the grid. Glass bottom dishes are recommended because the grid can be gently pushed up against the lip at the glass/plastic interface, making it easier to work the forceps under the grid. It is worth noting that glass bottom dishes may occasionally leak at the seam of the plastic and glass so it is worth checking for this during the poly-lysine coating step, before putting the grids in the dish. To do this, touch the bottom of the plate to a paper towel multiple times and see if there is any persistent liquid leaking.

4. Placing a sterile circle of filter paper below the slide is helpful for retrieving grids with forceps, in the case that some of the grids fall off the slide into the bottom of petri dish.

5. Glow discharging grids make them hydrophilic and aid in placing them in aqueous media. To ensure that the glow discharging protocol is working appropriately, pipette ~5 μL of water onto a freshly glow discharged grid. On a properly discharged grid, the water will spread evenly across the entire surface. If it is still hydrophobic, the water will ball up and avoid spreading.

6. Make sure that the grids are well separated so the entire surface is treated. If they are overlapping, use the fine tips of the forceps to gently push against the side of the grids until they separate.

7. To maximize cell count, triturate until as much of the visible particles as possible have been broken up after treating the tissue with the dissociation media. This may take longer or shorter than the 1 min suggested in the kit protocol. Additionally, the BrainBits protocol suggests spinning cells for 1 min, but extending the final spin to 2 min leads to a better cell recovery.

8. Manual or "hand" blotting from behind avoids applying pressure directly to the cells by the paper and any mechanical stress it would induce. Cells like to hold water so it is easier to under-blot than it is to over-blot. Even if parts of the grid are dry, areas containing cells or cell processes will tend to hold enough media to encapsulate them. The holes in the carbon support help to allow water trapped between cells to be removed.

9. To ensure vitrification, the ethane needs to be below $-160\ ^{\circ}C$. Since the Vitrobot does not have a thermal sensor in the ethane, seeing frozen ethane forming around the inside of the brass cup is a good visual indicator. With the outer portion of the cryo cup filled with liquid nitrogen, and the Styrofoam floating ring in place, it should take 5–10 min to see frost building up. It is useful to keep a small copper rod around for melting ethane if the whole cup freezes.

10. The grid will disassociate from the paper once most of the media has been wicked away and there is no more surface tension holding the grid and paper together. With enough practice this can be observed during blotting and used as a cue to progress with plunge freezing.

11. Make sure that any grids you load for tomography are clearly flat as seen by eye. If there are large bends or creases in the grid, it will be difficult to maintain eucentric height while moving around the grid.

12. To use the K2 in live mode, the camera will need to be controlled by the Gatan Digital Micrograph Software (GMS). Currently, there is no way to use the live mode directly through the Tomography software. It is important not to dwell on one spot for long periods of time during live viewing to avoid burning through the sample. If ice is bubbling very quickly, expand the beam or increase the spot size to minimize electron dose.

13. The defocus value is entirely up to the user, but 6–8 µm under focus is a good range. For more detail, collect closer to -6 µm and for more contrast collect closer to -8 µm. For thicker objects (>600 nm), where contrast may already be poor, it is better to shoot near -10 µm to increase contrast. As a point of reference, -8 µm is required to resolve the leaflets of a membrane bilayer.

14. 10 nm gold fiducials are acceptable for the magnification range where most tomography is done. 5 and 20 nm fiducials can be useful when collecting data at higher or lower magnifications, respectively. As long as the fiducials can be auto-tracked by the software, it is better to stay on the smaller side, simply to minimize the artifacts fiducials add to the reconstruction. While it is not covered in this protocol, Etomo can remove the gold fiducials from the tilt-series, post-alignment.

15. As long as they are tracking well, select a high number of fiducials to track in the tomogram (upwards of 20). If there are fewer than 10 points available, the tracking will still be adequate if they are distributed evenly across the field of view or around the object of interest. If there are fewer than 3–5 well-distributed fiducials, using the "patch tracking" method to generate a fiducial model may be the best option. In this situation, there must be good contrast for the algorithm to align to.

16. Not every fiducial must be tracked completely through the tilt-series. If some do not track completely or go outside of the field of view, it is ok, as long as there are multiple fiducials that do track across the entire tilt-range or multiple that overlap in their tracking across the whole tilt-series.

17. Until you are sure you are done processing a tomogram completely, do not delete intermediate files. It may be that you will want to come back and reconstruct at 1 × binning instead of 2 × binning, remove gold, or perform CTF correction. To do this, navigate to the directory the reconstruction files are in using the command terminal and run the command: etomo <filename.edf>. This will launch your reconstruction workflow just as you left off, allowing alterations to be made easily.

18. This segmentation feature of EMAN2 is likely to upgrade quickly, so it is recommended that these instructions be checked against the newest documentation on the EMAN2 website if they are not working (https://blake.bcm.edu/emanwiki/EMAN2/Programs/tomoseg). If your AI segmentations are very noisy, it may be worth exploring different frequency bands for training. Also, unlike the workflow in the original publication of this approach [7], our workflow treats all membranes as the same, and uses Amira to separate them into different materials later. We find that the fidelity of the segmentation is higher if the neural networks only have to distinguish between membranes and non-membranes, versus distinguishing between the membranes of different organelles.

19. In this workflow, pen size 2 was used for membranes and pen size 3–4 for microtubules and their lumen, but this number will vary depending on the pixel size in the tomogram.

20. The option "from_trained" will be used if you are refining a training set against a previously trained network. While not included in this protocol, it can improve the accuracy of segmentations in some cases.

Acknowledgments

The authors would like to acknowledge the Penn State Cryo-Electron Microscopy Core Facility and funding provided by TSF CURE award 4100079742.

References

1. Gan L, Jensen GJ (2012) Electron tomography of cells. Q Rev Biophys 45:27–56
2. Beck M, Baumeister W (2016) Cryo-Electron tomography: can it reveal the molecular sociology of cells in atomic detail? Trends Cell Biol 26:825–837
3. Ladinsky MS (2010) Micromanipulator-assisted vitreous cryosectioning and sample preparation by high-pressure freezing. Methods Enzymol 481:165–194
4. Rigort A et al (2012) Focused ion beam micromachining of eukaryotic cells for cryoelectron tomography. Proc Natl Acad Sci 109:4449–4454
5. Rigort A et al (2012) Automated segmentation of electron tomograms for a quantitative description of actin filament networks. J Struct Biol 177:135–144
6. Lucić V, Fernández-Busnadiego R, Laugks U, Baumeister W (2016) Hierarchical detection and analysis of macromolecular complexes in cryo-electron tomograms using Pyto software. J Struct Biol 196:503–514
7. Chen M et al (2017) Convolutional neural networks for automated annotation of cellular cryo-electron tomograms. Nat Methods 14:983–985
8. Thermo Fisher Scientific's CryoEM University. https://www.thermofisher.com/blog/micros copy/cryo-em-university-an-online-electron-microscopy-curriculum/
9. Kremer JR, Mastronarde DN, McIntosh JR (1996) Computer visualization of three-dimensional image data using IMOD. J Struct Biol 116:71–76
10. Schrod N et al (2018) Pleomorphic linkers as ubiquitous structural organizers of vesicles in axons. PLoS One 13:e0197886
11. Tao C-L et al (2018) Differentiation and characterization of excitatory and inhibitory synapses by Cryo-electron tomography and correlative microscopy. J Neurosci 38:1493–1510
12. Fischer TD, Dash PK, Liu J, Waxham MN (2018) Morphology of mitochondria in spatially restricted axons revealed by cryo-electron tomography. PLoS Biol 16:e2006169

Practical Approaches for Cryo-FIB Milling and Applications for Cellular Cryo-Electron Tomography

Vinson Lam and Elizabeth Villa

Abstract

Cryo-electron tomography (cryo-ET) is a powerful technique to examine cellular structures as they exist in situ. However, direct imaging by TEM for cryo-ET is limited to specimens up to ~400 nm in thickness, narrowing its applicability to areas such as cellular projections or small bacteria and viruses. Cryo-focused ion beam (cryo-FIB) milling has emerged in recent years as a method to generate thin specimens from cellular samples in preparation for cryo-ET. In this technique, specimens are thinned with a beam of gallium ions to gradually ablate cellular material in order to leave a thin, electron-transparent section (a lamella) through the bulk material. The lamella can be used for high-resolution cryo-ET to visualize cells in 3D in a near-native state. This approach has proved to be robust and relatively simple for new users and exhibits minimal sectioning artifacts. In this chapter, we describe a general approach to cryo-FIB milling for users with prior cryo-EM experience, with extensive notes on operation and troubleshooting.

Key words Cryo-focused ion beam milling, Cryo-electron tomography, Sample preparation, Lamella, Mammalian cells, Yeast, Bacteria

1 Introduction

A major goal in biological imaging is to visualize macromolecular complexes and the intricate networks they form at high-resolution in situ with minimal artifacts. However, most commonly accessible techniques lack sufficient resolution (such as fluorescence light microscopy) or sufficient cellular context (such as single particle cryo-EM). Cryo-electron tomography is a high-resolution imaging modality that can visualize macromolecular structures in situ in a near-native state in their cellular context [1]. However, specimen thickness is a major limitation with cryo-ET and other electron microscopy techniques. Due to multiple electron scattering, specimen thickness is limited to less than 400 nm at 300 kV [2], which is well below the size of eukaryotic cells and most bacteria.

One approach to overcome this limitation involves sectioning cells under cryogenic conditions using a microtome [3]. However,

Tamir Gonen and Brent L. Nannenga (eds.), *CryoEM: Methods and Protocols*, Methods in Molecular Biology, vol. 2215,
https://doi.org/10.1007/978-1-0716-0966-8_3, © Springer Science+Business Media, LLC, part of Springer Nature 2021

this approach is technically demanding and introduces visual and structural artifacts that arise from the sectioning process [4]. Recent work has adapted the versatile focused ion beam (FIB) instrument for cryo-electron microscopy applications as a method to generate thin sections through cellular material, yielding unprecedented insight into eukaryotic and prokaryotic cell biology [5–24]. This technique offers significant advantages including relative ease of use, minimal artifacts, and the ability to target specific cells for high-resolution cryo-ET.

Focused ion beam (FIB) microscopy and milling has been used in materials science as a method to thin samples for analysis and to create micro/nanoscale patterns [25]. The operation uses a tightly focused (5–10 nm) beam of ions, typically gallium, that rasters over the sample [26], similar to the beam of electrons in conventional scanning electron microscopy (SEM). Secondary electrons are generated by the ion beam's interaction with the sample, which can be detected to form images. Furthermore, the relatively large mass of gallium ions can be used as a tool for micro/nano machining by selectively ablating material from a specimen. In this manner, the FIB can be used to precisely remove material above and below a region of interest to leave a thin section—a lamella—that remains supported by the unmilled material around it. FIB milling sample preparation is usually performed in a "Dual-Beam" instrument with both FIB and SEM columns for simultaneous milling and multi-perspective imaging. These instruments are also typically equipped with gas-injection systems that allow targeted metal deposition or modification of milling characteristics [27]. For cryo-FIB applications, these instruments are equipped with a stage cooled to liquid nitrogen temperatures to maintain samples at cryogenic temperatures and an airlock quick-loading system to introduce cryogenic specimens into the chamber under vacuum. This allows cold samples to be transferred into and out of the instrument while avoiding atmospheric ice contamination and to be milled at liquid nitrogen temperatures for several hours.

This chapter will briefly describe practical principles and procedures for cryo-FIB milling including considerations for upstream sample preparation, downstream TEM tomography, and evaluation of sample quality as part of a typical cryo-ET workflow (Fig. 1). A closely related technique using cryo-FIB lift-out can be used for much thicker samples such as tissue blocks [28–30], but will not be discussed in this chapter. At this time of writing, commercial dual-beam instruments can be outfitted with third party cryostages/quickloader systems such as the Quorum PP3006 cryo-stage, the Leica EM VCT500, and the Hummingbird Scientific cryotransfer system. In this chapter, we will describe a typical FIB milling protocol using the Thermo Fisher Scientific (TFS) Aquilos, a dual-beam FIB/SEM platform that includes a cryo-stage/quickloader system. However, many of the general principles should be applicable across instruments.

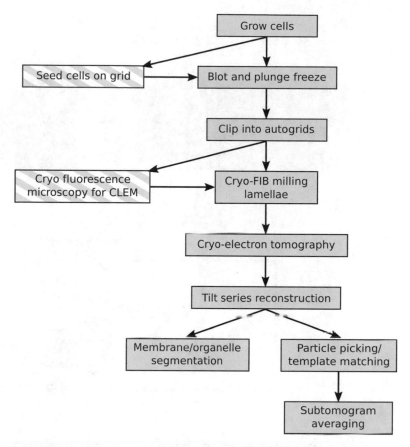

Fig. 1 Typical cryo-FIB-ET workflow. Cryo-FIB-ET is a vertically oriented work-flow in which the same sample is processed throughout. This technique is sensitive to the success rate of each step and is limited by low throughput during lamella milling. This chapter seeks to provide practical guidance on increasing success at cryo-FIB milling, thereby increasing throughput for cryo-ET. Typical workflow steps are highlighted with a solid background. Optional steps are highlighted with a striped background

1.1 Principles of Operation

Cryo-FIB milling is most easily performed using a dual-beam focused ion beam/scanning electron microscope (FIB/SEM) that allows simultaneous monitoring and processing of the sample. The SEM column is mounted vertically onto the chamber and the FIB column is mounted at a 52° angle relative to the SEM such that the beams intersect. This intersection point, the beam-coincidence point, is defined as the working distance of the microscope. In a well-aligned system, the beam-coincidence point should be located at the stage eucentric height (Fig. 2).

Samples for cryo-FIB milling are typically prepared on TEM grids by plunge-freezing (*see* Subheading 3.1). The grid is positioned at the beam-coincidence point and oriented such that its surface forms a low incident angle (typically 5°–10°) relative to the FIB column (Fig. 2c). Milling thus results in lamellae that are nearly

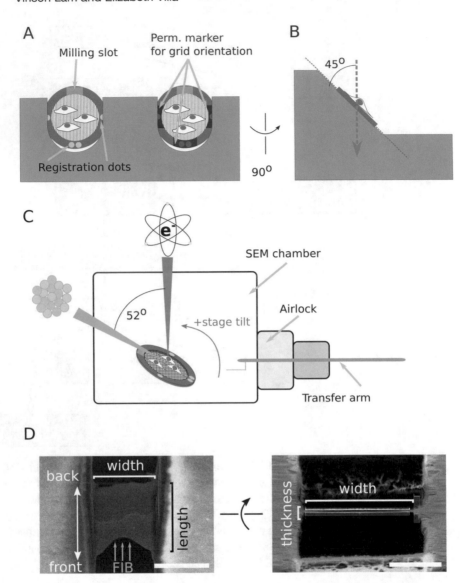

Fig. 2 Schematic of grid and FIB/SEM geometry. (**a**) Schematic of two grids clipped into a cryo-FIB-autogrid support and loaded into the shuttle, with cells on the top surface. The cryo-FIB-autogrid has a milling slot that enables lower angle milling and registration dots for aligning the grid when loading at the SEM or TEM. Additional markings with permanent marker can be made to enhance visibility under liquid nitrogen. Compared to Fig. 3e. (**b**) Side view of grid loaded into shuttle, illustrating the grids are held at a 45° angle relative to vertical. (**c**) Schematic of relative orientation of the SEM column, FIB column, sample stage, and transfer arm. The grid is positioned at the beam-coincident point, and the milling slot is oriented towards the FIB column. The stage is arranged to have positive tilt towards the FIB column. This results in FIB-sample incident angle 7° smaller than the reported stage angle. SEM beam is indicated in blue, FIB in green, and the specimen in copper. (**d**) Schematic of terminology used in this chapter to refer to lamella dimensions and orientation. Scale bars: 5 μm [3]

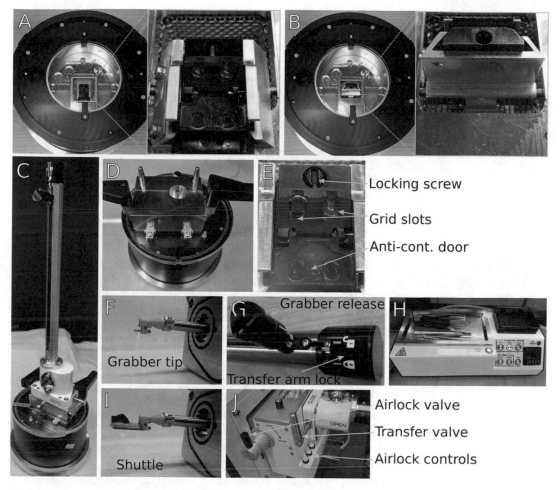

Fig. 3 Equipment for FIB/SEM sample preparation and loading. (**a**) Loading station with shuttle inserted and in the loading position. Inset shows detailed view of shuttle position. (**b**) Same as (**a**) but with the shuttle rotated vertically into the transfer position. Inset shows detailed view of shuttle in vertical transfer position. (**c**) Transfer arm attached to the transfer lid and positioned over the loading station. (**d**) Loading station with transfer lid, showing the transfer lid airlock chamber and the registration pins to place the transfer arm. (**e**) Detail of shuttle showing a loaded grid in the appropriate location and orientation, the locking screw to secure grids, and the anti-contamination door that covers the grids during transfer steps. Users should take care to keep the milling slot unmarked due to potential interactions with FIB imaging. (**f**) Detail of the transfer arm grabber tip, showing the registration pin and the grabber locking mechanism. (**i**) Detail of shuttle and grabber together. (**g**) Detail of transfer arm locking mechanism. The transfer arm lock on the right releases the rod to insert or retract. The smaller grabber release knob on the left opens and closes the grabber. (**h**) Preparation controller with tools: fine handling tweezers, grid box opening tool, screwdriver. The preparation controller is used to pump or vent the transfer lid and transfer arm assembly when moving specimens to and from the microscope. (**j**) transfer arm attached to airlock of FIB/SEM quickloader system. The transfer and airlock valves control access to the transfer arm chamber and the microscope chamber, respectively. Airlock pumping and venting actions are controlled by the green "P" and "V" buttons, respectively

parallel to the substrate and long enough to contain enough material of interest as detailed in Subheading 1.2. The FIB is used to mill away material by continually rastering over the sample in a user-specified pattern to gradually ablate material. When starting a lamella, the milling patterns are set ~2 μm apart, and are gradually brought closer together to reach the target thickness, <200 nm. Additionally, the ion-beam current is reduced as the lamella becomes thinner in order to afford more control over the milling process and to minimize beam damage. The final target lamella thickness should be informed by the biological question and any downstream analysis (e.g., subtomogram averaging or membrane segmentation) (*see* Subheading 3.3). Typical lamellae thicknesses may range from 85 to 250 nm, and typical TEM pixel sizes may range from 0.2 to 2 nm (*see* Subheading 3.3).

1.1.1 Platinum Sputtering and GIS Deposition

Vitrified biological samples are not conductive, and thus often result in beam-induced charging. This charging effect can be detrimental during milling and also during TEM imaging that leads to low quality images. A platinum sputtering system can be used to reduce charging by depositing a thin conductive layer on the sample. The sputtering system may be built into the chamber or immediately outside the chamber as part of a cryo-stage/quickloader installation, both enabling sputtering to be performed on cryogenic samples without excess contamination.

For cryo-FIB milling, an organo-platinum gas injection system (GIS) is used to coat the surface of the cold grid before lamella milling. The gas is released into the chamber close to the sample and immediately condenses on the cold grid. When exposed to the ion beam, the organic component is partially sublimated, leaving a metallic/organic film behind [31]. This residual platinum is essential for protecting the leading edge of the lamella during the milling process to prevent uncontrolled milling that will form "curtain" artifacts [26]. Additionally, the platinum layer acts as structural support that prevents lamella cracking during subsequent handling.

1.1.2 Stage and Shuttle

Up to two grids previously clipped into an autogrid support (*see* **Note 3**) can be loaded into the specimen shuttle (Figs. 2a, b and 3e) for insertion into the microscope. The grids are held in place by a metal spring flap that contacts the edge of the autogrid and maintains thermal contact. In order to minimize atmospheric ice contamination, the shuttle is transferred from the loading station to the microscope under vacuum using a sealed transfer arm (Fig. 3a), via a pumped airlock system (the quickloader, Fig. 3j). Additionally, the shuttle has a spring-loaded "door" that covers the grids whenever the shuttle is removed from the loading station or the microscope stage to minimize atmospheric ice contamination.

The shuttle holds the grids at a 45° angle relative to the vertically mounted SEM column (Figs. 2b and 3e). During operation, the stage may be tilted anywhere from 0° to 45° relative to the horizontal stage position. The FIB column and platinum GIS needle are mounted 52° relative to the SEM (Fig. 2c). The stage is capable of XYZ positioning as well as rotation about the Z-axis, and tilt about the X-axis (the direction of sample insertion). Consequently, there is one XYZ position that brings the milling target to the beam-coincidence point. The rotation angle is determined by the mounting position of the FIB column, and tilt is determined by the specific sample and lamella requirements.

1.1.3   Stage Temperature Control

The stage is actively cooled by nitrogen gas flowing through a heat exchanger (Fig. 4a). Nitrogen gas from the source is split into two lines, each passed through a flow regulator, then into a hollow copper coil (Fig. 4b) immersed in a liquid nitrogen dewar (Fig. 4c) to reach liquid nitrogen temperature. The cooled nitrogen gas is then passed to the microscope chamber through a vacuum-isolated tube to minimize thermal loss. Nitrogen gas exits the chamber through return lines through the vacuum tubing and is

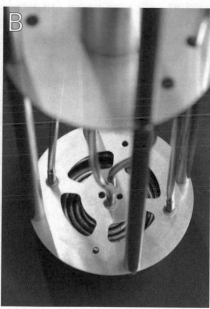

Fig. 4 Heat exchanger assembly. (**a**) Heat exchanger in a stand when not in use. During cryogenic operation, the heat exchanger assembly is immersed into a liquid nitrogen dewar. The heat exchanger assembly is kept under partial vacuum through the clear tubing attached near the top to maintain thermal isolation of cold gas nitrogen. (**b**) Details of heat exchanger coils and nitrogen gas lines. Warm nitrogen gas is passed from the blue gas lines into the copper coils and cooled to liquid nitrogen temperatures. Cold gas is routed through the body of the heat exchanger into the microscope stage. (**c**) Image of heat exchanger assembly inserted into a liquid nitrogen container during cryogenic operation. The thick clear tubing near the top of the heat exchanger unit is part of the vacuum isolation system

vented into the liquid nitrogen dewar to prevent frost formation that may potentially block gas flow and to prevent depositing liquid nitrogen in the lab environment.

Within the chamber, one of the lines of cooled nitrogen gas is passed through the lower part of the stage. The specimen is located on the upper part of the stage and is cooled by thermal conduction with the lower stage. The second line is passed through the cryo-shield, a separate anti-contamination fixture that has a large surface area to adsorb residual water molecules in the chamber. The stage and shield nitrogen flow rates are regulated separately from the input nitrogen line by the dedicated flow controller. Temperatures are monitored and recorded by computer software through thermocouples embedded in the stage and shield.

1.2 Considerations for Sample Milling Angle and Orientation

The target milling angle depends on the specimen. Typically, a large milling angle relative to the grid surface will result in shorter lamellae (i.e., measured from the front edge to the back edge) due to reduced cross-sectional area (Fig. 2d). High milling angles may be useful to create lamellae in thicker specimens or in specimens where it is difficult to identify a cell at low angles. Conversely, low milling angles result in longer lamellae, but may be more prone to obstruction from surrounding material or may make it difficult to identify a cell due to the nearly parallel viewing angle. Practically, it is best to mill lamellae at as low an angle (i.e., as parallel to the substrate as possible) for two reasons:

1. The lamella angle limits the tilt range of tomography in the TEM, contributing to resolution anisotropy [32].

2. Milling at low angles results in long lamellae, maximizing the usable area for tilt-series acquisition.

The lower limit to milling angle is set by sample geometry–at low angles, the edge of the autogrid or the grid bars will block the beam from reaching the grid square surface. This lower limit is about 11° stage tilt, corresponding to a beam incident angle of about 4° with respect to the grid surface. The upper limit to milling angle is set by the TEM stage. Due to the pre-tilt of the lamella, in one tilt direction on the TEM stage, the apparent sample thickness will be greater than compared to the other direction, corresponding to a greater relative tilt between the lamella and the beam. The apparent thickness of the sample increases proportionally to 1/cosine of the stage angle and at high TEM stage tilt angles, the sample will become too thick to image. If the initial lamella angle is too high, it will unnecessarily limit the range of TEM stage angles for tilt series acquisition. This upper limit for lamella milling is typically around 22° on the SEM stage, corresponding to an incident angle of about 15° (*see* **Note 1**). In general, "tall" cells such as yeast and many eukaryotic cells may be milled at a higher angle than unicellular bacteria, which require low milling angles to generate enough usable area.

In order to mill at low angles, special autogrids are available with a cutaway notch that increases accessibility at low stage tilts. The milling notch on the autogrid should be oriented to face towards the FIB source–the clipped grid should be rotated so that the notch is at the top position in the shuttle. To help with orientation, we routinely mark the autogrids with colored permanent marker to create a highly visible reference point (Fig. 2a).

2 Materials

2.1 Equipment

1. Dual beam (FIB/SEM) instrument equipped with a temperature controlled stage capable of reaching −175 °C or lower and a vacuum transfer system. This chapter is focused on preparing specimens using the Thermo Fisher Scientific (TFS) Aquilos system with a built-in cryo stage and quickloader, but general principles should be applicable across instruments.

2. Sample loading station/equipment and preparation tools compatible with the intended dual-beam instrument (Fig. 3).

3. Grid boxes.

4. Grid box opening tools.

5. Fine manipulation tweezers (e.g., Dumont #5).

6. Hot plate or hair dryer for warming/drying tools.

7. Liquid nitrogen transport dewar to hold grid boxes.

8. 4-L liquid nitrogen storage containers.

9. Personal protective equipment (PPE) including cryo-gloves and goggles.

2.2 Consumables

1. Clean liquid nitrogen for sample preparation (NF grade; less than 5 ppm moisture).

2. Pressurized liquid/gas nitrogen (to be regulated to 80 psi with a capacity of 30 L/min, sufficient for 12 h use). This can be from either an in-house line or from a separate tank, e.g., a 230 L capacity, 230 PSI tank.

3. EM grids, quantifoil type or with other film substrate (*see* **Note 2**).

4. TFS regular autogrids or cryo-FIB-autogrids.

5. TFS c-clips, also called clip-rings or clamp rings.

3   Methods

On-grid cryo-FIB milling is suitable for many types of biological specimens including bacteria, yeasts, and mammalian cells. This section describes generalized methods suitable for these three common specimen types with considerations for different approaches required for each specimen. *These instructions assume that the reader is familiar with manipulating cryogenic samples to minimize atmospheric ice contamination and risk of sample devitrification.* **Always use appropriate PPE when working with cryogenic liquids,** and always pre-cool tools in liquid nitrogen before handling the specimens.

3.1   Sample Type Considerations

Cryo-FIB milling is adaptable to a wide variety of cell types (Fig. 5). Here, we offer some considerations on sample preparation and FIB milling for three broad categories of cells. This protocol is not applicable for bulk tissue samples due to major differences in vitrification procedures and FIB milling workflow [20, 28–30]. The categories are not strictly limited to only those cell types described, and we list some exceptions within the categories. In all cases, grid preparation is critical and users should optimize the following for their particular cell type:

1. Specimens should be thick enough to provide sufficient material to create a lamella. If the sample is too thin, the final generated lamella will be very short and contain almost no cellular features of interest. For these cases, milling wedges can provide a longer electron-transparent window, albeit of varying thickness, on which tomograms can be acquired [21, 33].

2. Specimens should be thin enough to ensure proper vitrification and minimize crystalline ice domains.

3. An individual milling target (whether a single cell or clump of cells) should be large enough to support a 5–10 μm wide lamella suitable for cryo-ET.

4. Appropriate cellular density for the cell type. Note that cell density also affects sample thickness and vitrification.

3.1.1   Mammalian and Flat Eukaryotic Cells

These cells are generally flat and extend several tens of microns in diameter (Fig. 5d, g). Typically, the nucleus appears as a small hill in the center part of the cell. Each lamella made in these samples will be a partial section of a single cell. Examples include fibroblasts and other cells that are cultured on substrates, and some amoebae.

Grid Preparation

Cells may be grown directly on the quantifoil grid by seeding cells and allowing time for them to adhere. The grids should be made of non-cytotoxic material such as gold. Additionally, these grids may be treated with extracellular matrix such as fibronectin or poly-L-

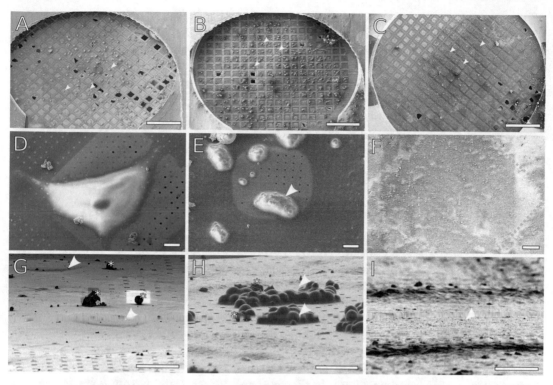

Fig. 5 Examples of ideal specimen concentrations. SEM images of (**a**) Adherent mammalian cells (NIH 3T3) grown on a grid. The cells density is low enough that there are only 1–2 cells per grid square. The surface is sufficiently blotted to reveal the carbon film, while enough liquid remains to cover the cells (some holes remain hydrated). (**b**) Yeast cells deposited on a grid. The cells form clumps consisting of several cells, with no more than 1–2 large clumps per grid square. As with (**a**), the surface is blotted to reveal the carbon film, while enough liquid remains to cover each clump. (**c**) Planktonic bacterial cells deposited on a grid. A monolayer of cells covers each grid square. Enough liquid has been removed to expose the grid bars, but the carbon film is not visible due to coverage by cells and media. Arrowheads indicate examples of cells/grid squares suitable for lamellae milling (**d–f**). Close up views of individual cells/grid squares for (**a–c**), respectively. In (**f**), the cells are obscured by the GIS platinum layer. The approximate location of the grid bars is outlined in blue (**g–i**). Representative FIB views of individual targets for lamella milling, all imaged at 13° stage tilt. Asterisks mark out atmospheric ice contamination. (**g**) NIH 3T3 cells. Nuclei are indicated by arrowheads. (**h**) Clumps of yeast, indicated by arrowheads. Notice that yeast clumps have a much taller profile compared to the 3T3 cells or the bacteria. (**i**) Monolayer of bacteria in a grid square. The darker horizontal ridges are the grid bars. Individual bacteria are not visible as they are embedded in an ice layer and overlaid with platinum from the GIS coating. Scale bars: **a–c** 500 μm; **d–i** 10 μm

lysine to facilitate cell adhesion. If sterility is important, grids may be sterilized under a germicidal UV lamp prior to coating and seeding with cells. The appropriate seeding concentration and recovery periods should be determined empirically.

Some mammalian cells may be deposited on grids immediately before plunging, similarly to protein solutions for single-particle cryo-EM. In this case, adherent cells may be dissociated from the culture surface (using trypsin/EDTA) with the aim to create single cells with minimal clumping. Cell clumps will lead to poor vitrification. Cells should be concentrated or diluted to a predetermined concentration prior to plunge-freezing.

Evaluating Grid Quality and Milling

In both seeded and deposited cells, the ideal specimen would have on average 1–2 cells per grid square (on a 200 mesh grid, ~6000 μm area) with minimal clumps (Fig. 5a). At the SEM, cells should appear well hydrated, and covered in a thin layer of ice (Fig. 5d, g). In some cases, the outline of the nucleus may be visible. Cells that are overblotted (i.e., too dry) may appear to be shrunken, starting to lift off from the substrate (for adherent cells), or have craters appearing on the surface.

Flat eukaryotic cells are typically milled at moderate to high stage tilt angles, from 15° up to 22°. FIB scanning patterns should be positioned to mill through the bulk of the cell in order to have sufficient support material to hold the lamella. Lamellae from cells may extend as far as 15 μm in length and may be thicker at the rear due to FIB beam spreading. The lamella thickness can be made uniform through a final milling step with additional stage tilt as described in Subheading 3.2.

3.1.2   Yeast and Tall Eukaryotic Cells

These cells are smaller than flat cells, typically encompassing less than 10 μm in diameter and have a roughly round shape (Fig. 5e, h). This is likely due to the presence of a rigid cell wall that holds the cell shape. Each lamella made in these samples will have sections of multiple cells appearing side by side. Examples include *Saccharomyces cerevisiae*, *Chlamydomonas*. In some cases, large bacteria clusters such as filamentous *Anabaena* [9] lend themselves to milling in this manner.

Grid Preparation

These cells are typically deposited on the grid before plunging, rather than cultured directly on the grid. Cells should be diluted or concentrated to an empirically determined concentration prior to plunge-freezing. In some cases (such as yeast), the cells are more tolerant of drying due to secreted extracellular polymers or the presence of a cell wall.

Evaluating Grid Quality and Milling

Contrary to the case with flat cells, the ideal yeast specimen *should* have clumps of 5–10 cells, with 1–2 clumps per grid square (on a 200 mesh grid, Fig. 5b). Due to the relatively small size of yeast, these clumps are necessary to provide enough material to make reasonably sized lamellae. At the SEM, cells should appear relatively dry, but with enough ice to cover the cells and to bridge the gap between neighboring cells (Fig. 5e, h). As a rule of thumb, the clumps should look like steep hills. If there is negative curvature of the clumps where they touch the grid surface, the cells are likely overblotted. If needed, it is possible to mill individual yeast cells, but with a narrow lamella about 3 μm wide instead of 10 μm [34].

Yeast may be milled at moderate to high stage tilt angles, from 15° up to 22°. Milling patterns should be positioned to cut through the center of the clump of cells. As in the case with flat cells, a final milling step with additional stage tilt is recommended to make uniform lamellae.

3.1.3   Bacteria

These cells are generally too small (less than 5 μm in diameter) to mill individually. During blotting, rod-shaped bacteria (e.g., *E. coli*, *B. subtilis*) will tend to lie down on the grid such that the long axis of the cell is parallel to the grid surface. Examples include most unicellular planktonic bacteria.

Grid Preparation

Bacteria are almost always deposited on the grid immediately prior to plunging. Cells should be diluted or concentrated to an empirically determined concentration prior to plunge-freezing with the aim to make a monolayer of cells embedded in media. The optimal concentration is highly dependent on the specific cell shape and the specific grid hole pattern used because the cells may slip through or become trapped by the holes. Bacteria are generally tolerant of drying because of the large number of closely packed cells and secreted extracellular polymers [35]. However, the relatively large total cell mass can lead to poor vitrification. This issue can be alleviated through the addition of cryoprotectants such as trehalose immediately prior to plunge-freezing.

Evaluating Grid Quality and Milling

The ideal bacterial specimen should have a uniformly flat monolayer of cells over each grid square (Fig. 5c), with the cells packed side by side. This arrangement will allow sufficiently long lamella with cells throughout. At the SEM, the ice should be just enough to cover the layer of bacteria, but without excessive hills or valleys (Fig. 5f, i). Generally, if each bacterium is well defined, the sample is overblotted.

Bacterial grids should be milled at low to moderate stage tilt angles, from 11° to 15°, in order to maximize the number of cells captured in each lamella. This will also result in a long GIS platinum leading edge. If needed, the GIS deposition time can be decreased slightly. A final milling step with additional stage tilt may be useful for some bacterial samples, but is not always necessary for samples milled at moderate stage tilts.

For FIB milling, the lamella should be targeted near the center of the grid square for maximum tilt range at the TEM. Unlike the case with mammalian cells and yeast, it is difficult to target one individual bacterium for milling. Instead, the milling strategy relies on having a near-complete grid square coverage to capture as many cells as possible in one single lamella.

3.1.4 Cryo-CLEM

Correlated light and electron microscopy (CLEM) of the same specimen is a powerful technique that allows targeting labeled cellular structures of interest for high-resolution EM [28, 36]. Cryo-CLEM combines the advantage of fluorescence microscopy to specifically locate organelles or proteins of interest and the advantage of TEM to visualize *the same object* at high resolution in its native cellular context. A full description of CLEM is outside the scope of this chapter. For FIB milling, correlation can provide great advantages in determining areas to mill, but also provides additional challenges for sample preparation including reduced throughput and increased atmospheric ice contamination due to additional imaging and transfer steps.

For typical mammalian and yeast samples, the cell density described above (about 1–2 cells or clumps per grid square) is typically compatible with cryo-fluorescent microscopy. For bacterial specimens, the cell density should be reduced to somewhat less than full coverage of the grid square in order to be able to resolve individual cells.

During FIB milling, the fluorescent data can be used to guide targeted milling of mammalian and yeast samples. For bacteria, milling is still done in non-targeted manner—while the fluorescent information is useful to screen individual grid squares for suitable cell coverage, ice thickness, and overall quality, the fluorescent data is more critical at the TEM to determine which specific cells should be targeted for tilt series acquisition.

There are commercially available cryo-fluorescence microscopes such as the Corrsight and the Leica Cryo-CLEM system. Additionally, aftermarket cryo-stage additions such as CMS-196 Linkam stage are available in addition to a number of custom-made cryo-light microscopes [37, 38]. Recent work has also demonstrated cryo super-resolution microscopy to be compatible with a typical cryo-CLEM workflow [28, 37, 39–42]. Key points to consider when acquiring cryo-fluorescence microscopes include stage temperature stability, anti-contamination features, and compatibility with existing EM instruments and workflows.

3.2 General FIB Milling Protocol

3.2.1 Before Starting

These steps should be done before the FIB session.

1. Culture and plunge-freeze specimens at the appropriate density (*see* Subheading 3.1) using your selected grid type (*see* **Note 2**). *Cells should be frozen on the carbon side of quantifoil grids.*

2. We recommend marking the autogrids with permanent marker before clipping to make sample more visible when under liquid nitrogen (Fig. 2a). Avoid marking the milling notch of the cryo-FIB-autogrid due to potential beam interactions with the marker residue.

3. While working at liquid nitrogen temperatures, clip specimen grids into a cryo-FIB autogrid support. The grid should be clipped so that the cell side (the carbon side) is facing towards the flat surface of the autogrid so that cells are visible during milling (*see* **Note 3**). Clipped grids may be kept for several weeks in a liquid nitrogen storage dewar before subsequent processing.

4. If fluorescent correlation is required for targeted FIB milling, it should be collected beforehand.

   When idle, our microscope is routinely kept in the following manner:

   - At room temperature, the microscope chamber is pumped down to about 2×10^{-6} mbar.

   - Heat exchanger flow controller set to 1 L/min for both stage and shield. This keeps the gas lines free of moisture.

   - Preferably, the sample transfer arm (Fig. 3c) is attached to the microscope quickloader and pumped. This minimizes dust and moisture contamination inside the transfer arm and the quickloader.

   - Typically, we keep the FIB source heating is turned off if the system is idle for more than a day, to conserve gallium. Otherwise, the source may be left on for the next user. However, practices may vary among facilities depending on instrument usage frequency.

3.2.2 Day of Milling

Microscope Startup
and Cooldown

1. Turn on the FIB and SEM beams, typically done by "waking" the microscope. Start the user interface and inspect the system status as described in the following **steps 2–4**.

2. Check that the chamber base pressure is about 2×10^{-6} mbar (*see* **Note 4**)

3. Check that the stage is empty.

4. Home the stage to reset stage coordinates and verify range of motion.

5. Check gas nitrogen source pressure and adjust to 80 psi. If the system is attached to a tank, check that the tank contains sufficient nitrogen for the session. Replace if needed.

6. Begin purging the heat exchanger gas lines by opening the flow controller to the maximum. Purge for at least 10 min.

7. Begin venting the transfer lid assembly (Fig. 3d). This step is necessary to remove residual moisture inside the lines.

8. Prepare for loading: turn on the hot plate on the preparation controller (Fig. 3h) and place required tools on it.

9. Fill a 4-L dewar with clean liquid nitrogen for sample preparation. Make sure the nitrogen hose attached to the source is dry before filling to minimize ice contamination. Be sure to use appropriate PPE when handling cryogenic liquids.

10. Fill the heat exchanger tank with liquid nitrogen.

11. Start the temperature logging software to monitor and record stage and shield temperature.

12. Slowly insert the heat exchanger into the heat exchanger tank and ensure that it is seated appropriately. Monitor the stage and shield temperature until it stabilizes around $-180\,^{\circ}\mathrm{C}$. Reduce the flow rate to 8.5 L/min to further reduce temperatures by about 3 °C. Allow the system to cool for an additional 30 min (*see* **Note 5**) When the stage is at cryogenic temperatures, the chamber vacuum should improve considerably compared to the room temperature base pressure. On our system, the chamber pressure is around 7×10^{-7} mbar when the stage is at cryogenic temperatures.

Sample Preparation and Loading

1. Retrieve grid boxes containing the frozen, clipped specimens, and have them ready in liquid nitrogen.

2. Insert the shuttle into the loading position in the loading station (Fig. 3a). Fill the loading station with liquid nitrogen and allow to cool until the liquid is no longer boiling. Refill as needed. Keep the loading station covered with the standard lid to minimize atmospheric contamination while cooling.

3. Transfer the grid box containing the grids to the loading station and unscrew the lid to access the grids.

4. Place one grid into each slot of the shuttle and rotate the grid in position so that the milling slot is at the top (Figs. 2a and 3e).

5. Secure the grids in the slots by turning the locking screw (Fig. 3e) at the top of the shuttle. Verify that the grids are secured by turning the shuttle to the vertical position and check that the grids remain in place.

6. Flip the shuttle into the vertical transfer position (Fig. 3b) and replace the standard lid with the transfer lid (Fig. 3d).

7. Press "V" to vent the quickloader airlock (Fig. 3j) and retrieve the sample transfer arm. Ensure that the transfer arm valve is closed.

8. Attach the transfer arm to the transfer lid on the loading station. If present, secure the transfer arm with the locking clamps (Fig. 3c). Otherwise, securely hold the base of the transfer arm. Check that the shuttle grabber is in the "open" position (Fig. 3g).

9. Pump the transfer lid airlock for 35–40 s, then open the transfer arm valve.

10. Vent the transfer lid airlock for 3 s, then close the venting line.

11. Open the sliding valve on the transfer lid to allow access to the shuttle.

12. While observing the shuttle, insert the transfer arm and close the grabber to hold the shuttle. Verify the hold by pulling the shuttle back slightly by a few millimeters. **The next two steps should be done quickly in succession to minimize contamination.**

13. Quickly and smoothly withdraw the shuttle completely to the back position, lock the transfer arm, and close the sliding valve on the transfer lid.

14. Pump the transfer lid airlock for 35–40 s to evacuate both the transfer lid airlock and the transfer arm, then close the transfer arm valve. Check that the transfer arm is in the "locked" position.

15. Vent the transfer lid airlock, then release and lift the transfer arm from the transfer lid.

16. Attach the transfer arm to the microscope airlock and pump the airlock by pressing the "P" button. This will pump the airlock and move the stage to the loading position.

17. Wait for the airlock vacuum to reach an acceptable level–the "OK" button will turn on and the quickloader valve will unlock (*see* **Note 6**). Listen for a click.

18. Open the microscope airlock gate valve, then open the transfer arm valve.

19. Unlock the transfer arm and insert it all the way to dock the shuttle into the stage. Release the shuttle grabber and retract the arm. Check that the shuttle is no longer attached to the arm. Lock the transfer arm.

20. Close the microscope gate valve to isolate the microscope chamber. Leave the transfer arm attached to the quickloader. The transfer arm valve may be left open.

21. Verify that the stage temperature is around $-180\ ^\circ\text{C}$ and that the chamber pressure is similar to the value before the transfer (within 2–3×10^{-7} mbar).

Sample Inspection and Milling Preparation

While milling, the user should periodically check the stage temperature and chamber pressure for appropriate values. If there is any large deviation, the user may need to respond rapidly to salvage the sample. The heat exchanger dewar will last about 8 h after initial cooldown, depending on gas flow rate. If the instrument needs to be operated for longer, the dewar can be refilled by slightly propping the heat exchanger aside and using a funnel to pour liquid nitrogen into the tank.

Table 1
Typical imaging conditions for SEM and FIB for frozen hydrated biological specimens

	Voltage (kV)	Current (pA)	Detector	Mode
	2–5	25	ETD	Secondary electrons
FIB	30	1.5[a]; 10–500[a]	ETD	Secondary electrons

[a]1.5 pA for live imaging. 10–500 pA for active milling

Table 2
Typical scan settings for SEM and FIB views for frozen hydrated specimens

	Live (FIB/SEM)	Snapshot (SEM)	Photo (SEM)
Dwell time	200 ns	200 ns	2 μs
Line integration	1	1	1
Resolution	1536 × 1024	3072 × 2048	6144 × 4096
Bit depth	8	8	8–16

"Live scan" and "snapshot" are used to monitor lamella milling. The "photo" preset is typically used to generate a low magnification grid overview due to the high electron dose. Snapshots and photos are not used in FIB view due to the potential for beam damage

See Tables 1 and 2 for FIB/SEM imaging and scanning conditions. The preset positions for mapping, sputtering, and GIS deposition should be determined beforehand (e.g., during initial system installation) and kept for all users.

1. Switch on the SEM and FIB beams. If the FIB was previously turned off, it may require 10–15 min to start up.

2. Set scan rotation to 180° for both SEM and FIB (*see* **Note 7**).

3. On the SEM view, change to the lowest magnification and find the grids to ensure that they are present. Check that the grids are seated appropriately in the shuttle.

4. Set the working distance: Start live SEM imaging and roughly focus the sample. Increase magnification to 5000× and refine focus and astigmatism. While actively imaging at focus, click "Link Stage to Z" to set the working distance in the microscope coordinate frame.

5. Move the stage to the "Mapping Position" for one of the grids—at 45° tilt at about 7.5 mm working distance. This will place the sample perpendicular with respect to the SEM column.

6. Acquire and save an SEM "Photo" preset image at a low enough magnification (80×) to *see* the majority of the grid. Evaluate the grid quality and pick out appropriate target cells or squares for lamella milling (*see* Subheading 3.1, Fig. 5, and **Note 8**).

7. If the samples are good, discard the nitrogen from the loading station and let the station dry. Otherwise, you may unload the shuttle and load new specimens.

8. Move the stage to the sputtering position and activate the platinum sputtering feature (*see* **Note 9**). When finished, click "Recover from Sputtering" to return to the mapping position.

9. Move the stage to the platinum deposition position (GIS position), insert the GIS needle, and deposit the organo-platinum compound for a predetermined amount of time (typically 5–10 s, *see* **Note 10**). When finished, ensure the GIS needle is retracted and return to the mapping position.

Lamella Milling

Lamella milling can broadly be divided into three phases: initial rough milling, thinning to about 400 nm thickness, and final milling to target thickness (Fig. 6).

In the rough milling phase, the goal is to develop a lamella that has been completely cleared of material throughout its length by the bottom and top milling patterns, respectively. We commonly refer to this as the lamella "breaking through" and is characterized by the appearance of empty space bordering the front and back of the lamella. This step is crucial because the lamellae may otherwise be obscured at high tilts in the TEM. Additionally, if the lamella has not broken through, the sample is likely too thick. A recent report has demonstrated that milling "expansion joints" adjacent to the lamella improves milling stability and quality [43]. In our hands, the use of expansion joints has increased overall lamella quality and is now a routine part of our cryo-FIB-ET workflow.

In the thinning phase, the goal is to gradually mill the lamella without introducing curtaining artifacts (*see* Subheading 3.4.3). Before reducing the milling pattern gap, each step should be allowed to continue until no biological material appears in the FIB milling pattern with increased software contrast. Additionally, the lamella surface should be monitored at regular intervals using the SEM to check for even texture.

In the final milling phase, the FIB current should be reduced to 30 pA or 10 pA. The milling patterns should be gradually moved closer to reach the target thickness. The lamella should be monitored more frequently by SEM to check that milling is proceeding evenly and that the GIS platinum layer is still sufficient to protect the leading edge of the lamella. The SEM can also be used to estimate lamella thickness by comparing the relative transparency of the lamella at 5 kV and 3 kV [13]. During this phase, the user

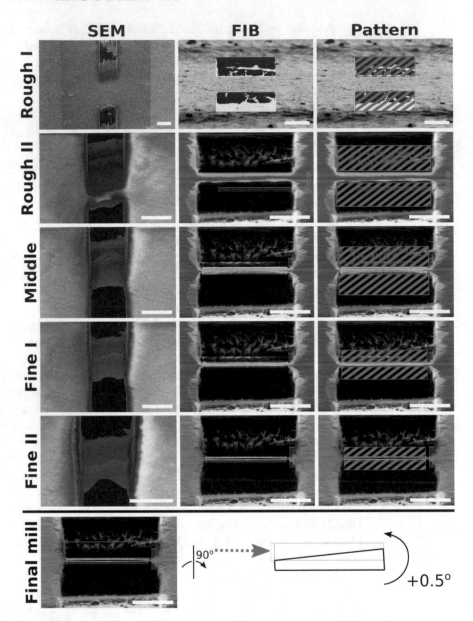

Fig. 6 Overview of typical FIB milling progression. SEM (left column) and FIB (middle column) views of one lamella milled from beginning to end. As material is removed, the lamella becomes more uniform and electron transparent seen from the SEM view. Additionally, the leading edge of the lamella becomes thinner seen from the FIB view. Third column: Schematic of milling patterns overlaid onto lamella FIB views (Table 4). Bottom row: Schematic of final milling step with a +0.5° stage tilt to make lamellae of uniform thickness using only the top milling pattern and an additional +0.5° stage tilt. Green arrow represents the direction of the ion-beam milling. For the final step, only the top pattern is activated to selectively mill the thicker portion in the back. The milled area is indicated in grey. Scale bars: SEM images 10 μm; FIB images 5 μm [4]

should monitor the FIB milling pattern for any image drift and compensate accordingly with the X/Y shift knobs. Continue milling until the lamella is judged to be thin enough (*see* **Note 11**). Sometimes, the lamella will not be able to survive continued milling

at which point milling should be stopped (e.g., lamella bending/
bulging, loss of GIS platinum layer, or holes in the lamella from
uneven milling).

1. Bring a selected milling target to the beam-coincidence point
   (*see* **Note 12**). Make sure sample is at desired tilt (*see* Subheadings 1.2 and 3.1) to determine appropriate milling angle).

2. Import the saved milling pattern into the FIB view (*see* Table 3
   for typical parameters). Adjust the milling position to create a
   lamella through the desired target. If needed, change the
   dimensions and gap between the milling patterns.

3. Change the FIB current to the desired beam current (*see*
   Table 4).

Table 3
Milling pattern parameters

Milling pattern basic properties	
Application	**Si**
X size	12–10 μm
Y size	[a]
Z size	~10 μm
Scan direction	[b]Outside to inside
Dwell time	1 μs
Beam	Ion

[a]The Y size for both patterns should be adjusted throughout milling to become smaller,
just enough to encompass the remaining unmilled material
[b]The scan direction for each pattern (i.e., top-to-bottom or bottom-to-top) should be set
so that the milling direction is towards the lamella. For very sensitive samples, the dwell
time may be reduced

Table 4
A typical scheme for performing step-wise milling starting from rough to final polish

	Rough I	Rough II	Middle	Fine I	Fine II	Final Polish
FIB current	0.5 nA	0.3 nA	0.1 nA	30 pA	10 pA	10 pA
Pattern separation	3 μm	1.5 μm	750 nm	400 nm	200 nm	n/a[a]
Pattern X size	12 μm	11.5 μm	11 μm	10.5 μm	10 μm	10 μm
Typical time	10 min	10 min	15 min	20 min	20 min	10 min

FIB current and pattern X size are gradually reduced as the lamella becomes thinner. Note that the pattern separation
does not correspond exactly to lamella thickness due to FIB beam spread. At low currents, milling time increases but there
is greater control over lamella quality
[a]For final polishing, the stage is tilted by an additional 0.5° and only one pattern is used to mill

4. After changing to a new FIB current, you may need to run the auto contrast routine for the FIB view. Additionally, take several brief FIB images to allow any image/beam drift to settle and to focus at the lamella position (*see* **Note 13**).

5. Start rastering the patterns. Monitor the milling progress regularly with the FIB and SEM. If desired, you may save SEM snapshots of intermediate milling steps to judge progress.

6. When milling at this current is finished, switch to a lower current, reduce the milling pattern separation, and adjust the pattern dimensions (*see* Table 4 and Fig. 6)

7. Continue milling while reducing pattern separation and FIB current until you reach a pattern separation of ~200 nm (*see* **Note 14** and Table 4)

8. On some specimens, it may be necessary to perform a final milling step with a stage tilt of $\pm 0.5°$ from the desired angle to compensate for the slightly diverging FIB beam (Fig. 6 "Final Mill"). In this case, only enable milling on the side corresponding to the tilt direction (i.e., only enable the top pattern if milling with a $+0.5°$ tilt. *See* **Note 15**) [13].

9. Continue on to other lamella locations as desired.

Sample Retrieval

About 1 h before retrieval, begin purging the loading station with dry nitrogen gas. If the loading station is at liquid nitrogen temperatures, increase the purge time to 2 h. The sequence of events for sample retrieval is essentially the reverse of sample loading, but it is reproduced here for the sake of completeness.

1. After completing all lamellae, perform a final cleaning step to remove amorphous ice buildup (*see* **Note 16**). When finished, close the SEM and FIB column valves.

2. Make sure the shuttle holder in the loading station is in the vertical position.

3. Stop purging the loading station and fill with liquid nitrogen. Allow to cool until the liquid is no longer boiling. Refill as needed. Place the transfer lid on the loading station.

4. Open the transfer arm valve, then press the "P" button to pump the airlock and move the stage to the loading position.

5. Wait for the airlock gate valve to unlock. Open the airlock gate valve.

6. Insert the transfer arm fully and close the grabber. Test the hold by pulling back slightly.

7. Fully retract the transfer arm and lock the arm in place. **Close the transfer arm valve and the airlock gate valve.**

8. Press "V" to vent the quickloader airlock. Retrieve the transfer arm and place it onto the transfer lid.

9. Pump the transfer lid airlock for 35–40 s (make sure transfer lid sliding valve is closed), then open the transfer arm valve. **The next three steps should be done quickly in succession to minimize contamination.**

10. Open the transfer arm valve, then vent the transfer lid airlock for 3 s. Stop venting.

11. Unlock the transfer arm.

12. Open the transfer lid slider and quickly and gently insert the transfer arm to dock the shuttle into the loading station.

13. Release the shuttle grabber and retract the arm. If desired, pump down the transfer arm as described above.

14. Remove the transfer lid from the loading station and replace with the standard lid.

15. Using pre-cooled tools, flip the shuttle into the horizontal loading position, turn the locking screw to release the grids, then transfer the grids to a pre-cooled, labeled grid box.

16. Store the grid boxes with your specimens in liquid nitrogen.

End of Session Tasks

1. Remove heat exchanger from dewar and allow the stage and shield to warm to room temperature (30–60 min)

2. Home the stage for the next user.

3. If the microscope will not be used in the next day, sleep the system to conserve gallium. Otherwise, keep it on but ensure the SEM and FIB column valves are closed.

4. Turn off hot plates and store tools.

5. When the stage and shield are near room temperature, turn the flow rates to 1 L/min to prevent moisture accumulation inside the heat exchanger gas lines.

3.3 Considerations for TEM

The specimen should be loaded such that the FIB milling axis is perpendicular to the TEM tilt axis (Fig. 7). This will allow maximum tilt range for tomography without the unmilled material blocking the field of view at high tilts. Due to FIB milling geometry, the lamella will have a measurable pre-tilt at the TEM ranging from 5° to 15° difference compared to reported stage goniometer angle. This pre-tilt will influence the tilt series acquisition scheme because in one tilt direction the specimen will appear to be much thicker at high tilt angles compared to the other direction (*see* Subheading 1.2). Additionally, sample orientation is important for accurate focus and tracking when acquiring data with low-dose techniques. TEM data acquisition is mostly handled through automated routines from software packages such as SerialEM [44, 45].

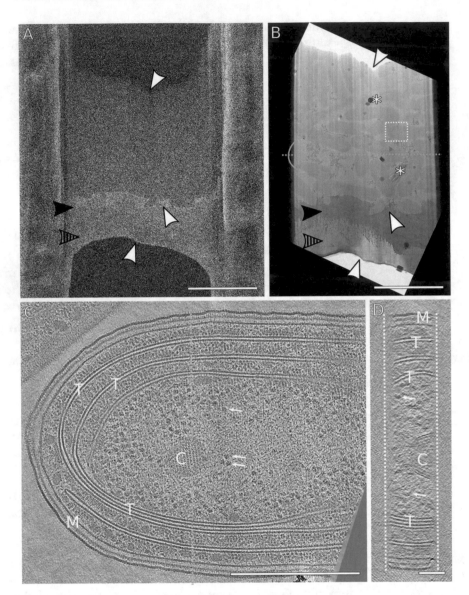

Fig. 7 Representative lamella and reconstructed tomogram. (**a**) SEM image of a finished lamella through a lawn of cyanobacteria. The milling direction was from bottom to top. The lamella was milled with expansion joints to reduce buckling (not visible here) [43]. (**b**) Corresponding TEM image of (**a**) showing a field of bacteria. White arrowheads mark areas of correlation between the SEM and TEM image. The solid and striped arrowheads point out the platinum sputtering and GIS platinum layers, respectively. Asterisks mark areas of atmospheric ice contamination and poor vitrification. The white box indicates the field of view for tilt-series acquisition. Green dotted line represents the TEM tilt axis. (**c**) XY view of a reconstructed tomogram of a single cyanobacterium from the lamella. *M* Inner/outer membrane. *T* Thylakoid membranes. *C* Carboxysomes. Arrows: ribosomes. (**d**) YZ view of reconstructed tomogram corresponding to the slice indicated by the blue line in (**c**) with structures labeled identically. The lamella thickness is approximately 150 nm. The surfaces of lamella are indicated by the dashed white lines. Scale bars: **a**, **b** 5 μm; **c** 500 nm; **d** 100 nm

When beginning a new project, it will be necessary to iteratively optimize samples between the TEM and FIB. For example, the final lamella thickness should be guided by the intended TEM data acquisition pixel size. This is important in order to limit the total dose applied to a specimen during tilt series acquisition. Doubling the TEM magnification while maintaining a similar beam intensity on the camera will result in four times the applied dose (in electrons per square angstrom) to the sample. A higher desired TEM magnification will therefore require thinner lamellae in order to reduce the dose applied to the specimen while maintaining an appropriate beam intensity on the camera. Conversely, thick specimens will require additional radiation, resulting in lower signal-to-noise ratio and lower resolution, but allow imaging of a larger volume for increased cellular context.

For reasons described above, it is important to track and orient grids during FIB milling and through to TEM data acquisition. This is most easily accomplished by using marked autogrid supports that allow the user to orient the sample at the aquilos loading station and at the TEM loading station. Examples of marked autogrids include the commercially available cryo-FIB autogrids that have laser-etched dots in the surface of the ring to indicate sample orientation. These marks may be difficult to see under liquid nitrogen, so we routinely mark the autogrids with a colored permanent marker.

While cryo-FIB milling exhibits fewer artifacts compared to sections from microtomy, the ion beam will damage the surface of the lamella as it ablates material [46]. Additionally, repeated use of the SEM for monitoring milling progress also contributes electron damage. In particular, sensitive structures such as bacterial polyphosphate bodies, accumulated damage from the FIB/SEM can be seen as localized bubbling confined to the surface of the specimen to several nanometers deep. However, previous work has demonstrated that FIB milled samples do not exhibit heat-induced devitrification beyond the immediate interacting surface [46].

3.4 Common Troubleshooting Items

3.4.1 System Maintenance

In addition to manufacturer suggested regular instrument upkeep, we recommend following a regular maintenance schedule for cryo-FIB/SEM systems focused on checking stage temperature, cleaning the stage of platinum buildup, and cleaning seals/o-rings of dust. These tasks should be done every 3–4 months. Additionally, the stage components should be inspected for wear and defects regularly due to the large temperature cycles that these parts experience.

System consumables include the gallium liquid-metal ion--source and the FIB aperture strip, both of which may need replacing every 6 months if the instrument is used routinely.

3.4.2 Equipment Troubleshooting

Poor Chamber Vacuum

The system should reach better than 2×10^{-6} mbar at room temperature and better than 7×10^{-7} mbar at cryogenic temperatures. If these are not satisfied, the chamber seals especially those on the front door and the quickloader should be inspected and cleaned. Poor vacuum may lead to an increased rate of water vapor accumulation on the sample resulting in an increase in the lamella thickness that reduces imaging quality.

Frost on Heat Exchanger Tubing

Frost buildup on the heat exchanger tubing is due to loss of vacuum insulation. The typical causes are pump failure, a leak in the outer tubing, or a leak in the internal nitrogen lines. Continued use of the system with frost buildup may lead to poor stage cooling.

Amorphous Ice Contamination

From our experience, poor stage cooling will lead to accumulation of amorphous ice or even "leopard skin" ice when observed at the TEM. The temperature of the top of the stage (where the shuttle is inserted) should be measured using a thermocouple. Starting from room temperature, the stage top should be able to reach below -170 °C within 30 min. If this is not the case, the stage should be disassembled and thoroughly cleaned. We have found that buildup from the sputter coater and GIS can be responsible for stage temperature issues.

Crystalline/Atmospheric Ice Contamination

Check that the o-rings on the transfer arm and transfer lid are clean and in good condition. Ensure that all tools and loading equipment are dry before handling samples. Ensure that the venting line on the transfer lid has been purged for at least 10 min, and keep purging continuously for the duration of the session. Ensure that any liquid nitrogen having direct contact with samples is free of frost.

Inaccurate Ion Beam Currents

When the beam is blanked, the FIB is deflected into a faraday cup to measure the beam current. If the reported value is much higher (10–15%) than the nominal value, the ion beam aperture strip is likely worn out (Fig. 8). Typical FIB milling may use one particular aperture (e.g., 10 or 30 pA) much more than the others, which may cause them to wear out earlier and become much larger than the initial size. This leads to drastically increased ion beam currents and poor control over milling. On certain microscopes, it may be necessary to switch to an unused "parking position" FIB aperture when the microscope is idle but the ion source is left on, due to the continual exposure of the aperture strip to the ion beam.

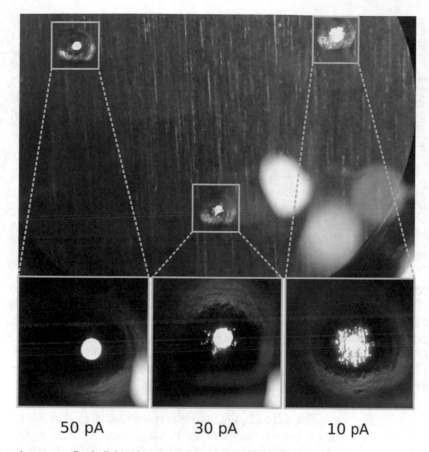

50 pA   30 pA   10 pA

Fig. 8 FIB aperture wear. Basic light microscope image of an FIB aperture strip retrieved after about 6 months of 4 sessions/week use. The smallest apertures used for milling (10 and 30 pA) have suffered extensive damage to the edge of the aperture hole leading to high beam currents, compared to an infrequently used aperture (50 pA)

3.4.3 Sample Troubleshooting

FIB Image Instability

The image may appear to drift or distort in the FIB view, especially at high beam currents or magnifications. Check that there are no ice or autogrid obstructions near the field of view that may be deflecting the ion beam. Check that the FIB accelerating voltage is at 30 kV. Increasing platinum sputtering time may help with drift.

Lamella Appears Skewed Relative to Milling Axis

The grid surface in the FIB view is not perfectly flat. Use additional scan rotation to make the FIB image appear horizontal (*see* **Note 7** and Fig. 9).

Curtaining Artifacts

These artifacts manifest as ridges and grooves on the surface of the lamella parallel to the milling axis. The FIB current is too high for the lamella thickness or milling has not completed before narrowing the patterns. Reduce the current and extend the milling time to ensure a flat lamella surface. Monitor the quality of the lamella at regular intervals with the SEM. If the GIS platinum layer is wearing away too quickly, the initial deposition time may be increased.

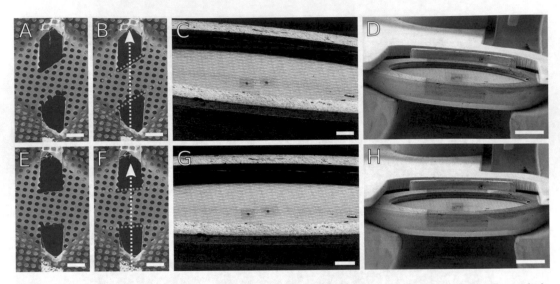

Fig. 9 Correcting skewed FIB milling with scan rotation. (**a, b**) SEM images of a skewed lamella made in carbon foil. Dotted green lines mark the approximate geometry of the lamella relative to the milling axis as indicated by the white arrow. (**c, d**) FIB views of the clipped grid at low magnifications, showing an apparent tilt of the grid relative to the horizontal. (**g, h**) By adjusting the apparent tilt using additional scan rotation applied to the FIB view, the lamella skew angle can be corrected (**e, f**) to become nearly perpendicular to the milling axis. These demonstrative lamellae were milled on a room temperature carbon film grid. Results are directly applicable for cryogenic samples. Scale bars: **a, b, e, f** 10 μm; **c, g** 100 μm; **d, h** 500 μm

For tall cells, the profile is presented such that the cell surface is nearly perpendicular to the beam. This is not the case for flat cells or bacteria, where the surface is much more oblique (compare Fig. 5h, i). The tall profile can lead to issues with insufficient GIS platinum deposition that would otherwise be acceptable. These cells may require a longer GIS platinum deposition time and rotating the stage to optimize deposition thickness.

Broken Lamellae at the TEM

The finished lamellae are relatively fragile and are susceptible to damage from rough handling. Slightly increasing the GIS platinum deposition time may prove extra support and make handling easier. It is important to only handle lamellae grids by the very edge of the autogrid support. Another option is to reduce the width of the lamellae to make them less susceptible to breakage.

Lamella Appears to Bend during Milling

Recent reports have suggested that lamella bending can be attributed to tension in the bulk material arising from the vitrification process. The proposed solution is to mill additional "micro-expansion joints" near the lamella to reduce the impact of surrounding material motion [43]. In our hands, the micro-expansion joints also reduce bending and subsequent breaking at the TEM. Silicon dioxide grids may also offer a stiffer support that may reduce lamella bending or breaking [47].

4   Notes

1. Because the grid is held at 45° and the FIB column angle relative to the SEM column is 52°, there is a 7° difference between the reported stage angle and the incident milling angle. To estimate the incident milling angle, subtract 7° from the reported stage angle.

2. The EM grid should be chosen such that each grid square is large enough to contain the widest desired lamella, but small enough to have sufficient sample rigidity so that the lamella will not bend and break during subsequent handling. We routinely use 200 mesh quantifoil grids for FIB milling. The specific pattern of holes will determine blotting speed and sample dryness. We are routinely using R1/4 grids for mammalian cells and R2/1 grids for bacteria. Grid material should be gold for cells cultured directly on the grid, copper for cells deposited immediately before plunge freezing. *See* Subheading 3.1 for sample type considerations.

3. The autogrid is a rigid ring support in which TEM grids are placed and secured with a spring clip. This arrangement allows improved handling and stability of samples. Grids may be clipped either using the provided loading station, or in any other suitable device that will minimize atmospheric contamination. Grids should be clipped so that the cell side is facing towards the flat side of the autogrid. We have found clipping grids under a very low level of nitrogen to be most efficient, with minimal grid square breakage. Once clipped, the grid is permanently mounted in the autogrid and is essentially impossible to remove at cryogenic temperatures without damaging the grid.

4. On our system, the chamber base pressure at room temperature is typically 2×10^{-6} mbar or better. If the base pressure is too high, the chamber seals, airlock seals, and nitrogen lines should be cleaned and inspected (*see* Subheading 3.4.2).

5. The reported stage temperature is measured from the lower stage, which is quickly cooled by cold nitrogen gas. However, the upper stage (where the sample sits) usually lags behind due to its thermal mass and reduced contact area with the lower stage. Waiting at least 30 min after full cooldown is important to ensure that the upper stage is well below devitrification temperature.

6. The microscope quickloader gate valve is kept locked to prevent accidental opening and venting of the chamber. The valve will only unlock if the user has issued the "pump" command through the button interface **AND** if the pressure in the quickloader airlock is low enough. Additionally, the gate valve

is only unlocked for about 90 s before the action times out and relocks. The valve must be opened within this timeframe. When closing the valve, wait for it to lock before reopening or re-issuing the pump command. Otherwise, we have found that this may lead to the gate valve being stuck in the open state.

7. Setting the scan rotation to 180° is optional, but it orients the SEM and FIB views to appear more natural such that the FIB view shows cells "sitting" on the grid, rather than "hanging" from the grid. In some cases, the scan rotation of the FIB column is not perfectly parallel to the grid surface. In this case, the resulting lamellae will be tilted relative to the milling axis (Fig. 9a–d). Such tilted lamellae become more difficult for tomography due to limited focus and tracking area (*see* Subheading 3.3). This issue can be corrected by rotating the scan direction of the FIB column until the grid surface appears horizontal (Fig. 9e–h).

8. In addition to sample specific considerations for milling, users should consider the accessible area of the grid at the TEM for tilt series acquisition. This depends on the specific TEM instrument, but is typically limited to the grid squares that are in the center 1 mm of the grid. For a 200 mesh grid, this is about 10-by-10 grid squares centered over the grid. Lamellae should be milled within this region. To minimize occlusion from the grid bars at high tilt at the TEM, it is generally best to make lamellae that are centered in the grid square.

9. Sputter coating is used to improve the image quality of non-conductive samples by depositing a layer of metal to allow charge to quickly dissipate. The Aquilos has a built-in sputter coater. Set the number of purge cycles to 0 (which would otherwise lead to sample contamination) and click "Prepare for Sputtering." The microscope will enter a low vacuum state (about 0.1 mbar) by partially venting with dry argon, and the stage temperature will increase slightly to about −170 °C. The stage will move to the sputtering position. When the preparation actions have finished, select the sputtering parameters or the preset values and run the process. Typical sputter conditions for initial rough sputtering are 1 kV, 30 mA, for 15 s at 0.10 mbar chamber pressure. You can see the plasma glow using the chamber camera. When sputtering is complete, click "Recover from Sputtering" to return the stage to the mapping position and return to high vacuum

10. The GIS deposition time depends on the sample type, the intended deposition thickness, and the preset deposition stage coordinate. If the stage is closer to the tip of the GIS needle, a shorter deposition time will yield the same thickness

of GIS platinum as a longer deposition time at a further distance. The deposition stage position/coordinates can be adjusted to allow long/short deposition times or can be adjusted to be centered between the two grids such that both the left and right grids are coated at the same time.

11. We have found that judging appropriate lamella thickness and when to stop milling requires some experience and iterative comparison between the SEM and TEM to "calibrate" the user. One helpful exercise is to make a lamella, estimate its thickness at the SEM, then measure the actual thickness from a tomogram.

12. To manually find the beam-coincidence point, tilt to the desired milling angle and go to a low mag view for both SEM and FIB. Find your target in both SEM and FIB, and center the target in the FIB view using XY moves. Switch to SEM view and move the stage in Z so that the target is about halfway closer to the center of the SEM field of view (e.g., if the target was 100 μm away from the center of the SEM, move the stage in Z so that the target is now about 50 μm away). Switch to the FIB view and recenter the target using XY moves. Switch to SEM and check positioning. Reiterate between XY moves in FIB view and Z moves in SEM view until the target is centered in both SEM and FIB. Alternatively, the proprietary MAPS software package from TFS allows correlation between fluorescent, SEM, and TEM images. Additionally, MAPS may be used as a microscope control program. This software package has the ability to calculate eucentric height for a given point on a sample, which should be very close to the beam-coincidence point. To do so, add a lamella object, then go to its mapping position. Select "Calculate Eucentric Height" which will ask you to keep the object centered at several tilts. The specific internal calibration for accurate eucentricity should be determined during software installation and kept for all users.

13. This apparent drift is not due to physical specimen movement, but due to slight movement in the beam due to sample interaction. Especially at high FIB currents, the sample may acquire charging that influences the FIB [48].

14. Step milling improves lamella stability by forming a gradual transition from the unmilled material to the lamella. Without step milling, the lamellae are more prone to developing cracks at the edge where they connect to the bulk material.

15. The FIB view has a shallow depth of field, which can be observed when the front of the lamella is in sharp focus, but the back of the lamella is out of focus. This leads to heterogeneity in milling efficiency over the surface of the lamella. At the TEM, this can be observed where the back of the lamella is

thicker than the front. To thin only the back portion of the lamella, the stage should be tilted approximately +0.5° which will expose the back for milling without disturbing the front edge of the lamella [13].

16. Given the large chamber volume, poor vacuum compared to the TEM, and cryogenic temperatures, amorphous ice buildup is inevitable inside the cryo-FIB/SEM. This can be minimized with proper instrument maintenance, but in our experience, long operating times of more than 6–8 h lead to ice buildup from the chamber and from milling redeposition. Before sample retrieval, it is recommended to perform a short 5-min lamella cleaning step with a low FIB current for all lamellae immediately prior to sample retrieval (Fig. 6). This step is essentially identical to the fine milling step, but with the pattern set to skim over the surface of the lamella without removing too much material.

Acknowledgments

We would like to acknowledge the following people: Reika Watanabe and Miles Paszek for providing images of NIH3T3 and yeast grids, respectively. Reika Watanabe, Kanika Khanna, and Sergey Suslov for fruitful discussion of FIB/SEM. Mario Aguilera for photography of microscope equipment.

This work was supported by an NIH Director's New Innovator Award 1DP2GM123494-01 (to E.V.) and NIH 5T32GM7240-40 (to V.L.). Images of cyanobacteria are from projects supported by NIH R35GM118290 awarded to Susan S. Golden. This work was performed in part at the San Diego Nanotechnology Infrastructure (SDNI) of UCSD, a member of the National Nanotechnology Coordinated Infrastructure, which is supported by the National Science Foundation (Grant ECCS-1542148).

References

1. Koning RI, Koster AJ, Sharp TH (2018) Advances in cryo-electron tomography for biology and medicine. Annal Anatomy Anatomischer Anzeiger 217:82–96

2. Russo CJ, Passmore LA (2016) Progress towards an optimal specimen support for electron cryomicroscopy. Curr Opin Struct Biol 37:81–89

3. Al-Amoudi A, Chang JJ, Leforestier A, McDowall AW, Michel Salamin L, Norlén L, Richter K, Sartori Blanc N, Studer D, Dubochet J (2004) Cryo-electron microscopy of vitreous sections. EMBO J 23:3583–3588

4. Al-Amoudi A, Studer D, Dubochet J (2005) Cutting artefacts and cutting process in vitreous sections for cryo-electron microscopy. J Struct Biol 150(1):109–121

5. Mahamid J, Tegunov D, Maiser A, Arnold J, Leonhardt H, Plitzko JM, Baumeister W (2019) Liquid-crystalline phase transitions in lipid droplets are related to cellular states and specific organelle association. Proc Natl Acad Sci 116(34):16866–16871

6. Chaikeeratisak V, Khanna K, Nguyen KT, Sugie J, Egan ME, Erb ML, Vavilina A, Nonejuie P, Nieweglowska E, Pogliano K,

Agard DA, Villa E, Pogliano J (2019) Viral capsid trafficking along treadmilling tubulin filaments in bacteria. Cell 177(7):1771–1780. e12

7. Khanna K, Lopez-Garrido J, Zhao Z, Watanabe R, Yuan Y, Sugie J, Pogliano K, Villa E (2019) The molecular architecture of engulfment during *Bacillus subtilis* sporulation. elife 8:e45257

8. Rast A, Schaffer M, Albert S, Wan W, Pfeffer S, Beck F, Plitzko JM, Nickelsen J, Engel BD (2019) Biogenic regions of cyanobacterial thylakoids form contact sites with the plasma membrane. Nat Plants 5(4):436–446

9. Weiss GL, Kieninger AK, Maldener I, Forchhammer K, Pilhofer M (2019) Structure and function of a bacterial gap junction analog. Cell 178(2):374–384.e15

10. Lopez-Garrido J, Ojkic N, Khanna K, Wagner FR, Villa E, Endres RG, Pogliano K (2018) Chromosome translocation inflates bacillus forespores and impacts cellular morphology. Cell 172(4):758–770.e14

11. Noble JM, Lubieniecki J, Savitzky BH, Plitzko J, Engelhardt H, Baumeister W, Kourkoutis LF (2018) Connectivity of centermost chromatophores in rhodobacter sphaeroides bacteria. Mol Microbiol 109(6):812–825

12. Mosalaganti S, Kosinski J, Albert S, Schaffer M, Strenkert D, Salomé PA, Merchant SS, Plitzko JM, Baumeister W, Engel BD, Beck M (2018) In situ architecture of the algal nuclear pore complex. Nat Commun 9(1):2361

13. Schaffer M, Mahamid J, Engel BD, Laugks T, Baumeister W, Plitzko JM (2017) Optimized cryo-focused ion beam sample preparation aimed at in situ structural studies of membrane proteins. J Struct Biol 197(2):73–82

14. Chaikeeratisak V, Nguyen K, Khanna K, Brilot AF, Erb ML, Coker JKC, Vavilina A, Newton GL, Buschauer R, Pogliano K, Villa E, Agard DA, Pogliano J (2017) Assembly of a nucleus-like structure during viral replication in bacteria. Science 355(6321):194–197

15. Rosenzweig ESF, Xu B, Cuellar LK, Martinez-Sanchez A, Schaffer M, Strauss M, Cartwright HN, Ronceray P, Plitzko JM, Förster F, Wingreen NS, Engel BD, Mackinder LC, Jonikas MC (2017) The eukaryotic co2-concentrating organelle is liquid-like and exhibits dynamic reorganization. Cell 171(1):148–162.e19

16. Bäuerlein FJ, Saha I, Mishra A, Kalemanov M, Martínez-Sánchez A, Klein R, Dudanova I, Hipp MS, Hartl FU, Baumeister W, Fernández-Busnadiego R (2017) In situ architecture and cellular interactions of polyq inclusions. Cell 171(1):179–187.e10

17. Mahamid J, Pfeffer S, Schaffer M, Villa E, Danev R, Kuhn Cuellar L, Förster F, Hyman AA, Plitzko JM, Baumeister W (2016) Visualizing the molecular sociology at the hela cell nuclear periphery. Science 351 (6276):969–972

18. Zhang J, Ji G, Huang X, Xu W, Sun F (2016) An improved cryo-fib method for fabrication of frozen hydrated lamella. J Struct Biol 194 (2):218–223

19. Engel BD, Schaffer M, Kuhn Cuellar L, Villa E, Plitzko JM, Baumeister W (2015) Native architecture of the *Chlamydomonas* chloroplast revealed by in situ cryo-electron tomography. elife 4:e04889

20. Harapin J, Börmel M, Sapra KT, Brunner D, Kaech A, Medalia O (2015) Structural analysis of multicellular organisms with cryo-electron tomography. Nat Methods 12:634–636

21. Villa E, Schaffer M, Plitzko JM, Baumeister W (2013) Opening windows into the cell: focused-ion-beam milling for cryo-electron tomography. Curr Opin Struct Biol 23 (5):771–777. Protein-carbohydrate interactions/biophysical methods

22. Wang K, Strunk K, Zhao G, Gray JL, Zhang P (2012) 3d structure determination of native mammalian cells using cryo-fib and cryo-electron tomography. J Struct Biol 180 (2):318–326

23. Rigort A, Bäuerlein FJB, Villa E, Eibauer M, Laugks T, Baumeister W, Plitzko JM (2012) Focused ion beam micromachining of eukaryotic cells for cryoelectron tomography. Proc Natl Acad Sci 109(12):4449–4454

24. Marko M, Hsieh C, Schalek R, Frank J, Mannella C (2007) Focused-ion-beam thinning of frozen-hydrated biological specimens for cryo-electron microscopy. Nat Methods 4:215

25. Rajput NS, Luo X (2015) Chapter 3: FIB micro-/nano-fabrication. In: Qin Y (ed) Micromanufacturing engineering and technology, Micro and Nano Technologies, 2nd edn. William Andrew Publishing, Boston, pp 61–80

26. Volkert CA, Minor AM (2007) Focused ion beam microscopy and micromachining. MRS Bull 32(5):389–399

27. Chyr I, Steckl AJ (2001) Gan focused ion beam micromachining with gas-assisted etching. J Vacuum Sci Technol B 19(6):2547–2550

28. Schaffer M, Pfeffer S, Mahamid J, Kleindiek S, Laugks T, Albert S, Engel BD, Rummel A, Smith AJ, Baumeister W, Plitzko JM (2019) A cryo-fib lift-out technique enables molecular-resolution cryo-et within native caenorhabditis elegans tissue. Nat Methods 16(8):757–762

29. Zhang J, Zhang D, Sun L, Ji G, Huang X, Niu T, Sun F (2019) VHUT-cryo-FIB, a method to fabricate frozen-hydrated lamella of tissue specimen for in situ cryo-electron tomography. bioRxiv. https://doi.org/10.1101/727149

30. Mahamid J, Schampers R, Persoon H, Hyman AA, Baumeister W, Plitzko JM (2015) A focused ion beam milling and lift-out approach for site-specific preparation of frozen-hydrated lamellas from multicellular organisms. J Struct Biol 192(2):262–269

31. Stevie FA, Griffis DP, Russell PE (2005) Focused ion beam gases for deposition and enhanced etch. Springer US, Boston, MA, pp 53–72

32. Diebolder C, Faas F, Koster A, Koning R (2015) Conical fourier shell correlation applied to electron tomograms. J Struct Biol 190(2):215–223

33. Strunk K, Wang K, Ke D, Gray J, Zhang P (2012) Thinning of large mammalian cells for cryo-TEM characterization by cryo-FIB milling. J Microsc 247(3):220–227

34. Medeiros JM, Böck D, Weiss GL, Kooger R, Wepf RA, Pilhofer M (2018) Robust workflow and instrumentation for cryo-focused ion beam milling of samples for electron cryotomography. Ultramicroscopy 190:1–11

35. Guo YS, Furrer JM, Kadilak AL, Hinestroza HF, Gage DJ, Cho YK, Shor LM (2018) Bacterial extracellular polymeric substances amplify water content variability at the pore scale. Front Environ Sci 6:93

36. Arnold J, Mahamid J, Lucic V, de Marco A, Fernandez JJ, Laugks T, Mayer T, Hyman AA, Baumeister W, Plitzko JM (2016) Site-specific cryo-focused ion beam sample preparation guided by 3D correlative microscopy. Biophys J 110(4):860–869

37. Nahmani M, Lanahan C, DeRosier D, Turrigiano GG (2017) High-numerical-aperture cryogenic light microscopy for increased precision of superresolution reconstructions. Proc Natl Acad Sci 114(15):3832–3836

38. Carlson DB, Evans JE (2011) Low-cost cryolight microscopy stage fabrication for correlated light/electron microscopy. JoVE 52:e2909

39. Tuijtel MW, Koster AJ, Jakobs S, Faas FGA, Sharp TH (2019) Correlative cryo super-resolution light and electron microscopy on mammalian cells using fluorescent proteins. Sci Rep 9(1):1369

40. Schellenberger P, Kaufmann R, Siebert CA, Hagen C, Wodrich H, Grünewald K (2014) High-precision correlative fluorescence and electron cryo microscopy using two independent alignment markers. Ultramicroscopy 143:41–51. SI: Correlative microscopy

41. Chang YW, Chen S, Tocheva EI, Treuner-Lange A, Löbach S, Søgaard-Andersen L, Jensen GJ (2014) Correlated cryogenic photoactivated localization microscopy and cryo-electron tomography. Nat Methods 11(7):737–739

42. Kaufmann R, Schellenberger P, Seiradake E, Dobbie IM, Jones EY, Davis I, Hagen C, Grünewald K (2014) Super-resolution microscopy using standard fluorescent proteins in intact cells under cryo-conditions. Nano Lett 14(7):4171–4175

43. Wolff G, Limpens RWAL, Zheng S, Snijder EJ, Agard DA, Koster AJ, Bárcena M (2019) Mind the gap: micro-expansion joints drastically decrease the bending of FIB-milled cryo-lamellae. J Struct Biol 208:107389

44. Mastronarde DN (2005) Automated electron microscope tomography using robust prediction of specimen movements. J Struct Biol 152(1):36–51

45. Schorb M, Haberbosch I, Hagen WJH, Schwab Y, Mastronarde DN (2019) Software tools for automated transmission electron microscopy. Nat Methods 16(6):471–477

46. Marko M, Hsieh C, Moberlychan W, Mannella CA, Frank J (2006) Focused ion beam milling of vitreous water: prospects for an alternative to cryo-ultramicrotomy of frozen-hydrated biological samples. J Microsc 222(1):42–47

47. Toro-Nahuelpan M, Zagoriy I, Senger F, Blanchoin L, Théry M, Mahamid J (2020) Tailoring cryo-electron microscopy grids by photo-micropatterning for in-cell structural studies. Nat Methods 17:50–54

48. Stokes DJ, Vystavel T, Morrissey F (2007) Focused ion beam (FIB) milling of electrically insulating specimens using simultaneous primary electron and ion beam irradiation. J Phys D Appl Phys 40(3):874–877

Chapter 4

In Situ Imaging and Structure Determination of Biomolecular Complexes Using Electron Cryo-Tomography

Mohammed Kaplan, William J. Nicolas, Wei Zhao, Stephen D. Carter, Lauren Ann Metskas, Georges Chreifi, Debnath Ghosal, and Grant J. Jensen

Abstract

Electron cryo-tomography (cryo-ET) is a technique that allows the investigation of intact macromolecular complexes while they are in their cellular milieu. Over the years, cryo-ET has had a huge impact on our understanding of how large biomolecular complexes look like, how they assemble, disassemble, function, and evolve(d). Recent hardware and software developments and combining cryo-ET with other techniques, e.g., focused ion beam milling (FIB-milling) and cryo-light microscopy, has extended the realm of cryo-ET to include transient molecular complexes embedded deep in thick samples (like eukaryotic cells) and enhanced the resolution of structures obtained by cryo-ET. In this chapter, we will present an outline of how to perform cryo-ET studies on a wide variety of biological samples including prokaryotic and eukaryotic cells and biological plant tissues. This outline will include sample preparation, data collection, and data processing as well as hybrid approaches like FIB-milling, cryosectioning, and cryo correlated light and electron microscopy (cryo-CLEM).

Key words Cryo-ET, In situ structural biology, Subtomogram averaging, FIB-milling, High-resolution tomography, Cryosectioning

1 Introduction

The continuous interaction of proteins with each other, and with other molecules, in the crowded environment of the cell generates life. Disentangling the structure and dynamics of proteins while they are in this native milieu is crucial to structural biology. To achieve this, multiple imaging and spectroscopic techniques have been developed in recent years that aim at investigating biomolecules at ever higher resolution inside the cell (see, e.g., Refs. 1–4). Cryo-ET has now become an indispensable tool to investigate molecular complexes in intact cells in a frozen hydrated state at

Tamir Gonen and Brent L. Nannenga (eds.), *CryoEM: Methods and Protocols*, Methods in Molecular Biology, vol. 2215, https://doi.org/10.1007/978-1-0716-0966-8_4, © Springer Science+Business Media, LLC, part of Springer Nature 2021

nanometer resolution [5, 6]. In favorable cases, atomic models of these nanomachines can be built using known high-resolution structures obtained by other methods like NMR, X-ray crystallography, and single particle analysis (SPA) which provides new mechanistic insights [7]. Moreover, comparing the structures of different nanomachines or of the same nanomachine in different species can shed light on how these biological machines have evolved (and continue evolving) over time [8, 9]. In addition, cryo-ET is very powerful in studying how large complexes assemble and disassemble in the cell and is capable of capturing and revealing the structures of intermediate and transient states during such processes which is crucial in deciphering such pathways [10–13].

Cryo-ET is one of the essential modes of Cryo-EM (together with SPA and Micro-Electron Diffraction (MicroED)) in which a 3D reconstruction of the sample is produced by back-projecting images of the sample taken at different tilt angles. Usually, the sample is tilted from $-60°$ to $+60°$ every $1°–3°$. As biological samples are sensitive to radiation, different tilt-schemes have been developed like the bidirectional tilt-scheme and the dose symmetric tilt-scheme [14] and the selection of a specific tilt-scheme is dependent on the sample and the question to be answered. In many cases, single tomograms may not have sufficient signal to provide a detailed knowledge of the molecular complex of interest. Therefore, to enhance the signal and the resolution of tomography data, subtomogram averaging can be performed whereby the subvolumes in the tomograms that have the complex of interest are cropped, aligned, and then averaged to produce an improved signal-to-noise 3D average of the complex of interest [15, 16]. The final resolution of subtomogram averaging depends on many factors with some of them being related to how the data was collected and others determined by inherent characteristics of the complex of interest (flexibility, thickness heterogeneity, etc.). Currently, by averaging many subvolumes the macromolecular structures of many biological machines have been solved (*see*, e.g., Refs. 8, 10, 12, 17–26). For some samples, atomic models have been generated (e.g., Refs. 27, 28).

Hybrid approaches that combine cryo-ET with other methods have recently extended its applicability and usefulness. This includes cryo-CLEM which allows the localization of specific protein complexes and targeted data collection on only the cells or parts of cells that have the complex of interest. Moreover, as the quality of tomography data depends on the sample thickness, cryo-ET has been limited for a long time to thin samples like bacterial cells or the thin edges of eukaryotic cells. Thicker samples required the cumbersome task of cryosectioning. However, the recent development of FIB-milling has extended the realm of tomography to targets located deep in the thick regions of eukaryotic cells [29]. Finally, recent hardware developments like phase plates

[30], direct electron detection cameras with high quantum efficiency and the development of fast-tilt-series schemes [31] (which decrease the time of data collection significantly) have further extended the reach, power, and resolution of cryo-ET.

In this chapter, we present a general outline of the workflow that we use in our laboratory to investigate different kinds of biological samples with cryo-ET, starting from sample preparation and proceeding through data collection and processing. This discussion will include prokaryotic as well as different types of eukaryotic samples (like plant tissues), hybrid approaches like cryo-ET with cryo-CLEM or FIB-milling or both, and cryosectioning of thick biological samples like plant tissues. Finally, we will comment on the requirements for high-resolution tomography in terms of data processing and data collection, including rapid tilt-series schemes.

This chapter is organized into two major sections which are sample preparation and data collection and processing. In the sample preparation section, we will discuss bacterial samples as an example of single small cells, isolated eukaryotic cells and plant tissues in separate subsections. Finally, one should bear in mind that the steps outlined here represent a general protocol and specific tweaks and modifications might be applied to each sample specifically.

2   Sample Preparation

2.1   Single Small Cells

This section is applicable to many types of unicellular small organisms. We will discuss bacterial samples as an example. For a detailed discussion of how to perform tomography on a specific large bacterial protein complex, we refer the reader to our recently published chapter [32].

2.1.1   Bacterial Cultures

In general, we start from a $-80\ ^\circ C$ glycerol stock of the bacterial strain which is cultured on a suitable agar plate under the required conditions (e.g., temperature, antibiotics, and CO_2). Subsequently, a single colony is inoculated into suitable liquid medium and allowed to grow to the desired OD_{600}. Note that this general workflow can differ for different conditions/strains. For example, it might be sometimes better to grow the cells on an agar plate then resuspend them in a liquid medium just prior to freezing. Therefore, one needs to optimize the growth conditions for the investigated bacterial species and for the specific aim of the experiment.

2.1.2   Grids and Gold Fiducials

For many bacterial samples destined for tomography, we use copper R2/2 200 Quantifoil holey extra thick carbon grids. However, if the cells need to be incubated on the grids for a significant amount of time, we recommend gold grids with Quantifoil holey carbon as copper is toxic to the cells. When correlated work between different kinds of microscopes is required, finder grids may be used to facilitate the process of correlation. Prior to use, the grids are glow-discharged to increase their hydrophilicity. In our lab, the grids are glow-discharged for 60 s with 15 mA negative discharge at 1×10^{-1} bar using an Emitech K100× glow-discharging device. Gold fiducials are coated with bovine serum albumin (BSA) to secure a homogenous dissemination of the beads. In a 1.5 mL Eppendorf tube, 1 mL of gold nanoparticles (generally 10–20 nm in diameter) is mixed with 5% BSA (two such tubes are usually prepared). After that, they are spun down at $14,000 \times g$ for 30 min. The pellets of the two tubes are resuspended in ~80 μL of the supernatant. This solution is subsequently diluted (7–8 times) with the bacterial sample prior to freezing. The remaining solution of gold nanoparticles can be kept for many weeks at 4 °C. If the fiducial markers aggregate around the cells because of the inherent characteristics of the bacteria (e.g., S-layer or extracellular matrix), they can be laid down on the grids prior to freezing. To do so, proceed as described above but dilute with PBS instead of the bacterial sample. From this diluted solution, pipette 3 μL onto each glow-discharged grid held by autoclosing forceps. Incubate the drop for approximately 1 min and carefully back-blot the drop on the other side of the grid, making sure the grid is not damaged. Leave to dry for several minutes. Optional screening under a binocular loupe can ensure the overall integrity of the grid. The grids are then glow-discharged again with the same parameters.

2.1.3 Plunge Freezing of Bacterial Cells

Samples are blotted either manually or automatically in the blotting chamber of the Vitrobot. Temperature is usually kept at 22 °C with 100% humidity. The blotting conditions (e.g., humidity, blotting force, and time) have to be identified and optimized for each sample. Once plunge-frozen, samples can be kept in liquid nitrogen dewars for long-term storage.

2.2 Eukaryotic Cells

The choice of cell type is a critical consideration for cryo-ET. For successful cryo-ET imaging of peripheral cellular thin edges and extensions, some types of mammalian cells are better suited and reveal extensively larger imageable areas of cytoplasm. In our experience, U2OS, 3T3, and fibroblasts spread optimally on EM grids to reveal peripheral thin edges that are electron transparent. Otherwise, additional thinning steps, such as FIB-milling (*see* below) may be required to gain access to the thicker regions of the cell body and nucleus.

Choice of grid support is critical. Standard copper EM grids are toxic to cells; therefore, an inert grid material such as gold should be used. For cryo-CLEM work, it is recommended that finder grids with reference markers such as letters, symbols, and numbers be used to easily locate target cells imaged between light microscope (LM) and EM modalities. Here, we include all the steps for a cryo-CLEM workflow.

2.2.1 Preparing Cells

Sample preparation depends on the cell type. Here we give an example for U2OS cells.

U2OS cells should be grown in a T75 flask containing McCoy's 5a Medium Modified (*see* **Note 1**). Typically, U2OS cells are passaged every 2 days and new media replaced every day. At all times, the cells are incubated at 37 °C with 5% CO_2. The U2OS cells are grown to 80% confluency in a T75 flask before being detached by trypsin and seeded on EM grids. Usually, 1 mL of trypsin is added to the flask and left to incubate at 37 °C for 2 min. Next, 6 mL of media is added to the flask to neutralize the trypsin (*see* **Note 2**). To remove clumps and achieve a single-cell suspension, the cells are pipetted up and down 20 times using a 10 mL pipette. A cell strainer can also be used to obtain a uniform single-cell suspension from mammalian cells. The suspended cells are then transferred to a 10 mL Falcon tube.

2.2.2 Preparing EM Grids for Tissue Culture

1. London finder grids with Quantifoil holey carbon are UV treated for 1 h before plating experiment (*see* **Note 3**). Specific extracellular matrix proteins (ECMs) can be used to improve attachment and spreading of various cells. The most commonly used ECMs are collagen, fibronectin, poly-L-lysine, and laminin. For U2OS cells, we use collagen. Several drops of human collagen diluted ten times in H_2O are pipetted onto parafilm.

2. A UV-treated grid is then placed carbon-side down on the drop for 2 min. The grid is then flipped so that the grid sinks to the bottom of the drop with the carbon-side now facing up (*see* **Note 4**). The grid is left in the drop for 20 min.

3. Once the grid has incubated in the drop, it is picked up with tweezers and the excess collagen solution on the grid is blotted away with Whatman 40 blotting paper and the grid is left to dry.

2.2.3 Growing Mammalian Cells on Grids

1. Glass slides are placed in a 10 mL petri dish and UV treated for 20 min. 20 mLs of media is then added to the petri dish to cover the glass slides. Alternatively, when smaller volumes of samples are required a 35 mm imaging dish with coverslip bottom can be used.

2. The pre-coated EM grids are then submerged into the media of the petri dish with the carbon side of the grid facing upwards and placed on the glass slide.

3. The detached cells are then added to the petri dish. The optimal density for growing cells on R2/2 grids is 1 or 2 cells per grid square. If there are more than three cells per grid square, there will be little room for the cells to spread and grow flat. Plating cells on grids so that roughly one or two cells are present in most of the grid squares can be achieved by

measuring a set density of cells in suspension using a cell counter. Or the trypsinized cells in the 10 mL Falcon tube can simply be pipetted onto the medium above the grids and left to settle into the grid squares. A light microscope can be used to evaluate if a suitable density has been achieved.

4. U2OS cells are grown on the grids for 15–20 h before plunge-freezing.

2.2.4 Freezing Grids, Including Adding 20 nm Gold Beads and Fluorescent Microspheres for Cryo-CLEM

1. The EM grids are taken out of the media with Vitrobot tweezers, which are placed into the Vitrobot (*see* **Note 5**). Blot the excess media from the grid before adding 3 μL of 20 nm gold solution and 0.5 μm TetraSpeck Microspheres (*see* **Note 6**).

2. We recommend manual blotting from the gold side of the grid with all Vitrobot parameters set to 0. It is recommended that the humidity of the Vitrobot chamber be at 95% when the grid is blotted (*see* **Notes 7** and **8**).

3. After plunge-freezing, the grids are stored within grid boxes cooled in clean liquid nitrogen.

4. Before cryo-LM or cryo-EM imaging, EM grids are clipped into FEI autogrids.

2.2.5 Cryo-LM Imaging for Cryo-CLEM Using the FEI Cryostage

1. In our setup, an FEI cryostage modified to hold Titan Krios cartridges is used, and liquid nitrogen is continuously pumped into the stage by a Norton cryo-pump to maintain the temperature at 80 K. Wait approximately 40 min for the stage to equilibrate and maintain its cryogenic temperature.

2. Load one grid into the stage at a time.

3. Collect a montage or a series of individual phase contrast and fluorescent Z-stack images of mammalian cells growing flat within the grid square and near the center of the grid (Fig. 1). If your target protein is tagged with a green fluorescent protein (GFP) (*see* **Note 9**), collection of red, blue, and green fluorescent images is recommended (*see* **Note 10**).

4. Evaluation of ice thickness should be made when imaging mammalian cells using cryo-LM. Ice cracks and thick ice around the circumference of the square are indications that the ice is too thick around the cell. In addition, obvious cracks in the carbon support close to target cells are not ideal and will affect the stability of the cell when being tilted, which can lead to suboptimal tilt-series alignment.

5. Align the fluorescent images to the fluorescent microspheres observed in the phase contrast image (*see* **Note 11**).

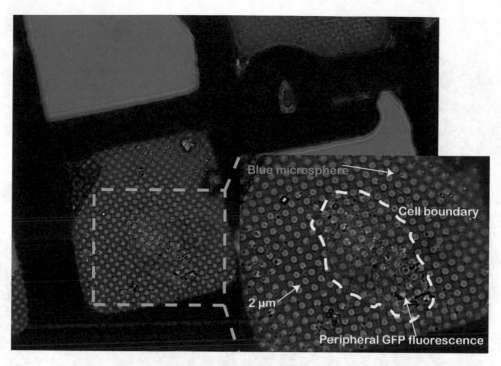

Fig. 1 Deconvolved cryo-LM image (composite of bright field and epifluorescence in FITC and DAPI channels) of U2OS cells, grown on London finder gold Quantifoil grids and stably expressing a GFP-tagged protein. Note the letter P which can be used for orientation. Relevant features are noted in the enlarged box

2.2.6 FIB-Milling for Imaging Thicker Regions of the Cell

In cryo-ET, sample thickness is a critical factor for data collection. Because of inelastic scattering, the application of cryo-ET is usually restricted to samples thinner than 600 nm, like small bacterial cells or the thin peripheral regions of eukaryotic cells. However, most regions of eukaryotic cells and many bacteria are thicker than that; thus, additional thinning procedures are required for cryo-ET to gain access to these samples (Fig. 2a). In recent years, cryo-FIB-milling has been developed as a tool to make thin lamellae from a thick sample while preserving most of the native state of the molecules. The concept of FIB-milling is to use high-energy ions such as gallium to remove materials above and below a targeted area. When milling at a low angle, this will create a lamella, usually a few hundred nanometers thick amenable for cryo-ET (Fig. 2b–d). Here, we describe how to do cryo-FIB-milling on an FEI Versa FIB/SEM equipped with a Quorum cryo-transfer system.

1. To prepare the Quorum transfer system for cryo work, mount the Quorum cryo-stage on the SEM stage.

2. Start the Quorum system in manual mode to pump vacuum in the chamber and scaffold. Prep-chamber vacuum should reach less than 6×10^{-6} before cooling to avoid contamination.

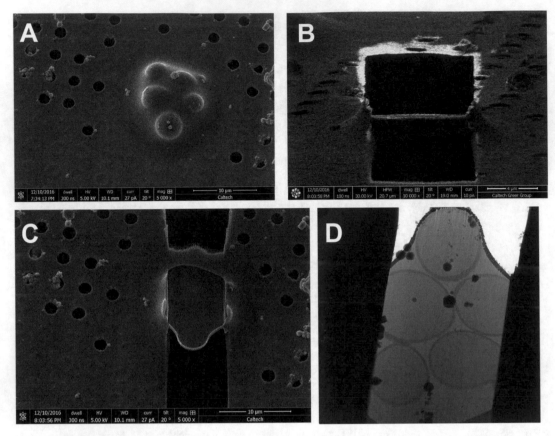

Fig. 2 Preparation of a lamella by focused ion beam milling. (**a**) SEM image (top view) of a cluster of yeast cells. (**b**) FIB image (oblique side view) of the lamella after milling and polishing both the top and bottom of the cell. (**c**) SEM image (top view) of the lamella at high magnification. (**d**) Lamella imaged by cryo-TEM

3. Pump the nitrogen gas isolated line to reach a vacuum lower than 5×10^{-2} mbar. This is important for the liquid nitrogen cooled gas to remain cold before it reaches the stage.

4. Set the temperature of the cryo-stage and anti-contaminator to desired number. Note that the temperature of the anti-contaminator is usually 20° lower than the stage. Set gas flow to auto. If using a self-pressured tank, make sure the pressure is higher than 30 psi.

5. Put the cooling rod into the liquid nitrogen dewar. Wait until the cryo-stage and anti-contaminator have reached their desired temperatures (usually -168 °C and -188 °C respectively). This usually takes 30–40 min.

6. To begin the sample transfer to the SEM cryo-stage, the shuttle and loading box should be pre-cooled with liquid nitrogen. Grids need to be clipped into autogrids with the cell side (the flat side) facing down before transfer to the loading box.

7. All tools used should be pre-cooled before contacting the sample.

8. Place the autogrids into the shuttle with the cell side facing up. Depending on which type of shuttle is used, you can load 1–2 autogrids at a time.

9. After attaching the transfer-unit's insertion rod to the shuttle, pump the vacuum in the slush chamber until liquid nitrogen freezes. Lift out the shuttle into the transfer unit and vent the slush chamber.

10. Attach the transfer unit to the airlock of the prep-chamber and press "pump vacuum." The vacuum has to reach a certain level to safely open the airlock valves. A layer of platinum can be sputter-coated in the prep-chamber to increase conductivity and provide additional protection to the sample.

11. After the vacuum in the prep-chamber reaches the desired level, open the valve between the prep-chamber and SEM. Make sure the SEM cryo-stage is in the loading position. Transfer the shuttle onto the cryo-stage with the insertion rod and then retract the rod. The transfer unit should be removed and the valves should be closed.

12. To begin milling lamellae, the electron and ion beams should be in the operational state.

13. Move stage to image position and adjust focus and astigmatism for the electron beam. Link Z to FWD after doing so.

14. Find a marker on the grid and adjust eucentric height so the ion and electron beams point to the same location.

15. An additional layer of organometallic platinum can be deposited using the gas ingestion system (GIS) in the SEM. A detailed description of the operation of the GIS can be found in a protocol developed by the Baumeister lab [33].

16. The milling process usually starts with a higher current like 0.3 nA or 100 pA depending on the settings your FIB/SEM has. Two rectangular areas several microns in width (in mammalian cells, usually 10 μm) are drawn above and below the target region. Leave 3–5 μm in between to avoid damage by high-current ions.

17. Reduce the current to 30 pA and redraw the pattern of milling to remove any material left. Leave ~1 μm between the two patterns.

18. Use a current of 10 pA to do the final cleaning and reduce the thickness of the lamella to 200–500 nm. For this last step of milling, an additional 1° tilt can be added so the final thickness of the lamella is more homogenous.

19. Multiple places on the grid can be milled. When done, transfer the shuttle back to the transfer unit, and close the valve.

20. After venting the airlock, place the transfer unit on the slush chamber and pump. When the vacuum has reached 10^{-2}, open the valve on the transfer unit and place the sample in liquid nitrogen. All steps in this part should be performed gently to avoid any damage to the lamella. The slush chamber can be pre-pumped beforehand so it can reach the desired vacuum faster.

21. The sample can then be stored in clean liquid nitrogen for cryo-ET.

2.3 Biological (Plant) Tissues

The following will focus on plant tissues; however, the workflow of High Pressure Freezing (HPF)—cryosectioning—cryo-ET can be adapted to other types of tissues suitable for HPF. Preparation of plant tissues for HPF and cryosectioning are discussed in further detail in references [34, 35]. We will focus here on two plant tissues: in vitro grown 6-day-old root tips and PSB-D *Arabidopsis* Landsberg *erecta* cells grown in liquid medium.

2.3.1 Tissue Cultivation

1. Liquid cultured PSB-D cells are grown in Murashig and Skoog with Minimal Organics (MSMO, *Sigma-Aldrich*) with sucrose pH 5.7 with added Kinetin (*Sigma-Aldrich*) and 1-Naphthaleneacetic acid (NAA) (*Sigma-Aldrich*). Dispatch 75 mL of media in four 250 mL flasks (*Pyrex No 4980*) and cultivate at 25 °C, continuous illumination (20 µE/m/s) with 120 rpm shaking.

2. Seeds are sterilized in bleach. 100 mL of commercial bleach is placed in a beaker in a large plastic Tupperware, in a chemical fumehood (*see* **Note 12**). Open 1.5 mL Eppendorf tubes containing the seeds are placed in a tube rack inside the plastic Tupperware. 3 mL of 100% HCl are added and the Tupperware is closed and left for 3–4 h. After treatment, the Tupperware is opened, and tubes are closed before transferring to a sterile hood. The seeds are aired out in the sterile hood by opening the tubes for 30 min. The bleach-HCl mix can be left in the chemical fumehood to evaporate.

3. Arabidopsis seedlings are grown in Murashig and Skoog medium + vitamins (*Duchefa Biochemie*), with 2-(N-morpholino)-ethanesulfonic acid (MES) buffer (*Euromedex*) and 3.5 g of plant agar (*Duchefa Biochemie*), pH 5.8 in square plastic petri dishes (*VWR*).

4. Sterilized seeds are resuspended in autoclaved DI water, then pipetted in horizontal lines onto the plates. The plates are then set vertically in a greenhouse at 22 °C, on a long day photoperiod (16 h, 100 µE/m/s) for a week to produce young root tips.

2.3.2 High Pressure Freezing of Plant Tissues

The size of plant cells (ranging from 15 to 50 μm), their large aqueous vacuoles and their pecto-cellulosic cell wall make them challenging to cryo-preserve. Optimal vitrification of whole plant cells can only be achieved by HPF and the samples must be less than 200 μm thick. In practice, any plant sample that can fit in the freezer carriers used for HPF can be vitrified. However, experience shows that some freeze better than others. In general, the following materials are required for a HPF session of plant tissues:

1. Dextran cryoprotectant solution: 25% and 40% solutions are needed for the freezing of *Arabidopsis thaliana* root tips and cultured cells. 5 mL of each is sufficient.

2. Benchtop mini-centrifuge.

3. Dissection stereomicroscope.

4. Single edged carbon steel razor blades (*EMS*).

5. *X-Acto* knife set (for decapsulating tool) (EMS) or hemostatic forceps (*EMS*).

6. Tooth picks.

7. Various EM grade precision forceps (EMS style 2, 5, 5×, and Dumont style 7).

8. Freezing carriers: two types can be used for cryosectioning of plant tissues: the 200 μm deep brass carriers described in more detail elsewhere [35] or the 100 μm carriers composed of Type A gold-plated copper dishes (3 × 0.50 mm with 0.1/0.2 mm cavities, *Technotrade International*) and Type B aluminum hats (3 × 0.50 mm with a 0.3 mm cavity, *Technotrade International*).

9. Liquid nitrogen.

10. High Pressure Freezer: Bal-Tec HPM 010.

    After preparing the above-mentioned materials, the process of HPF can be performed as follows:

2.3.3 Dissection and Freezing of Plant Tissues

1. Under a binocular loupe, put a drop of water on a glass slide.

2. Mount the bottom part of the type A carrier onto the Bal-tec HPM 010 freezing rod.

3. With a sharp razor blade, dissect the plant tissue of interest. The only requirement is that it is small enough to fit in the freezing planchette.

4. Quickly transfer the freshly dissected tissue to the type A carrier using the appropriate tools (forceps or toothpicks) and add a few μL of 25% Dextran until the liquid forms a shallow dome over the surface of the carrier.

2.3.4 Liquid Cultured
Plant Cells

Cell pellets should be compact and dense. Hence, cells grown in liquid culture must be concentrated, usually by centrifugation before freezing. The sample should resemble a thick paste.

1. Spin down a 1.5 mL aliquot of the liquid culture using a tabletop centrifuge for 1 min at $1000 \times g$.

2. Discard the supernatant and resuspend in a small volume of 40% dextran.

3. Mount the bottom part of the brass carrier onto the Bal-tec HPM 010 freezing rod.

4. With a toothpick transfer the cells into the freezing carrier and add a drop of 40% dextran.

2.3.5 Sample Position
Strategy

How the sample is positioned in the carrier will determine the cutting angle of the tissue (transversal or longitudinal) and can make the trimming process easier and more efficient.

Because each project and sample have their own requirements, here we will detail the advantages and disadvantages of two types of carriers used in our laboratory.

1. **200 μm brass carrier (used for PSB-D cell pellets)**

    In this system, the sample + cryoprotectant filler are laid in the bottom part of the carrier to form a prominent bubble, which is then enclosed by the concave top of the carrier (Fig. 3a). This system allows rapid trimming because the sample is more prominent; however, vitrification is only optimal in the edges of the sample, with the deeper parts usually forming hexagonal ice. This type of holder is appropriate for dense pellets of cells, where direction and area do not matter since the cells are randomly oriented in the pellet.

2. **100 μm type A gold-plated copper carrier (used for *Arabidopsis* root tips).**

    In this carrier, the sample + cryoprotectant filler forms a flat film within the rim of the carrier (Fig. 3b). This provides the advantage of homogenous vitrification throughout the thin sample. The disadvantage is that the carrier needs to be trimmed from the side until the sample is reached (Fig. 3b). This system is appropriate for tissues less than 100 μm thick that require sectioning in a particular orientation.

    After positioning the sample in the desired way inside the carrier, place the lid of the carrier on top of the sample, and secure it on the freezing rod and subsequently put it into the slot of the Bal-tec HPM 010 and perform HPF. From this point on, the samples should never be taken out of liquid nitrogen.

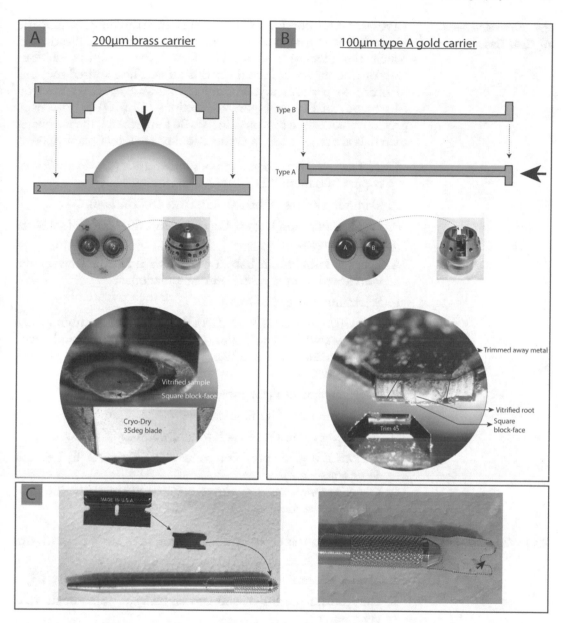

Fig. 3 (a) Top: schematic of a 200 μm brass carrier. Dashed arrows indicate how the system closes and red arrow indicates direction of sectioning. Green to yellow gradient indicates vitrification efficiency of sample (green = vitreous, yellow = hexagonal ice). Middle: top views of the carrier and its corresponding specimen holder. Half #2 lays flat in the specimen holder (dashed circle and arrow). Bottom: carrier containing a vitrified cell pellet mounted in the specimen holder in the cryo-ultramicrotome chamber facing the cryo-dry 35° knife. The block-face is visible on the cryo-planed surface. (**b**) Top: schematic of type A carrier. Dashed arrows indicate how the system closes and red arrow indicates direction of sectioning. Middle: top views of the type A and B halves and corresponding specimen holder. The carrier is mounted vertically in the clamp of the specimen holder (dashed circle and arrow). Bottom: trimming through the carrier to reach the sample. (**c**) Home-made decapsulating tool used to open the brass carriers after freezing. The screw tightening pen is outlined in green and the lever, carved from a razor blade in red. The groove where the carrier is placed is indicated by a black arrow

2.4 Cryosectioning of Plant Tissues

Cryosectioning consists of two steps: (a) trimming, where the block-face is carved out [35], and (b) cryosection collection in which the block-face is sectioned and the resulting sections and/or ribbons are collected on EM grids. The required material in order to perform cryosectioning followed by Cryo-Electron Microscopy of Frozen Hydrated Sections (CEMOVIS) has been previously addressed thoroughly [35, 36]. In general, the following materials are required for a cryosectioning session of plant tissues:

1. 200 mesh carbon-coated copper finder grids (*Ted Pella Inc*) or copper 200 mesh R2/2 NH2 Quantifoil grids.

2. Trimming knives: Trim 20 and Trim 45 (*Diatom*).

3. Section collection knives: Cryo-dry 25°, 35°, or 45° (*see* **Note 13**) (*Diatom*).

4. Hair (eyelash, blonde baby, or Dalmatian) glued to the tip of a wooden stick for cryosection manipulation.

5. Static line ionizer (*Diatom*).

6. Leica *Ultracut UCT* with *Leica EMFCS* cryo-chamber (*Leica Microsystems*) or equivalent cryo-ultramicrotome and associated microtools (pen-like pressing tool, freezing carrier holders, etc.).

7. Micromanipulator (for more stable ribbon collection).

8. Large amounts of liquid nitrogen.

9. Hair dryer or heating bench to warm and dry the tools.

10. Various EM grade precision forceps (EMS style 2, 5, 5×, and Dumont style 7).

After preparing the above-mentioned materials, the process of cryosectioning can be performed as follows:

2.4.1 Preparation

1. Cool down the cryo-ultramicrotome to −165 °C (*see* **Note 14**).

2. Mount the carrier in the cryo-ultramicrotome (*see* **Note 15**).

3. Decapsulate the lid of the carrier using a home-made tool (Fig. 3c).

4. Install and cool the knives, setting them as far as possible from the specimen holder to prevent accidental heat transfer.

2.4.2 Trimming

1. Light the system from below in order to highlight the knife-to-sample distance.

2. Bring the knife close to the sample, using the command panel and/or sliding the knife platform forward. Use the command panel only in the final stages of approach.

3. If the block has already been started, line up the knife with the cryo-planed surface by rotating the knife holder until the thin light sheet emerging below is of homogenous thickness.

4. Introduce the cryostat into the chamber and set to 25–30%.

2.4.3 Trimming Domes from the 200 μm Carrier

1. Manually advance the block for a few microns and then set it on automatic mode with a feed <100 nm and a cutting speed between 80 and 100 mm/s until reaching a depth of approximately 10 μm. The circular cryo-planed area should have a very smooth, reflective surface without any irregularities or holes. If not, vitrification is not good, and the sample should be discarded.

2. Cut 50–100 μm more on the left and right sides of the cryo planed circle to generate a prominent thin vertical rectangle.

3. Rotate the specimen holder by 90° in order to make a square block-face.

4. Finish block-face by repeating **step 2** above.

2.4.4 Trimming the Type a 100 μm Deep Carrier

This method entails cutting through metal and sample, two materials with different cutting properties. Trimming steps should favor as much as possible cutting through only one material at a time. When cutting through both cannot be avoided, the cutting speed should be reduced.

1. Orient the specimen holder so that the carrier is presented vertically to the knife.

2. Bring the knife to the carrier.

3. Trim through the metal at 70 nm feed and 60–80 mm/s speed. It will take approximately 800 μm of trimming to reach the sample.

4. When the sample is reached, reduce the trimming speed to 50 mm/s. At this point, you will be cutting through the dextran and the sample will look like a clear circular structure.

5. Continue cutting until the sample enlarges horizontally to fill the full 100 μm depth of the carrier. At this point, there should be a vertical separation between the sample and the bottom of the carrier.

6. Shift the knife in order to cut only through the vertical metal and not the sample.

7. Trim 200 μm deep, gradually approaching the sample at 100 mm/s (*see* **Note 16**).

8. When both the sample and metal are being trimmed, lower the speed to 50 mm/s.

9. When the vertically oriented sample is well cleared on both sides, rotate 90°.

10. Trim the metal on both sides at 100 mm/s to a depth of 200 μm.

11. Trim through the frozen dextran on each side in stages, each approximately one-fourth of the width of the diamond blade, until the final width of the block-face is reached.

The section collection technique is precisely described in reference [35] and thus will not be repeated here. The nominal cryosection thickness should be around 100 nm for tomography.

2.5 Other Samples

These include any purified organelle, compartment [37], in vitro assembly [38], or purified protein complex. The sample should be purified and concentrated as gently as possible to avoid damage [28, 39]. As many of these samples are thin, they can be suitable for high-resolution tomography. These samples are compatible with the different material sections discussed above. However, for high-resolution samples, we typically use Protochip C-flat 2/2 holey carbon grids with 300 mesh, and mix the sample with 10 nm gold fiducials.

3 Data Collection

A detailed protocol for data collection and processing is available in reference [40].

3.1 Data Acquisition Using SerialEM: Conventional Tilt-Series

After loading the samples to the microscope (in our lab a Titan Krios) and assuming that the microscope is already aligned, data collection can be started. Currently, we use SerialEM [41] to collect our data. We refer the reader to the tutorial of SerialEM for detailed discussion of what the software can do. In general, we start by making an atlas of the grid at low magnification (82×). This will give a general idea of how the grid looks and the ice gradient. Subsequently, a montage of some good squares (at magnification 2245×) is collected providing a better view of the cells on these squares. Subsequently, points are added on cells in the holes (or other interesting cells in the montage). Then, the stage is moved to each of these points and a preview image (with low dose) is taken for that cell. The position of the cell is optimized, for instance, centered or offset if one wants to collect data from a pole of the cell). After that, a view image (at medium magnification, 2245×) is taken and is saved and used as the anchor state. View images are used to help tracking and aligning the target during data collection. One can further proceed with the process for all the required targets by pressing the "Anchor map" button. At the end, click the "TS" (tilt-series) option for the targets where data will be collected and tilt-series parameters are filled in as required (e.g., starting and ending angles, increment degrees, defocus, how to change exposure time with tilting angle). Once that is done, click on "Navigator" in the main menu and go to "Acquire at points", where data collection can be started. Usually, we select "Rough eucentricity", "Autofocus", and "Autocenter beam" for batch collection. To calculate the exposure time, we use the following equation:

$$\text{Exposure time (s)} = \left[\text{Total Dose} \left(e^-/\text{Å}^2\right) \times \left(\text{Pixel size (Å)}\right)^2\right]$$
$$/\left[\left(\text{Dose measured } e^-/\text{pix/s}\right) \times \left(\text{Number of images in the tilt} - \text{series}\right)\right]$$

where

Total dose is the total dose desired, usually between 100 and 200 $e^-/\text{Å}^2$.

Pixel size depends on the magnification used to collect data.

Dose measured is the number of electrons ($e^-/\text{pix/s}$) passing through the hole. For K3 cameras, this can be up to 30 $e^-/\text{pix/s}$ on the sample area.

Number of tilt images in the tilt-series depends on the maximum and minimum tilt angles and the angular increment (e.g., a tilt-series from $-50°$ to $50°$ with $1°$ step will have 101 images).

3.2 Rapid Tilt-Series Acquisition

Two rapid tilt-series methods are currently being developed on a Titan Krios (Thermo Fisher Scientific) equipped with a single-axis holder: continuous-tilting and fast-incremental [31]. In continuous-tilting, the camera continuously records frames as the sample is tilted. For this reason, only unidirectional and bidirectional tilt-schemes are possible. In fast-incremental, even though the camera also records continuously, the beam shutter is used to rapidly expose the target at discrete tilt angles, resulting in blank frames while the stage is tilting, and allowing more complex tilt-schemes, such as dose symmetric [14]. Figure 4 shows examples of cellular features reserved in tomograms collected using both methods.

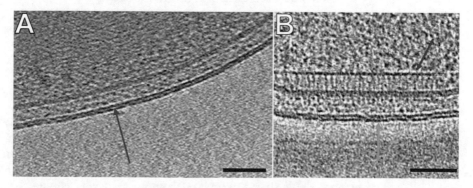

Fig. 4 Snapshots taken from tomograms acquired using the (**a**) continuous-tilting, and (**b**) fast-incremental methods display visible features. (**a**) The double leaflet of the outer membrane lipid bilayer (blue arrow). (**b**) Side view of a chemoreceptor array (blue arrow), and the outer membrane lipid bilayer (green arrow). Figure adapted from Ref. 31

3.2.1 Pre-data Collection Steps

Because the target is not tracked during tilting, both methods require accurate eucentric height measurement prior to collecting data to minimize field-of-view loss. While this is currently being optimized, we describe one method to measure eucentric height below.

1. Find and center the desired target in the field of view at "Record" magnification in low-dose mode.

2. Perform a rough eucentricity adjustment routine to adjust the Z-height to the eucentric point by clicking the "Eucentric–Rough" command in the Tasks menu.

3. Run the "Realign to item" command to recenter the target in the field of view after rough eucentricity.

4. Perform the fine eucentricity routine in the Trial area, at the same magnification as data collection. First, check the "Use Trial in LD Refine" command in the Tasks menu, then click the "Eucentric–Fine" command. Ensure the Trial area is unobstructed.

3.2.2 Continuous Tilting Data Collection in SerialEM

For a unidirectional tilt-scheme, from −60° to +60°:

1. Set the exposure time in the Record settings to be long enough to record the entire tilt-series.

2. Turn on dose fractionation and select an adequate time per frame.

3. Tilt to −60°.

4. Run the command "TiltDuringRecord #A #D" to record while tilting the stage to angle #A after a delay of #D in milliseconds. For instance, TiltDuringRecord 60 1000.
   For a bidirectional tilt-scheme, starting from 0°:

5. Set the exposure time in the Record settings to be long enough to record each direction of the tilt-series.

6. Turn on dose fractionation and select an adequate time per frame.

7. Tilt to 0°.

8. Run the command TiltDuringRecord −60 1000.

9. Return to 0°.

10. Run the command TiltDuringRecord 60 1000.

3.2.3 Fast-Incremental Data Collection in SerialEM

SerialEM version 3.7 includes the FrameSeriesFromVar command, which allows the collection of fast-incremental data given an input array variable that describes each step of the tilt-scheme.

1. Generate the array variable containing the following information for each tilt, separated by spaces: Tilt Angle, Shutter open time (exposure time), Wait time between step (before opening

the shutter), change in defocus, image shift in X, image shift in Y. Units for the last three entries are all in microns. By saving the array as a persistent variable in SerialEM, this step only needs to be done once per tilt-scheme.

2. A typical fast-incremental SerialEM script includes the following commands:

   (a) "FrameThresholdNextShot 0.1" automatically skips saving blank frames based on a mean value below a threshold of 0.1. This threshold can be adjusted as desired.

   (b) "FrameSeriesFromVar var 31 #d", where #d is an initial delay time, followed by the Record command to record movie frames.

   (c) ReportLastFrameFile, to assign the filename to reported Value1.

   (d) "WriteFrameSeriesAngles $reportedValue1.angles" writes the tilt angles collected into a text file appended with . angles for post-processing.

3.3 Data Collection for High-Resolution (Near-Atomic) Tomography

Data collection for high-resolution tomography begins with determining the desired resolution, and choosing parameters based on both the desired resolution and the features of the object of interest. Collection is typically optimized through one or more sessions, with alignment and subtomogram averaging resolution used to determine optimal settings.

For higher-resolution collection, samples must be thin (ideally less than 200 nm) and fiducials must be plentiful (typically 20 or more beads in the field of view). The data collection is typically reminiscent of a single-particle averaging collection, with a pixel size of less than 2 Å, defocus gradient of −1 to −5 μm, and dose-fractionated frame collection [28, 42]. If subtomogram averaging is intended, a phase plate is not recommended. The tilt-scheme begins at zero degrees and is acquired with symmetric dose alternating between positive and negative angles [28]. Optimal dose is determined based on the point at which the sample displays visible damage, but is often approximated at 120 electrons/$\mathrm{Å}^2$ with the highest accumulated doses at the highest tilt angles. Tilt increment and dose have frequently been set to −60° to +60° in 3° increments, but smaller tilt increments or a larger range may be beneficial for some samples.

Also, particular attention is paid to minimize the dose used to find and center the target and for setting the eucentric height and to minimize drift during exposure. Movies can be collected to motion-correct any residual drift. In addition, it is recommended to use off-targets focus spots to maintain focus throughout tilt-series collection. Finally, the beam has to be highly coherent and parallel to preserve high-resolution signals.

3.4 Imaging Fluorescent Targets in the Titan Krios Using Cryo-CLEM

1. Use SerialEM to collect a 200× magnification montage for each EM grid loaded into the Titan Krios (*see* **Note 17**).

2. Find a cell of interest imaged previously by cryo-LM using the letters, symbols, or numbers of the grid support.

3. Use the whole-grid 200× montage to generate a polygon montage of the thin cellular peripheral area of interest and surrounding area. It is advisable to capture fluorescent microspheres in the surrounding area in this montage for registration of the EM and LM images (*see* **Note 18**).

4. Using your polygonal montage, evaluate if the ice is too thick for tilt-series collection by evaluating the hole that is closest to the fluorescent puncta using the cryo-LM aligned phase contrast/fluorescence images for orientation.

5. If the ice is thin enough, overlay the fluorescence images with the polygonal montage images using the fluorescent microspheres seen in both the LM and EM modalities. Any image processing software can be used to roughly align the images.

6. Next, localize more precisely the location of the fluorescent puncta in the thin edge.

7. Collect a tilt-series in this location.

4 Data Processing and Tomographic Reconstruction

4.1 Conventional Tilt-Series

We reconstruct tomograms from raw tilt-series using the IMOD software [41, 43, 44]. We refer the reader to the useful IMDO tutorial (http://bio3d.colorado.edu/imod/doc/tomoguide. html). We also refer the reader to our recently published chapter [32] where a similar procedure is described for bacterial secretion systems.

The tilt-series is opened ("Build Tomogram") using the IMOD program Etomo's graphical user interface (GUI). Fill the diameter of the gold fiducials and specify the tilt-series axis type. Click "Scan header" to scan the pixel size and image rotation. Then, click "Create Computer Scripts". Following this, several steps will appear on the GUI that need to be performed. These steps include:

4.1.1 Pre-processing

During this step, pixels with very high or low values are removed using "Ccderaser" to circumvent the generation of artifacts. Use value 10 for "Peak criterion" and 8 for "Difference criterion". Generate a stack by clicking "Create Fixed Stack", then check the quality of the stack by clicking "View Fixed Stack" and remove any bad images from the stack then click "Use Fixed Stack".

4.1.2 Coarse Alignment

The aim of this step is to align the images of the tilt-series using cross-correlation before producing a finer-aligned stack using the fiducials. Click first "Calculate Cross-Correlation" and then "Generate Coarse Aligned Stack." Check the aligned stack by clicking "View Aligned Stack in 3dmod." Remove any bad images from the stack and create and use a new stack lacking the bad images.

4.1.3 Fiducial Model Generation

If the sample has gold fiducials, these can be tracked, usually using the "Make seed and track" or "Raptor" option. Generally, we select between 10 and 30 beads. We usually begin with "Make seed and track", activate the "Refine center with Sobel filter" provide a value of 8, and then click "Generate Seed Model" in the "Track beads" subwindow. After that, click "Fix Fiducial Model" to fix any untracked or badly tracked beads manually. If the tracking using "Make seed and track" is bad, one can try tracking the beads using "Raptor". Tracking with "Patch Tracking" is usually used for eukaryotic samples which lack gold beads in many cases.

4.1.4 Fine Alignment

Click "Compute Alignment" to correct the inappropriately placed fiducials by clicking "View/Edit Fiducial Model."

4.1.5 Tomogram Positioning

A tomogram thickness value is provided (usually a thickness of 600 is used for many single-small-cell samples). Click "Create Whole Tomogram" with a binning of three and after that "Create Boundary Model". Rotate the tomogram 90°, save the model, and click on "Create Final Alignment".

4.1.6 Final Aligned Stack

Here, we only perform "Create Full Aligned Stack". However, one can also apply "Contrast Transfer Function (CTF) correction", "Erase Gold", and/or "2D Filter".

4.1.7 Tomogram Generation

The aligned image stack is used to build the tomogram by using either: (a) Weighted Back Projection or (b) Simultaneous Iterative Reconstruction Technique (SIRT-like). The generated tomogram can be viewed by clicking "View Tomogram in 3dmod."

4.2 Fiducial-Less Patch Tracking Using IMOD

Sometimes it is difficult to add gold fiducial markers to a specimen, particularly FIB-milled lamellae or cryosections. In these cases, the alignment is usually done by image correlation within. Here, we briefly describe how to reconstruct data without fiducials by using the patch tracking method in IMOD. More detailed information can be found on the IMOD website: https://bio3d.colorado.edu/imod/doc/patchTrackExample.html

1. Raw tilt-series are usually dose-weighted to boost the contrast of higher angle tilt images.

2. Data are imported and pre-processed in IMOD as described before.

3. In the "Fiducial Model Generation" step, select "Use patch tracking to make fiducial model."

4. Change the patch size, for cryo-tomograms a larger patch size is recommended. Note that you can use the "Use boundary mode" option to only patch-track a subarea of the tile-series. This is often used to avoid areas with few features or with contamination.

5. The "Advanced" button at the bottom gives options to play with filter parameters, trimming pixels and range of views, etc.

6. Check "Break contours into pieces" to perform robust fitting with a tuning factor.

7. Click "Track Patches."

8. After this is done, go to "Fine Alignment." Make sure to check "Do robust fitting with tuning factor" and turn on "Find weights for contours, not point". Click "Compute Alignment." This will give you a mean residual in the log window.

9. Press "View/Edit Fiducial Model." All the tracks will be shown with overlapping segments. Tracks with obvious mistracking can be deleted. If needed, go back to **step 4** above to refine the region used for tracking until you reach the desired alignment quality.

4.3 Rapid Tilt-Series Processing

4.3.1 Continuous tilting data

Due to low signal-to-noise ratio (SNR) of individual images, some processing is required to enhance the contrast for fiducial tracking. One strategy we use is to sum neighboring images to enhance the contrast of fiducials at each tilt angle (Fig. 5). This requires neighboring images to first be stretched by an amount equal to the cosines of its tilt angle and of the tilt angle of the reference image, followed by motion correction. These functions have been coded in the Neighbor-enhance script, which requires the IMOD package.

1. Download the Neighbor-enhance script and the accompanying rotmagstr program (https://github.com/chreifi/fasttilt) and add both to your list of environment variables on your linux machine.

2. Run the Neighbor-enhance script to gain normalize, correct defects, remove deviant pixels, stretch, align, and sum images, and perform a rough alignment of each summed image of the tilt-series. Outputs are the raw data after gain normalization, defect correction, and deviant pixel removal (.st), a roughly aligned contrast-enhanced tilt-series (.preali file) and the transforms (.xf, .xg).

3. Run the eTomo software in the IMOD package, using the raw tilt-series as input. Enter values for the tilt axis angle, start angle, and an estimate of tilt increment.

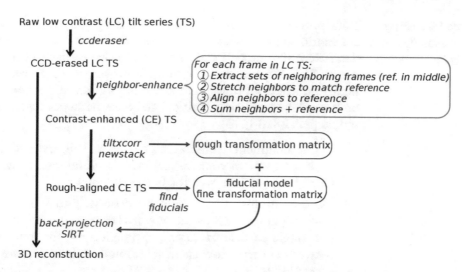

Fig. 5 Workflow for processing data acquired using the continuous-tilting method. Font color is black for tilt-series data, red for programs and scripts, and blue for alignment transforms. A general description of the Neighbor-enhance script is shown in the green box. Figure adapted from Ref. 31

4. Skip directly to seed model creation to generate a seed model and track the beads throughout the tilt-series. These steps are identical to conventional tilt-series alignment.

5. Using IMOD, the transforms generated by the fiducial model will automatically be applied to the raw tilt-series (original input). The reconstruction step is thus also identical to conventional tilt-series as shown below.

4.3.2
Fast-Incremental Data

The current SerialEM version (3.7) produces a single stack of movie frames containing the entire tilt-scheme data and skips saving blank frames based on a user-defined threshold. Nevertheless, the program automatically pads the tilt-series with one blank frame at the start, and one at the end, which first need to be removed. Frames that are part of each discrete tilt angle must then be extracted, motion-corrected, and placed into the final tilt stack in the correct order. We have written a bash script that performs these post-processing steps.

1. Download the Fast-incremental script (https://github.com/chreifi/fasttilt) and add it to your list of environment variables on your linux machine.

2. Execute the script, providing the following files for input: input tilt-series, angles.txt file obtained from running the WriteFrameSeriesAngles function in SerialEM, as well as gain reference and defects file. The script also provides options for doing motion correction using Alignframes or Motioncor2.

3. Once successfully processed, the output tilt stack from the Fast-incremental script can be processed in IMOD just as in conventional cryo-ET.

4.4 Data Processing for High-Resolution Tomography

Data processing for high-resolution tomography mirrors that of conventional data processing, but with additional steps to optimize data. Recently, user-friendly software packages have been developed (EMAN2, emClarity, relion); however, to date every novel high-resolution subtomogram averaging structure has been based on a workflow utilizing multiple software packages and personal scripts [27, 28, 38, 42].

1. Frame alignment: gain-correct, rotate, and align the tilt frames in the software of your choice. IMOD alignframes or Motion-Cor2 both perform well for frame alignment.

2. Dose-weighting: tilts should be dose-weighted prior to tilt stack alignment. All available software methods are based on the single-particle averaging estimation of the relationship between electron dose and final resolution from the Grigorieff group [45].

3. CTF estimation: the CTF can be estimated with any standard software package. However, CTF estimation software routinely underperforms at high tilt angles. It is common to test multiple packages for different data collections or to use homemade software to assess or improve outputs.

4. Reconstruction: the most common reconstruction method is weighted back projection, coupled with either two-dimensional CTF correction in IMOD or three-dimensional CTF correction in NovaCTF [46]. For low-defocus tomograms, an SIRT-like filter can be applied during reconstruction for visualization purposes.

5. Subtomogram averaging: for a low particle count (100s), high resolution is not expected and PEET or RELION may be used as in conventional tomography. For particle counts above 10,000, high-resolution subtomogram averaging often involves parallelized GPU processing and a highly sample-specific workflow. Dynamo and AV3 have both performed to high resolution on novel structures [28, 42] and are flexible for a customized workflow. emClarity and EMAN2 have performed to high resolution on a dataset previously solved in AV3, but are less flexible for advanced users.

4.5 Subtomogram Averaging

In subtomogram averaging, certain subvolumes are cropped from the tomogram, aligned, and then averaged to enhance the signal-to-noise ratio of a certain repetitive feature (e.g., a protein complex of interest) within a single tomogram or across different tomograms of the same species collected with similar pixel size. Many software packages are currently available to perform subtomogram averaging, including: Dynamo, PEET, RELION, EMAN2, and PyTOM (*see*, e.g., Refs. 47–52).

As an example, we will discuss how to perform subtomogram averaging using PEET, a routine procedure we do in our group. Open the tomogram using the command 3dmod and activate the "Toggle between regular and high-resolution image" and "keep current image or model point centered" buttons. Activate the "Model" mode in the 3dmod window and set the slicer thickness to ten and then go to "Edit" and select "Angles". A new window will appear where the coordinates of each one of the chosen particles can be viewed. Find the optimal view of the complex of interest by rotating the particle and moving through the different slices of the tomogram and then save these coordinates and angles. Create a model point for each particle with a middle mouse click and then save this model.

Before averaging, compute an initial motive list (MOTL) by running the command "stalkInit name.mod." Usually, we create a separate folder for the PEET run and then subfolders entitled "run1", "run2", etc. In "run1", load all the tomograms, the MTOL files, and the ".tlt" files (which have the tilt angles required for missing wedge compensation). In the first run, use one of the good particles as a reference without a mask. Subsequently, use the obtained structure from each run as a reference for the next run. Different masks can be constructed in IMOD to perform focused alignment on different parts of the protein complex. This process is repeated until the structure cannot be improved anymore. If the symmetry of the complex is known, this symmetry can be imposed on the structure; however, one should bear in mind that different parts of the complex might have different symmetries and imposing one specific symmetry might affect other parts of the complex that have different symmetries. If the symmetry is unknown, we usually apply two-fold symmetry along the mirror plane to improve the signal-to-noise of the side view (checking that no new densities appear compared to the non-symmetrized structure). Note that when two-fold symmetry is applied to a complex with unknown symmetry, no conclusions should be made on the top view of that average.

5   Notes

1. The use of phenol red-free media is advised to prevent the generation of unwanted autofluorescence at 80 K when imaging the cells using cryo-LM [53].

2. Over-digestion with trypsin may prevent the cells from adequately attaching and spreading on the grid.

3. Different types of films (silicon vs. carbon) and shapes of grids support (square vs. hexagonal) are available. For U2OS cells,

R2/2 square supports coated with Quantifoil holey carbon film are recommended. The R2/2 square support allows a sufficient area for the cells to adequately attach and spread on the grid.

4. To prevent the grid from sliding on the glass slide, both sides of the grid are coated with collagen.

5. We recommend touching the grid to the edge of the glass slide to gain optimal purchase.

6. Due to their increased contrast in low signal:noise tomograms of mammalian cells, 20 nm gold beads are used as fiducial markers to align the tilt-series. If performing cryo-CLEM, fluorescent microspheres are mixed and vortexed in 1:1 dilution with the 20 nm gold beads before being added to the grid. It is recommended that the gold beads be concentrated as much possible to ensure their presence around the thin edges of cells.

7. The grid is blotted from the side where the cells are not attached so as to prevent damaging the cells during the blotting process. Blotting is achieved by using a strip of blotting paper which has been folded 90° at its tip and held by tweezers which enter the Vitrobot chamber from the side port.

8. A liquid ethane/propane mixture is used to avoid solidification as when using ethane alone.

9. In our experience, the signal from GFP can be too weak to detect due to high background autofluorescence in cryogenic conditions. Using the brighter mNeonGreen fluorescent protein is recommended.

10. Collection of phase contrast and red, green, and blue fluorescence images is needed to distinguish autofluorescence from real fluorescent protein signal [53]. For example, in U2OS cells the blue channel can be used to distinguish the 0.5 μm Tetra-Speck Microspheres (which fluoresce bright red, green and blue, and are seen by phase contrast) from autofluorescence (bright red and green, dim blue), and real green fluorescent protein signal (bright green). We recommend plotting the intensities of puncta in the red, green, and blue channels in unlabeled cells vs. cells expressing the fluorescent protein to identify targets.

11. Aligning the fluorescent images to the fluorescent microspheres seen in the phase contrast image can help identify whether the target fluorescent protein signal is present in an area of the ice that has sufficient electron transparency.

12. Mixing HCl and bleach creates very hazardous chlorine gas therefore this should be done in a chemical fumehood with an exhaust leading to the exterior of the building rather than a hood which recycles air in the same room.

13. The lesser the angle of the knife, the less compression artifacts (crevassing) will occur on the cryosections, and the more fragile the blade will be. While a 25° cryo-dry knife will give the best results, it should be reserved for experts as it is fragile. The 35° knife is a good compromise between sturdiness and compression artifacts.

14. The devitrification temperature of water is −135 °C. Above this temperature, the sample will devitrify and cubic ice will nucleate and spread. The state of the ice in cryosections can be checked by cryo-electron diffraction.

15. For the cutting to be smooth, everything in the chamber needs to be tightened in order to eliminate sources of vibration during the process.

16. In order to optimize trimming time, it is good to trim as much metal as possible during the first trimming session. This way, when the sample block-face needs to be cleared after several section collection sessions, one will only need to trim 50 μm around the existing block-face, instead of going through the whole process again.

17. For high-resolution imaging and to prevent premature radiation damage, we recommend that each image of the montage be collected at a low-dose ($0.5 \ e^-/Å^2$) and far from focus (−40 μm) to maintain good contrast of the outline and thin edges of the cells.

18. Before collecting the polygon montage, align the whole-grid $200 \times$ montage with the microscope X and Y stage coordinates using the "shift to marker" function in SerialEM.

Acknowledgments

Cryo-ET work in the Jensen lab is supported by NIH grants R35 GM122588, RO1 AI27401, and P50 AI150464 and by the Howard Hughes Medical Institute and performed in the Beckman Institute Resource Center for Transmission Electron Microscopy. M.K. acknowledges a Rubicon postdoctoral fellowship from De Nederlandse Organisatie voor Wetenschappelijk Onderzoek (NWO). We thank Catherine M Oikonomou for reading and editing the manuscript.

References

1. Sigal YM, Zhou R, Zhuang X (2018) Visualizing and discovering cellular structures with super-resolution microscopy. Science 361:880–887

2. Hänsel R, Luh LM, Corbeski I, Trantirek L, Dötsch V (2014) In-cell NMR and EPR spectroscopy of biomacromolecules. Angew Chem Int Ed 53:10300–10314

3. Kaplan M et al (2015) Probing a cell-embedded megadalton protein complex by DNP-supported solid-state NMR. Nat Methods 12:649–652

4. Kaplan M et al (2016) EGFR dynamics change during activation in native membranes as revealed by NMR. Cell 167:1241–1251.e11

5. Oikonomou CM, Jensen GJ (2017) A new view into prokaryotic cell biology from electron cryotomography. Nat Rev Microbiol 15:128

6. Lučić V, Rigort A, Baumeister W (2013) Cryo-electron tomography: the challenge of doing structural biology in situ. J Cell Biol 202:407–419

7. Chang Y-W et al (2016) Architecture of the type IVa pilus machine. Science 351:aad2001

8. Kaplan M et al (2019) The presence and absence of periplasmic rings in bacterial flagellar motors correlates with stator type. elife 8: e43487

9. Chaban B, Coleman I, Beeby M (2018) Evolution of higher torque in campylobacter-type bacterial flagellar motors. Sci Rep 8:97

10. Kaplan M et al (2019) In situ imaging of the bacterial flagellar motor disassembly and assembly processes. EMBO J 38:e100957. https://doi.org/10.15252/embj. 2018100957

11. Ferreira JL et al (2019) γ-proteobacteria eject their polar flagella under nutrient depletion, retaining flagellar motor relic structures. PLoS Biol 17:e3000165. https://doi.org/10.1101/ 367458

12. Zhu S et al (2019) In situ structures of polar and lateral flagella revealed by Cryo-Electron tomography. J Bacteriol 201:13

13. Zhao X et al (2013) Cryoelectron tomography reveals the sequential assembly of bacterial flagella in Borrelia burgdorferi. Proc Natl Acad Sci 110:14390–14395

14. Hagen WJH, Wan W, Briggs JAG (2017) Implementation of a cryo-electron tomography tilt-scheme optimized for high resolution subtomogram averaging. J Struct Biol 197:191–198

15. Briggs JAG (2013) Structural biology in situ— the potential of subtomogram averaging. Curr Opin Struct Biol 23:261–267

16. Leigh KE et al (2019) Subtomogram averaging from cryo-electron tomograms. Methods Cell Biol 152:217–259

17. Murphy GE, Leadbetter JR, Jensen GJ (2006) In situ structure of the complete Treponema primitia flagellar motor. Nature 442:1062–1064

18. Chen S et al (2011) Structural diversity of bacterial flagellar motors: structural diversity of bacterial flagellar motors. EMBO J 30:2972–2981

19. Zhao X, Norris SJ, Liu J (2014) Molecular architecture of the bacterial flagellar motor in cells. Biochemistry 53:4323–4333

20. Gold VA, Salzer R, Averhoff B, Kühlbrandt W (2015) Structure of a type IV pilus machinery in the open and closed state. elife 4:e07380

21. Hu B, Lara-Tejero M, Kong Q, Galán JE, Liu J (2017) In situ molecular architecture of the salmonella type III secretion machine. Cell 168:1065–1074.e10

22. Ghosal D, Chang Y-W, Jeong KC, Vogel JP, Jensen GJ (2017) In situ structure of the legionella dot/Icm type IV secretion system by electron cryotomography. EMBO Rep 18:726–732

23. Chang Y-W, Shaffer CL, Rettberg LA, Ghosal D, Jensen GJ (2018) In vivo structures of the helicobacter pylori cag type IV secretion system. Cell Rep 23:673–681

24. Ghosal D et al (2019) Molecular architecture, polar targeting and biogenesis of the legionella Dot/Icm T4SS. Nat Microbiol 4:1173–1182

25. Yang W et al (2019) In situ conformational changes of the *Escherichia coli* serine chemoreceptor in different signaling states. MBio 10: e00973-19

26. Rapisarda C et al (2019) *In situ* and high-resolution cryo-EM structure of a bacterial type VI secretion system membrane complex. EMBO J 38:e100886

27. Wan W et al (2017) Structure and assembly of the Ebola virus nucleocapsid. Nature 551:394–397

28. Schur FKM et al (2016) An atomic model of HIV-1 capsid-SP1 reveals structures regulating assembly and maturation. Science 353:506–508

29. Rigort A et al (2012) Focused ion beam micromachining of eukaryotic cells for cryoelectron tomography. Proc Natl Acad Sci 109:4449–4454

30. Danev R, Buijsse B, Khoshouei M, Plitzko JM, Baumeister W (2014) Volta potential phase plate for in-focus phase contrast transmission electron microscopy. Proc Natl Acad Sci U S A 111:15635–15640

31. Chreifi G, Chen S, Metskas LA, Kaplan M, Jensen GJ (2019) Rapid tilt-series acquisition for electron cryotomography. J Struct Biol 205:163–169

32. Ghosal D, Kaplan M, Chang Y-W, Jensen GJ (2019) In situ imaging and structure determination of bacterial toxin delivery systems using electron cryotomography. In: Buchrieser C, Hilbi H (eds) Legionella, vol 1921. Springer, New York, pp 249–265

33. Schaffer M et al (2015) Cryo-focused ion beam sample preparation for imaging vitreous cells by cryo-electron tomography. Bio Protoc 5: e1575

34. Nicolas W, Bayer E, Brocard L (2018) Electron tomography to study the three-dimensional structure of plasmodesmata in plant tissues—from high pressure freezing preparation to ultrathin section collection. Bio Protocol 7:1

35. Ladinsky MS (2010) Micromanipulator-assisted vitreous cryosectioning and sample preparation by high-pressure freezing. Methods Enzymol 481:165–194

36. Ladinsky MS, Pierson JM, McIntosh JR (2006) Vitreous cryo-sectioning of cells facilitated by a micromanipulator. J Microsc 224:129–134

37. Iancu CV et al (2007) The structure of isolated Synechococcus strain WH8102 carboxysomes as revealed by electron cryotomography. J Mol Biol 372:764–773

38. Kovtun O et al (2018) Structure of the membrane-assembled retromer coat determined by cryo-electron tomography. Nature 561:561–564

39. Metskas LA, Briggs JAG (2019) Fluorescence-based detection of membrane fusion state on a cryo-EM grid using correlated cryo-fluorescence and cryo-electron microscopy. Microsc Microanal 25:942–949

40. Resch GP (2019) Software for automated acquisition of electron tomography tilt series. Methods Cell Biol 152:135–178

41. Mastronarde DN (2005) Automated electron microscope tomography using robust prediction of specimen movements. J Struct Biol 152:36–51

42. Hutchings J, Stancheva V, Miller EA, Zanetti G (2018) Subtomogram averaging of COPII assemblies reveals how coat organization dictates membrane shape. Nat Commun 9:4154

43. Kremer JR, Mastronarde DN, McIntosh JR (1996) Computer visualization of three-dimensional image data using IMOD. J Struct Biol 116:71–76

44. Mastronarde DN (2008) Correction for non-perpendicularity of beam and tilt axis in tomographic reconstructions with the IMOD package. J Microsc 230:212–217

45. Grant T, Grigorieff N (2015) Measuring the optimal exposure for single particle cryo-EM using a 2.6 Å reconstruction of rotavirus VP6. elife 4:e06980

46. Turoňová B, Schur FKM, Wan W, Briggs JAG (2017) Efficient 3D-CTF correction for cryo-electron tomography using NovaCTF improves subtomogram averaging resolution to 3.4 Å. J Struct Biol 199:187–195

47. Nicastro D (2006) The molecular architecture of axonemes revealed by cryoelectron tomography. Science 313:944–948

48. Hrabe T et al (2012) PyTom: a python-based toolbox for localization of macromolecules in cryo-electron tomograms and subtomogram analysis. J Struct Biol 178:177–188

49. Bharat TAM, Scheres SHW (2016) Resolving macromolecular structures from electron cryo-tomography data using subtomogram averaging in RELION. Nat Protoc 11:2054–2065

50. Castaño-Díez D, Kudryashev M, Arheit M, Stahlberg H (2012) Dynamo: a flexible, user-friendly development tool for subtomogram averaging of cryo-EM data in high-performance computing environments. J Struct Biol 178:139–151

51. Tang G et al (2007) EMAN2: an extensible image processing suite for electron microscopy. J Struct Biol 157:38–46

52. Galaz-Montoya JG, Flanagan J, Schmid MF, Ludtke SJ (2015) Single particle tomography in EMAN2. J Struct Biol 190:279–290

53. Carter SD et al (2018) Distinguishing signal from autofluorescence in cryogenic correlated light and electron microscopy of mammalian cells. J Struct Biol 201:15–25

Part II

Single Particle Analysis and Electron Cryomicroscopy

Chapter 5

Progress Towards CryoEM: Negative-Stain Procedures for Biological Samples

Shane Gonen

Abstract

In recent years, electron cryo-microscopy (CryoEM) has become a powerful method for the high-resolution studies of biological macromolecules. While CryoEM experiments can begin without additional microscopy steps, negative-stain EM can tremendously minimize CryoEM screening. Negative-stain is a quick method that can be used to screen for robust biochemical conditions, the integrity, binding, and composition of samples and to get an estimation of sample grid concentration. For some applications, the map resolutions potentially afforded by stain may be as biologically informative as in CryoEM. Here, I describe the benefits and pitfalls of negative-stain EM, with particular emphasis on Uranyl stains with the main goal of screening in advance of CryoEM. In addition, I provide a materials list, detailed protocol and possible adjustments for the use of stains for biological samples requiring imaging and/or diffraction-based methods of EM.

Key words Negative-stain, Uranyl formate, Uranyl acetate, CryoEM, Single-particle, Averaging, Electron crystallography, Crystals, Electron microscopy

1 Introduction

Electron cryo-microscopy (CryoEM) has gained tremendous momentum over the last few years with the advent of new microscopes with stable optics, more sensitive and fast cameras capable of taking movies and faster, more powerful and accessible software for data collection, correcting motion and processing [1–5]. Traditionally, a typical CryoEM workflow first involved screening using negative-stain electron microscopy (EM), which is a very quick, albeit low-resolution method to visualize biological samples [6] (Figs. 1, 2, and 3). Due to the powerful nature of modern EM, negative-stain is sometimes skipped in favor of direct CryoEM workflows. For some samples, such as very stable macromolecules [7] or those where precious little amount is available (and high-resolution paramount), directly going to CryoEM is likely a better choice. However, for many samples, especially those that can be

Tamir Gonen and Brent L. Nannenga (eds.), *CryoEM: Methods and Protocols*, Methods in Molecular Biology, vol. 2215, https://doi.org/10.1007/978-1-0716-0966-8_5, © Springer Science+Business Media, LLC, part of Springer Nature 2021

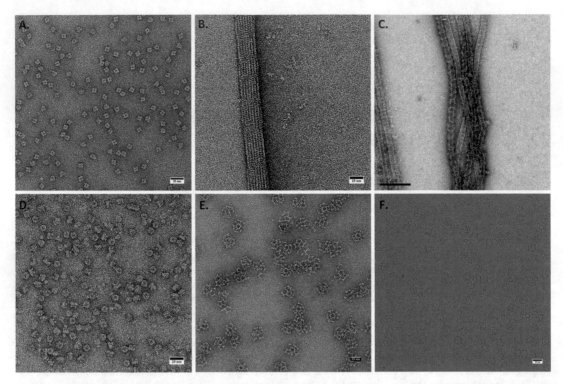

Fig. 1 Examples of different particles stained with UF and a comparison with CryoEM. (**a**) A designed protein tetrahedron. (**b**) Taxol-stabilized microtubule. (**c**) Microtubules with bound Dam1 rings. (**d**) Membrane protein TRPV1. (**e**) A designed icosahedrally symmetric protein cage highlighting the possible flattening effect of stain (**f**) Micrograph of the same particles in (**e**) frozen in ice for CryoEM without flattening

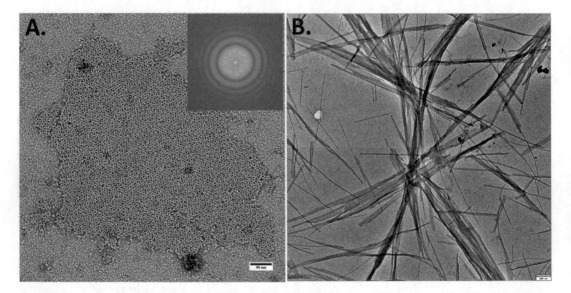

Fig. 2 UF Stained protein crystals. (**a**) A designed 2D protein crystal with a power spectrum (inset) showing reflections. (**b**) Three-dimensional crystals exhibiting needle morphology

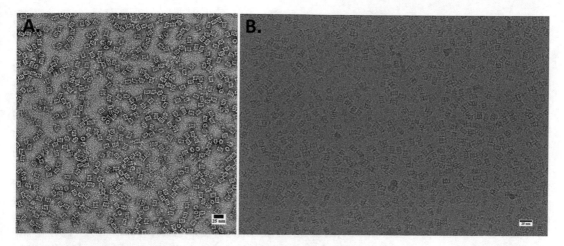

Fig. 3 Negative-stain and CryoEM micrographs of the Budding Yeast 20S proteasome complex (BY20S). (**a**) Negative stain micrograph of the BY20S proteasome complex stained using UF. (**b**) CryoEM micrograph of the BY20S proteasome complex

readily obtained or purified, negative-stain can not only significantly lower the barrier to a well-designed CryoEM experiment but can sometimes also serve to add biological information in cases when CryoEM is not viable [8].

While there are unique elements to screening in CryoEM, including minimizing grid contamination [9] and getting particles distributed in different orientations in thin layers of vitreous ice [10], some elements are common and can be quickly accessed using negative-stain. Additionally, negative-stain can afford tremendous benefits and advantages as both a prior and complementary tool for CryoEM, including:

The viability, stability, purity, and homogeneity of particles can be quickly accessed [6]; Samples can be stained within seconds after preparations are made and therefore can spend little time outside of their optimal biochemical conditions; Concentrations can be checked and an initial starting point for CryoEM determined (see **Note 1***); Greater contrast of the sample when compared to CryoEM allowing for smaller proteins to be imaged [11]; Initial datasets can be obtained for single particle averaging and electron crystallography and result in maps upwards of 14–20 Å in resolution allowing quick assessment [12]; Buffers, salts, and other conditions can be optimized; Cheaper than a CryoEM experiment; Once the grid is stained it can be left at ambient temperatures for years either for archiving or while waiting for microscope time; Grids can be checked using the majority of electron microscopes already available at most institutions (including those without direct electron detectors); Imaging and interpretation can rely less on perfect alignment of the beam; Grids can withstand more mechanical bending than in CryoEM; Technique is more accessible to non-experts and provides a good gateway to CryoEM* *and EM in general; Typically require less sample than in CryoEM, enabling the finding and visualization of rare/sparse particles; Does not have some of the drawbacks of CryoEM, for example, air-water interface and electron dose destruction of the particles* [13]*; Can be used more efficiently and can provide insight at any point of a biochemical experiment—including stages prior to purification; Can ascertain if all components of a macromolecule are present and at what orientations and stoichiometries and if any flexibility is present* [14]*; Can screen for binding of antibodies and other binders or co-complexes* [11]*; etc.*

In CryoEM, the sample appears darker than the surrounding ice while stained samples appear white giving rise to the term "negative-stain" [6]. The stain surrounds the sample, covering it, causing drying, and therefore likely flattening it (Fig. 1e). If the stain is too thick, it may also break the sample, so damage and particle breakage can be common. The most widely used stains are the heavy metal stains Uranyl Acetate (UA) and Uranyl Formate (UF), each having their own benefits and drawbacks [6]. Both stains are electron-dense and provide great contrast, are radioactive, toxic, light-sensitive, and low in pH. Solutions of UA can be stored for long periods of time ready to be used while UF needs to be freshly prepared every few days. UF, however, has a smaller grain size, which can help with the visualization of finer details—especially advantageous in the case of smaller proteins and thin crystals.

In the following sections, I give a detailed list of the materials and equipment required and a primer for a robust Uranyl-based negative-stain workflow, much of which is shared among other methods of staining. I also show examples of different macromolecules in stain (Figs. 1, 2, and 3), comparisons with CryoEM (Figs. 1 and 3) and provide a notes section which includes discussing changes users can make to optimize the protocol. One of the most powerful aspects of negative-stain is its versatility. The ideal result is to get thinly stained samples on a carbon support. As such, many microscopists have made tweaks and found their own methods of staining and some can vary wildly from the method I outline below. Once a user gets familiar with the basics of staining and relating that to results from the microscope, they can easily tailor their use of negative-stain to their own samples.

2   Materials

2.1   Grid Handling and Preparation

1. Carbon-coated grids (200–400 copper mesh) (*see* **Notes 2** and **3**).

2. Tweezers (e.g., Dumont high-precision) for general grid handling.

3. Anti-capillary tweezers (e.g., Dumont #5) for staining.

4. Glass slide wrapped in parafilm or aluminum grid block.

5. Grid storage holder.

6. Glow-discharge system.

2.2 Stain Solution: Preparation and Storage

1. Uranyl Acetate (UA) or Uranyl Formate (UF) powder.

2. Balance and weighing paper.

3. Milli-Q or double-distilled water.

4. Tubes wrapped with aluminum foil (for storage).

5. Bunsen burner, tongs, and conical flask (*to be used only for boiling water*).

6. 10 mL beaker for UF or appropriate volume beaker for UA.

7. Magnetic stir bars for beakers in #6.

8. 250 mL beaker wrapped with aluminum foil (or equivalent).

9. Stir plate.

10. Timer.

11. 5 M NaOH (filtered).

12. 0.22 μm filters and syringes.

2.3 Staining Workflow

1. Parafilm.

2. Filter paper (Whatman #1).

3. Pipettes (capable of dispensing 50 μL and 2–4 μL).

4. Milli-Q water.

5. (optional) Vacuum system with a fine tip and filtered to block the aspirated stain reaching the vacuum source.

---

3 Methods

Preparation of Uranyl Stains.

Important: UA and UF are light-sensitive, toxic, and radio-active. Adhere to appropriate health and safety procedures (including use of a fume hood due to the stain powder) and also separate stain equipment from other lab equipment.

3.1 Uranyl Acetate (2%) (Concentration Can be Adjusted)

Here, it is important to minimize light exposure of UA during the following steps.

1. Weigh UA for a final 2% (w/v) and add to Milli-Q or double-distilled water in a glass beaker with a stir bar.

2. Stir in the dark until fully dissolved.

3. Filter solution using a 0.22 μm filter and store in the dark. UA solution can be stored at room temperature and used for months.

**3.2 Uranyl Formate
(see Note 4)**

As with Uranyl acetate, it is important to minimize light exposure of UF during the following steps. Final UF solution is best used fresh and made fresh when required

1. Weigh 37.5 mg Uranyl Formate.

2. Place in 10 mL beaker (containing a stir bar) and cover with a larger beaker wrapped with aluminum foil.

3. Add 5 mL of just boiled Milli-Q or double-distilled water.

4. While covered, stir gently for 5 min.

5. Stop stirring and add 6 µL of filtered 5 M NaOH.

6. Wait for ~15 s and continue stirring in the dark for an additional 5 min.

7. Filter the solution using a 0.22 µm filter and 5 mL syringe.

8. Solution should be yellow in color without precipitation.

9. Store at room temperature in the dark for 3–7 days or until precipitation starts (*see* **Note 5**).

3.3 Staining Protocol

Protocol below is based on starting with a high concentration of protein.

The protocol can be adjusted for this and other types of samples (*see* **Note 6**).

1. Prepare all your starting material, including any sample dilutions, preparation of stain, and organization of the negative stain area before any of the next steps.

2. Place grids carbon-side up on a glass slide (wrapped in parafilm for easier handling) or aluminum grid block. Glow-discharge the grids at default vacuum for ~30 s at 15 mA (*see* **Note 6**). Stain grids within ~1 h of glow-discharging.

3. Place 3 drops of 50 µL Milli-Q water on the surface of a large sheet of parafilm for every grid to be stained. Keep the areas per grid separate and leave space between the drops for slight movement. For samples containing detergent, add 3–10 drops of Milli-Q water and adjust use accordingly (*see* **Note 7**).

4. Place 2 drops of 50 µL UF or UA on the parafilm for the current grid to be stained only. Perform **steps 5–10** below immediately following **step 4** to minimize the exposure of stain to light.

5. Using anti-capillary tweezers, pick up one grid from its edge being careful not to bend the grid. Place 3 µL of sample on the carbon side of the grid (*see* **Note 8**) and without the tip of the pipette touching the grid.

6. Wait for 5 s and blot the excess sample by touching the side of the grid to a piece of filter paper briefly (*see* **Note 6**).

7. Turn the grid upside town (so that the carbon and sample side are facing down) and quickly touch the grid on the surface of the first water drop. Quickly remove the excess as per the last step and continue for all water drops. For samples with detergent, stop after the detergent has been removed by this process—you can visually see when this occurs as the drops containing detergent will be more opaque than clean Milli-Q.

8. Repeat **step 7** with the first stain drop.

9. For the second stain drop, touch the grid on the surface of the stain but keep the grid touching the stain. Carefully lift the grid just enough as to not break the surface tension between the stain and grid and gently move the grid side to side and up and down. Do this step for 15–20 s and blot the excess as previously.

10. Air dry the grid and store in a grid holder. For best results, remove the remaining stain using a vacuum setup with a fine tip on the edge of the grid for a few seconds.

11. Once stained, grids can be stored for years at room temperature. Refer to standard procedures for imaging (*see* **Notes 9** and **10**).

4   Notes

1. Transitioning from negative-stain to CryoEM: This is largely sample dependent. First step is to access everything you know about the protein biochemically, for example, is it a multimeric protein and are all the components present? Is the protein homogeneous? Second is to estimate conditions for CryoEM from the negative-stain results. For example, concentration estimation: CryoEM requires more sample. For single proteins that are found to be highly concentrated, but still well dispersed, on a negative-stain grid, a good estimate to start CryoEM with is ~5–10× the concentration. This can vary between crystals, fibers, single-proteins, etc.

2. 200 vs. 300/400 copper-mesh grids: There are different aspects to consider when choosing between 200 mesh copper grids and 300 or 400, for example, the integrity of the carbon support. Thinner carbon can easily break on a 200-mesh grid when compared with 400, leading to less usable areas. Thick carbon will hold well on 200 mesh. Most single-particle experiments will work well with the smaller areas in a 400-mesh grid while larger samples such as fibers and crystals may benefit from the larger areas afforded by a 200-mesh grid.

3. Carbon-coated grids. Copper-mesh grids can be purchased with or without a pre-coated carbon support. For some

applications, it may be advantageous to coat grids on a per user basis—detailed protocols are available [15]. Coating grids allows the user to create thin carbon supports which can result in cleaner backgrounds and help for samples such as small proteins or thin two-dimensional (2D) crystals.

4. UF Solution: The volumes used in making UF can be adjusted accordingly along with the amount of 5 M NaOH.

5. Crystals of the stain can be observed on the microscope when precipitation of the stain occurs. This could hinder screening for crystalline samples and can affect the quality of the staining and usable areas for data collection.

6. The protocol can be optimized in many ways depending on the results seen on the microscope, for example:

    Concentration of sample on the grid: Adjust sample volume, adsorption time, glow-discharge time, add chemicals to change the surface properties of the grid, etc.

    Flattening: Adjust thickness of the staining, skip the water wash in order to stain faster, skip the vacuum aspiration.

7. While I recommend water as a default step for washing, some samples may need either a buffer wash (if water is not favored) or even skipping the wash step completely.

8. Grids with intact or mostly intact carbon will be able to contain all the solutions in the staining procedure to the carbon side of the grid. If a solution is observed on the opposite side, most or all of the carbon support has been broken and has allowed the solution to penetrate the grid. If the grid can be re-stained, it is recommended to do so, if not, check the grid on the microscope, there may be enough usable area to answer the questions required and/or collect data for processing.

9. While electron dose is not a major factor in a negative-stain experiment as in CryoEM, areas being imaged can burn and carbon supports can break. For data collection, it is best to image away from screening and beam-alignment areas.

10. When the concentration of the sample is too high, the particles can cluster and not stain well, resulting in a lack in definition. Low concentrations can result in large grid areas of only background carbon. A well-dispersed sample will fill each micrograph with many particles but still be individually well defined with space between each particle (Figs. 1 and 3). For cases involving fibers or crystals, the translation of concentration to a grid can be difficult to establish. Long fibers for example can bunch up in certain areas of the grid and may need to be broken up for dispersal and crystals can contain different amounts of repeat units, sometimes numbering in the thousands.

Acknowledgments

Thank you to Yifan Cheng, Zanlin Yu, and Kaihua Zhang for providing micrographs used in Figs. 1d and 3 and to Tamir Gonen for micrographs used in Figs. 1c and 2b.

References

1. Cheng Y (2018) Single-particle cryo-EM—how did it get here and where will it go. Science 361:876–880. https://doi.org/10.1126/science.aat4346

2. Glaeser RM, Hall RJ (2011) Reaching the information limit in Cryo-EM of biological macromolecules: experimental aspects. Biophys J 100:2331–2337. https://doi.org/10.1016/j.bpj.2011.04.018

3. Scheres SH (2016) Processing of structurally heterogeneous cryo-EM data in relion. Methods Enzymol 579:125–157. https://doi.org/10.1016/bs.mie.2016.04.012

4. Li X et al (2013) Electron counting and beam-induced motion correction enable near-atomic-resolution single-particle cryo-EM. Nat Methods 10:584–590. https://doi.org/10.1038/nmeth.2472

5. Schorb M, Haberbosch I, Hagen WJH, Schwab Y, Mastronarde DN (2019) Software tools for automated transmission electron microscopy. Nat Methods 16:471–477. https://doi.org/10.1038/s41592-019-0396-9

6. Ohi M, Li Y, Cheng Y, Walz T (2004) Negative staining and image classification—powerful tools in modern electron microscopy. Biol Proced Online 6:23–34. https://doi.org/10.1251/bpo70

7. Kim LY et al (2018) Benchmarking cryo-EM single particle analysis workflow. Front Mol Biosci 5:50. https://doi.org/10.3389/fmolb.2018.00050

8. Gonen S et al (2012) The structure of purified kinetochores reveals multiple microtubule-attachment sites. Nat Struct Mol Biol 19:925–929. https://doi.org/10.1038/nsmb.2358

9. Passmore LA, Russo CJ (2016) Specimen preparation for high-resolution Cryo-EM. Methods Enzymol 579:51–86. https://doi.org/10.1016/bs.mie.2016.04.011

10. Cheng Y, Grigorieff N, Penczek PA, Walz T (2015) A primer to single-particle cryo-electron microscopy. Cell 161:438–449. https://doi.org/10.1016/j.cell.2015.03.050

11. Wu S et al (2012) Fabs enable single particle cryoEM studies of small proteins. Structure 20:582–592. https://doi.org/10.1016/j.str.2012.02.017

12. De Carlo S, Harris JR (2011) Negative staining and cryo-negative staining of macromolecules and viruses for TEM. Micron 42:117–131. https://doi.org/10.1016/j.micron.2010.06.003

13. D'Imprima F et al (2019) Protein denaturation at the air-water interface and how to prevent it. elife 8:e42747. https://doi.org/10.7554/eLife.42747

14. Coscia F et al (2016) Fusion to a homo-oligomeric scaffold allows cryo-EM analysis of a small protein. Sci Rep 6:30909. https://doi.org/10.1038/srep30909

15. Booth DS, Avila-Sakar A, Cheng Y (2011) Visualizing proteins and macromolecular complexes by negative stain EM: from grid preparation to image acquisition. J Vis Exp 22 (58):3227. https://doi.org/10.3791/3227

Chapter 6

Setting Up Parallel Illumination on the Talos Arctica for High-Resolution Data Collection

Mark A. Herzik Jr

Abstract

Illuminating a specimen with a parallel electron beam is critical for many experiments in transmission electron microscopy as deviations from this condition cause considerable deterioration of image quality. Carefully establishing parallel illumination is particularly important on two-condenser lens transmission electron microscopes (TEMs) as the parallel illumination condition is limited to a single beam intensity value on these instruments. It was recently shown that a Thermo Fisher Scientific Talos Arctica, a two-condenser lens TEM operating at 200 kV, equipped with a Gatan K2 Summit direct electron detector is capable of resolving frozen-hydrated macromolecules of various sizes and internal symmetries to better than 3 Å resolution using single particle methodologies. A critical aspect of the success of these findings was the careful alignment of the electron microscope to ensure the specimen was illuminated with a parallel electron beam. Here, this chapter describes how to establish parallel illumination conditions in a Talos Arctica TEM for high-resolution cryogenic data collection for structure determination.

Key words Single-particle electron cryomicroscopy, Parallel illumination, Two-condenser lens electron microscope, Talos Arctica

1 Introduction

For both single particle analysis electron cryomicroscopy (cryo-EM) and electron cryotomography, it is of paramount importance to illuminate the specimen using a parallel electron beam so as to minimize deterioration in image quality [1, 2]. For modern transmission electron microscopes (TEMs), the strong pre-field of condenser-objective lenses requires a crossover in the front focal plane of the upper objective lens to ensure that the non-isoplanatism β angle, the angle with which the electron beam interacts with the specimen, is zero—this gives parallel illumination of the specimen [3]. Conditions in which the source image lies above the front focal plane of the objective lens (i.e., convergent illumination) or below it (i.e., divergent illumination) result in an electron beam with a non-zero β angle (Fig. 1) [3]. Due to the

Tamir Gonen and Brent L. Nannenga (eds.), *CryoEM: Methods and Protocols*, Methods in Molecular Biology, vol. 2215, https://doi.org/10.1007/978-1-0716-0966-8_6, © Springer Science+Business Media, LLC, part of Springer Nature 2021

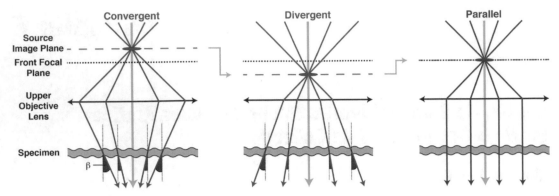

Fig. 1 Schematic ray diagrams portraying convergent, divergent, or parallel illumination of a specimen in a transmission electron microscope. A converging (left panel) or diverging (middle panel) electron beam on the specimen can be tuned to a parallel beam (right panel) by adjusting the height of the source image (shaded disc) in front of the upper objective lens. By adjusting the beam intensity (i.e., C2 lens strength) the source plane (dashed line) can be placed at the front focal plane of the objective lens (dotted line) to confer parallel illumination of the specimen. The optical axis is shown as a green line. Importantly, for symmetrical twin-objective lenses (such as those in the Talos Arctica), the height of the source image in front of the upper objective lens directly relates to the height of the diffraction pattern after the lower objective lens. The non-isoplanatism β angle for each ray in the convergent and divergent illumination diagrams are shown as red triangles

spherical aberration of the objective lens, this β angle varies across the area of illumination, and the amount of variation increases with the degree of convergent (or divergent) illumination (Fig. 1) [3]. Imaging specimens using non-parallel illumination conditions—non-zero β angle—can impede high-resolution structure determination due to two resulting phenomena: (a) local defocus variations across the specimen and (b) local variation of magnification resulting from points of the specimen locating to different planes within the objective lens (i.e., a tilted specimen or objects imaged at different under focus values) (Fig. 2) [1, 3]. Importantly, distortions in magnification become more pronounced at higher defocus values (Fig. 3). Although the magnification changes resulting from non-parallel illumination are subtle, they will nonetheless degrade the high-resolution information in a 3D reconstruction due to the averaging of thousands of differently sized particles.

Consider as example imaging a perfectly flat specimen, such as those typically desired for single particle cryo-EM. Illuminating the specimen with a divergent beam will result in a defocus gradient across the image, where the degree with which the recorded defocus deviates from the nominally set value increases with both distance from the optical axis and the degree of non-isoplanatism (Fig. 2). As a consequence, correcting the particle images extracted from the periphery of the micrograph using the contrast transfer function (CTF) parameters estimated from the entire aligned micrograph results in scrambling of the phases at higher spatial

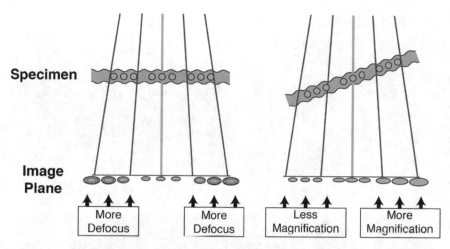

Fig. 2 Schematic ray diagram portraying the effects of illuminating a specimen with a divergent beam. When a specimen (blue circles) is illuminated with a divergent beam, the observed defocus of the specimen increases as the radius from the optical axis (green line) increases. A tilted specimen imaged with a divergent beam results in differently magnified specimen images, with the degree of apparent magnification related to the tilt axis and the radius of the specimen from the optical axis

frequencies due to differences between the CTF parameters used for imaging (experimental) and those applied during correction (estimated), resulting in a considerable loss in resolution. Although recent advances in data processing algorithms enable refinement of estimated CTF parameters on a per-particle basis [4–6], these methodologies are not foolproof, and care should be taken to illuminate the specimen under ideal conditions rather than to depend solely on computational correction of aberrations post-acquisition. The above example assumes all objects within the specimen lie on a single plane located at eucentric height within the objective lens. Due to variations present in nearly all EM specimens, this assumption is almost never satisfied and objects within the specimen inevitably reside at different Z-heights [7]. If these objects were imaged using a non-parallel electron beam, then each object will exhibit a change in magnification that is proportional to: (a) the distance the object lies away from eucentric height, (b) the degree by which the electron beam deviates from the parallel conditions (i.e., non-isoplanatism β angle), and (c) the distance the object lies away from the optical axis. For a tilted specimen, such as is required for tomography [8] or single particle specimens with pathologically preferred orientation [9], portions of the image distant from the optical axis would exhibit a change in magnification that is proportional to: (a) the angle of the specimen tilt, (b) the degree of deviation from the parallel condition, and (c) the distance the object lies away from the optical axis (Fig. 2). As a result, objects extracted from these images would differ in

0 μm underfocus   **10 μm underfocus**   **25 μm underfocus**

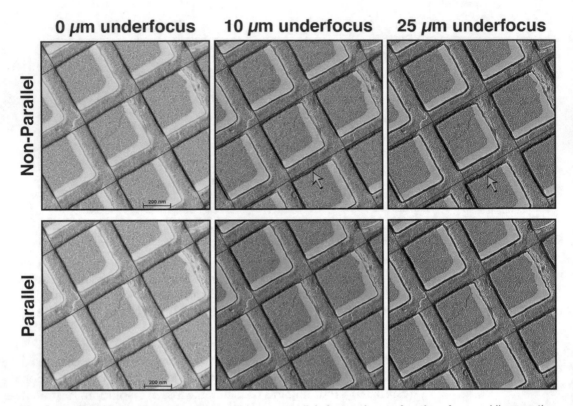

Fig. 3 Effects of illuminating a specimen with a non-parallel electron beam. A series of crossed lines grating replica calibration grid images taken at SA 45,000× magnification at different underfocus values using a non-parallel (divergent) beam (top row) or a parallel beam (bottom row). The outline of the waffling pattern of the crossed lines grating replica calibration squares are highlighted by red grid lines, generated using the in-focus images obtained with a divergent (top left) or parallel (bottom left) beam. Note: there is a negligible difference in size of the squares obtained between the non-parallel and parallel in-focus images. The square reference lines are transposed onto the underfocus images to serve as a visual reference. The blue arrows indicate significant changes in magnification of the specimen observed at higher defocus values as a result of imaging the specimen with a non-parallel (divergent) beam

magnification and the averaging of these images during generation of a three-dimensional (3D) reconstruction would degrade the available high-resolution information.

Three-condenser lens TEMs such as the Thermo Fisher Scientific (formerly FEI) Titan Krios utilize a double zoom condenser lens system that pairs neighboring lenses (e.g., condenser 1 (C1)—condenser 2 (C2) and C2—condenser 3 (C3)) with an image plane between them that can be moved up or down without changing the position of the image before the first lens or after the second lens (Fig. 4). Importantly, this configuration maintains a fixed image distance for the third condenser lens (C3) such that the crossover below C3 can be maintained in the front focal plane of the objective lens to produce parallel illumination over a wide range of crossover positions either between the C1–C2 lenses (i.e., spot size) and the

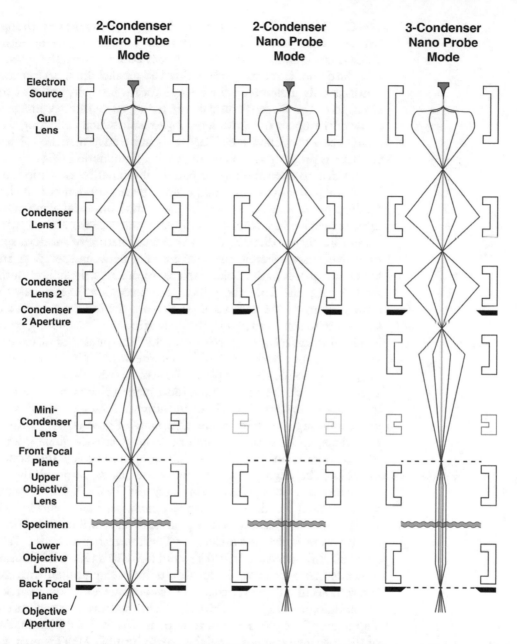

Fig. 4 Schematic ray diagrams detailing parallel illumination in modern transmission electron microscopes. Parallel illumination of a specimen (gray) in a 2 condenser lens system operating in micro probe mode (left panel), in a 2 condenser lens system operating in nano probe mode (Middle panel), and in a 3 condenser lens system operating in nano probe mode (Right panel). Switching micro probe mode (Left panel) to nano probe mode (Middle panel) turns off the mini-condenser lens (shown in gray) in front of the objective lens which increases the convergence angle, resulting in a smaller diameter beam. The third condenser lens ensures the beam is focused in the front focal plane of the upper objective lens over a wide range of crossover positions, either between the C1–C2 lenses (i.e., spot size) and the C2–C3 lenses (i.e., beam intensity). The unscattered beam is shown as a green line

C2–C3 lenses (i.e., beam intensity) (Fig. 4). Although two-condenser lens TEMs, such as the Thermo Fisher Scientific Talos Arctica, utilize a C1–C2 condenser zoom system, the absence of a third condenser lens means that the parallel illumination condition is only guaranteed for a single beam intensity value as the strength of the C2 lens controls the height of the source image in front of the upper objective lens. As a result, when collecting data using a two-condenser lens TEM, it is imperative that the C2 lens strength is properly set to confer parallel illumination (Fig. 4).

Critical to determining the parallel illumination condition in a TEM is aligning the source image after the last condenser lens (i.e., C2 for a two-condenser lens TEM) with the front focal plane of the upper objective lens (Fig. 4) [10]. Although the front focal plane cannot be viewed directly, the symmetrical nature of modern split objective lenses, wherein the specimen lies between the upper and lower pole pieces of the objective lens, means that the height of the front focal plane from the upper pole piece is *approximately* the same distance as the back focal plane from the lower pole piece of the objective lens. As a result, the procedure for determining parallel illumination relies on maximizing the "sharpness" of diffraction spots when the back focal plane of the lower objective lens is projected onto the viewing plane. To accomplish this, the objective aperture, which lies in the back focal plane of the objective lens, must first be brought into focus in diffraction mode. By adjusting beam intensity to maximize the sharpness of gold powder diffraction rings, such as those obtained from a crossed lines grating replica calibration grid containing sputtered gold, the source image will be aligned to the front focal plane. Importantly, for two-condenser lens TEMs, this singular beam intensity value must be used for data collection; otherwise, image quality will suffer from the effects of convergent or divergent illumination.

It was recently shown that the Thermo Fisher Scientific Talos Arctica TEM operating at 200 kV coupled with a Gatan K2 Summit direct electron detector is capable of resolving frozen-hydrated macromolecules to better than 3 Å resolution using single particle methodologies [11–13]. Critical to this success was the careful alignment of the electron microscope to ensure ideal illumination of the specimen using a parallel electron beam [12]. This protocol details how to establish parallel illumination conditions for a two-condenser lens TEM, such as the Talos Arctica, in preparation for high-resolution data collection using a K2 Summit direct electron detector. This protocol also includes instructions for determining the optimal exposure rate of the specimen while ensuring the parallel illumination condition is maintained.

2 Materials

1. 500 nm crossed lines grating replica calibration grid (*see* **Note 1**).

2. Autogrid C-clip rings (Thermo Fisher Scientific, Product #1036173) and Autogrid C-clips (Thermo Fisher Scientific Product #1036171) for Talos Arctica TEM autoloader.

2.1 Equipment

1. Thermo Fisher Scientific Talos Arctica (or Glacios) TEM (*see* **Note 2**) operating at 200 keV equipped with a Gatan K2 Summit direct electron detector and a 4 K × 4 K CETA 16 M CCD camera (or similar, *see* **Note 3**), operating with the following microscope settings:

   (a) Extraction voltage: 4150 V (*see* **Note 4**).

   (b) Gun lens setting: 8 (*see* **Notes 5** and **6**).

   (c) Spot size 3 (*see* **Notes 5** and **6**).

   (d) C1 aperture: 2000 μm.

   (e) C2 aperture: 70 μm (*see* **Note 6**).

   (f) Objective aperture: 100 μm (*see* **Note 7**).

2.2 Software

1. Tecnai User Interface (Thermo Fisher Scientific).

2. Tecnai Imaging & Analysis (Thermo Fisher Scientific).

3. Gatan Microscopy Suite® (Gatan, Inc.) which includes Digital Micrograph®.

4. Leginon [14], SerialEM [15], or EPU [16].

3 Methods

3.1 Basic Column Alignments

This protocol assumes the following: (a) that basic column alignments, including gun alignments (*see* **Note 8**), pivot points, beam and image shifts, and lens calibrations have already been performed, (b) the Tecnai User Interface/Tecnai Imaging & Analysis and Digital Micrograph® software packages are installed on the scope and camera computers, respectively, (c) a standard crossed lines grating replica calibration grid has been clipped and loaded into the compustage of the microscope (*see* **Note 9**), and (d) all SA mode magnifications (e.g., SA 45,000×, 73,000×, or 120,000×) use Nano probe mode while all M and LM mode magnifications use Micro probe mode (*see* **Note 10**).

1. Set the microscope to a lower magnification (e.g., LM-280×) and use the trackball to move the stage to the center of an intact square on the calibration grid that is devoid of obvious physical

imperfections (e.g., large cracks, broken areas, foil separated from the grid bar). This area will be used for all column alignments outlined below.

2. Switch to a higher magnification (e.g., SA 11,000×) in Nano probe mode and bring the specimen to eucentric height using any of the following methods:

   (a) Choose a recognizable image feature (e.g., large gold particle or obvious grid feature) and place that feature at the center of the fluorescent screen. Switch on the Alpha wobbler (±15° tilts) and minimize image movement by adjusting specimen height (Z-shift). Iterate until the image no longer moves (*see* **Note 11**).

   (b) Tilt the stage to a defined angle (e.g., −25°) and adjust the Z-height of the specimen until that feature is recentered on the fluorescent screen. Reset the stage back to 0° tilt and recenter the feature on the fluorescent screen. Iterate until the image no longer moves. Larger stage tilts can also be used to fine-tune the accuracy of eucentric height (*see* **Note 12**).

   (c) Press the Eucentric Focus button on the hand panel to bring the focus level to, or very close to, eucentric height. Focus the beam to a spot and adjust the height of the specimen (Z-height) until the caustic ring is minimized to a spot (*see* **Notes 13** and **14**).

   (d) Use the auto-eucentric feature built into an automated data collection software package (e.g., Leginon [14], EPU [16], SerialEM [15]).

3. Press the Eucentric Focus button on the hand panel. Switch to a higher magnification (e.g., SA 45,000×) in Nano probe mode and use the Intensity knob on the hand panel to spread the beam (right side of crossover or clockwise) until the diameter of the beam is slightly larger than the 40 mm circle on the fluorescent screen. With the objective aperture removed, use Digital Micrograph® to display the Fourier Transform (FFT) of a live image and determine proper focus of the specimen by adjusting the Focus knob until the observed Thon rings in the FFT are eliminated. Proper focus can also be determined using automated data collection software packages (e.g., Leginon [14], EPU [16], SerialEM [15]), but it is always best practice to manually confirm proper focus has been determined by viewing the live FFT in Digital Micrograph® (*see* **Notes 15** and **16**). After determining proper focus, press the "Reset Defocus" button on the hand panel.

4. Select and center the 70 μm C2 aperture by first focusing the beam to a spot and then using beam shift to center the focused beam on the fluorescent screen. Expand the beam to the 40 mm circle. Click on the "Adjust" tab for the C2 aperture

and center the beam by adjusting the C2 aperture position using the multifunction X and Y knobs. Focus the beam to a spot and recenter using beam shift. Expand the beam to the 40 mm circle and recenter the C2 aperture using the multifunction knobs. Iterate until minimal improvements are obtained. Turn off the "Adjust" button.

5. Select the "High-Resolution" setting of the fluorescent screen in the Tecnai User Interface and focus the beam to a spot. In the Tecnai User Interface, click on the "Condenser" tab in the Stigmator panel and use the multifunction X and Y knobs to adjust the currents through the condenser stigmators while cycling above and below the crossover point to make the beam circular. Iterate until the beam is properly stigmated (i.e., the beam shape is similar on both sides of crossover).

6. With the specimen at eucentric height and proper focus, focus the beam to a spot and align the beam tilt pivot points using the direct alignments feature in the Tecnai User Interface. For each direction (e.g., X or Y), use the multifunction X and Y knobs to center the bright spots of each beam on top of each other. The step size on the Focus knob can be used to adjust the frequency of tilting.

7. Insert and align the objective aperture. Press "Diffraction Mode" on the hand panel to switch to diffraction mode (D 850 mm) and center the central spot of the diffraction image by adjusting Diffraction shift X and Y (multifunction knobs X and Y, respectively). Insert the 100 μm (or larger diameter) objective aperture, click on the "Adjust" tab for the objective aperture within Tecnai User Interface, and use the multifunction X and Y knobs to center the aperture around the gold powder diffraction rings (see **Note 7**).

8. Exit Diffraction Mode and minimize the astigmatism of the objective lens. Manually set the defocus to ~1 μm using the Focus knob on the hand panels and use Digital Micrograph® to display the FFT of a live image (see **Note 17**). In the Tecnai User Interface, click on the "Objective" tab in the Stigmator panel and use the multifunction X and Y knobs to adjust the currents through the objective stigmators until the ellipticity of the Thon rings observed in the live FFT is minimized (see **Notes 18** and **19**).

9. Switch to a higher magnification (e.g., SA 120,000×) in Nano probe mode and center on an image feature (e.g., at the edge of a calibration grating waffle). Perform the rotation center alignment of the objective lens using the direct alignments feature within Tecnai User Interface (see **Notes 20** and **21**). This final step attempts to align the unscattered electron beam with the optical axis, minimizing image coma. Accurately performing this step will minimize time spent on Subheading 3.2.

3.2 Coma-Free Alignment

In order to avoid the introduction of large phase errors at higher resolution due to axial and off-axis coma, one must ensure that illumination of the specimen occurs at an axis that is parallel to the optical axis. This can be performed using the coma-free alignment procedures described in Glaeser et al. 2011 [2] as implemented within the Leginon [14] software (*see* **Notes 22** and **23**).

1. Move to an intact square on the calibration grid and ensure the specimen is at eucentric height and proper focus.

2. Create a preset within Leginon [14] that uses the same magnification (e.g., SA 45,000× in Nano probe mode) and beam intensity settings as the exposure preset (e.g., C2 strength 43.5%). Set this preset to use a defocus of -300 nm and acquire 924×924 (binned by 2) images with a two-second exposure (*see* **Note 24**).

3. Go to the "Beam Tilt Image" node in Leginon [14] and set these parameters in the settings tab.

   - Tableau Type: beam tilt series-power.
   - Beam Tilt: 0.005 radians.
   - Number of tilt directions: 4.
   - Start angle: 0.
   - Correlation type: phase.
   - Tableau binning: 2.
   - Beam tilt count: 1.

4. Click on the "Simulate target" icon in the Beam Tilt Image node. Leginon [14] will now acquire a series of 5 images—a zero-tilt image and 4 tilted images—with differing amounts of beam tilt—5 milliradians of beam tilt in each "minus" X, "plus" X, "minus" Y, and "plus" Y directions, respectively.

5. The FFTs of opposing images in the resulting tableau (i.e., "minus" Y and "plus" Y images) should have approximately the same amount of astigmatism and Thon ring spacing. The presence of slightly different astigmatism and/or number of Thon rings in the FFTs of the "plus" versus "minus" tilt angles demonstrates that the beam direction is not parallel to the coma-free axis. To make adjustments, click on the zero-tilt image in the direction towards the image(s) that is (are) further from focus (i.e., FFT images with more Thon rings). If significant adjustments are necessary, it is recommended to repeat the rotation center alignment (**step 9**, Subheading 3.1) and recenter the objective aperture (*see* **Note 20**). Repeat **step 8** (Subheading 3.1) to ensure the objective lens is stigmated.

3.3 Parallel Illumination

The parallel illumination condition for a two-condenser lens TEM requires the source image to lie at the front focal plane of the upper objective lens pole piece. As the objective lens is symmetrical, focusing the diffraction pattern in the back focal plane of the objective lens will ensure parallel illumination (Fig. 4).

1. Confirm that the specimen is at eucentric height and proper focus at a magnification of SA 45,000× in Nano probe mode. Switch to Diffraction Mode (D 850 mm) and insert the 100 μm objective aperture if it is not already inserted. In the Tecnai User Interface, click on the "Adjust" tab for the objective aperture and use the multifunction X and Y knobs to move the objective aperture off-center so that the edge of the aperture is overlapping the gold powder diffraction rings and partially occluding the central spot (Fig. 5a). If needed, adjust the brightness of the fluorescent screen camera until the edge of objective aperture is clearly visible. This setup provides maximal contrast for observing the edge of the objective aperture.

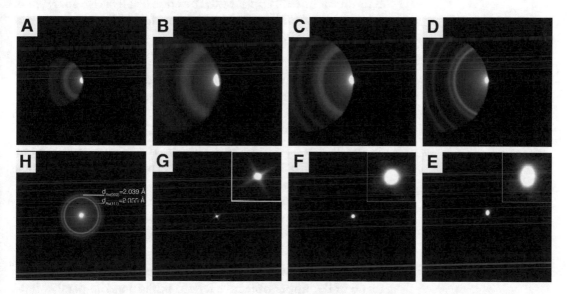

Fig. 5 Determining parallel illumination in Nano probe mode on a Talos Arctica. (**a**) 1× zoomed image of gold powder diffraction obtained from a crossed line grating replica calibration grid imaged at D 850 mm. A 100 μm objective aperture is intentionally displaced over the central spot to increase the contrast of the edge of the objective aperture. (**b**) 2× zoomed image of image shown in (**a**). Note blurriness of both the edge of the objective aperture and the gold powder diffraction ring. (**c**) Resulting image after the objective aperture has been focused into the back focal plane of the lower objective lens. Note the objective aperture is now crisp while the gold powder diffraction rings are still blurry. (**d**) Resulting image after the width of the gold powder diffraction rings have been minimized by adjusting the beam intensity (i.e., C2 lens strength). (**e**) 2× zoomed image of the central spot obtained at D 5.7 m with the objective aperture recentered so that the central spot is no longer occluded. Note the astigmatism of the central spot. (**f**) Resulting image after correcting the astigmatism of the diffraction lens. (**g**) Resulting image after the diameter of the central spot has been minimized by fine-tuning the beam intensity (i.e., C2 lens strength). (**h**) 1× zoomed image of the gold powder diffraction rings at parallel illumination on a Talos Arctica (Nano probe mode, D 850 mm, 70 μm C2 aperture, 100 μm objective aperture, gun lens 8, spot size 3, 2000 μm C1 aperture). For panels **e**, **f**, and **g**, inset images are 10× zoomed images of the central spot

2. Switch to "high-resolution" mode in the fluorescent screen viewer. If needed, adjust the brightness of the fluorescent screen camera until the edge of the objective aperture is clearly visible (Fig. 5b). Note that both the objective aperture and gold powder diffraction rings are blurry. Bring the back focal plane of the objective lens into view (i.e., to the object plane of the diffraction lens) by adjusting diffraction focus using the Focus knob until the edge of the objective aperture is as sharp/crisp as possible (Fig. 5c) (*see* **Note 25**). If the gold powder diffraction rings are not crisp, then the source image is not located in the front focal plane of the objective lens and the specimen is not illuminated using a parallel beam.

3. Adjust the beam intensity (i.e., C2 strength) until the width of the gold powder diffraction rings is minimized (Fig. 5d).

4. If the objective aperture is no longer crisp, repeat **steps 2** and **3** (Subheading 3.2) until both the edge of objective aperture is crisp, and the width of the gold powder diffraction rings are minimized (*see* **Note 27**).

5. Center the objective aperture (*see* **step 7**, Subheading 3.1).

6. Increase the diffraction camera length to D 5.7 m or until just the central spot is visible (*see* **Note 26**). If needed, adjust the brightness of the fluorescent camera screen until the shape of the central spot is clearly visible (*see* Fig. 5e).

7. In the Tecnai User Interface, click on the "Diffraction" tab in the Stigmator panel and use the multifunction X and Y knobs to adjust the shape of the central spot while cycling above and below the crossover point to make the beam circular. Iterate until the beam is properly stigmated (i.e., the beam shape is similar on both sides of crossover) (*see* Fig. 5f).

8. Adjust the beam intensity until the central spot is minimized (Fig. 5g). This will align the source image to the front focal plane of the upper objective lens, establishing the parallel illumination condition. **Make note of the beam intensity value, as this is the *only* value under the current microscope settings that confers parallel illumination, and should be used for exposure image acquisition (*see* Note 28). Important: adjustments to the exposure rate must be performed through changes in spot size, C2 aperture, gun lens, or magnification rather than changes in beam intensity.**

9. Decrease the camera length to D 850 mm and ensure that the central spot is centered, and the objective aperture is centered (Fig. 5h).

10. Exit Diffraction Mode and center the beam on the camera by performing beam shift alignment in the Tecnai User Interface Direct Alignments. Ensure that the objective lens is properly stigmated (*see* **step 7**, Subheading 3.1).

11. Measure the diameter of the beam using the measurement tool in the Tecnai User Interface or by using the calibration grid waffles as a reference. If the beam diameter is too large for the desired data collection strategy, then switch to a smaller C2 aperture (*see* **Note 29**). If the smallest C2 aperture is already selected, please adjust the data collection strategy accordingly to ensure areas of the grid will not be double-exposed during data collection. This is particularly important when using Micro probe mode or larger diameter C2 apertures (Fig. 4) (*see* **Note 10**). Experience has shown that the approximate diameter of a parallel beam at SA 45,000× (Nano probe mode, 70 μm C2 aperture, gun lens 8, spot size 3, 2000 μm C1 aperture, 100 μm objective aperture) is approximately 1.8 μm.

3.4 Determine and Set the Exposure Rate

1. Set the microscope to a lower magnification (e.g., LM-280×) and use the trackball to move the stage to the center of an empty area of the calibration grid (i.e., a large crack or empty square).

2. Insert the K2 camera. In Digital Micrograph®, click on the settings icon in the "Camera View" window, making sure that the binning is set to 1 and that the entire sensor is being used by clicking on "Full CCD". Set an exposure time of 1 second and click on "start view". The exposure rate (in e^-/pixel/s) will be displayed in the Camera Monitor window. A measured exposure rate of 2–5 e^-/pixel/s is optimal for the K2 (*see* **Note 31**).

3. If the measured exposure rate is not ideal for the current experiment and adjustments to the exposure rate are required, adjustments must be performed by changing spot size, C2 aperture, magnification, and/or guns lens rather than changing beam intensity. If changes in spot size, C2 aperture, and/or gun lens are made, then the above-outlined steps should be checked to ensure the scope has not deviated from the parallel condition.

   (a) Changes to spot size: Lowering the spot number (i.e., from 3 to 2) will, under most conditions, result in an approximate doubling of the measured exposure rate. On the contrary, increasing the spot number (i.e., from spot size 3 to 4) will result in an approximate two-fold decrease in measured exposure rate (*see* **Notes 28** and **32**).

   (b) Changing to C2 aperture: Using a larger C2 aperture will allow more electrons to illuminate the specimen and will therefore increase the exposure rate. Likewise, using a smaller C2 aperture will decrease the number of electrons illuminating the specimen and will therefore decrease the exposure rate (*see* **Note 29**).

(c) Changes in magnification: Increasing the magnification will decrease the measured exposure rate on the camera but the flux should not change (*see* **Note 30**). Changes in magnification only affect the projection system and should not alter conditions for parallel illumination.

(d) Changes in gun lens: Increasing the gun lens (e.g., from 2 to 3) results in increased demagnification of the source and to decreased current in the beam. In general, increasing the gun lens by 1 increment leads to a reduction in the beam current by approximately 35%.

4. Using the determined exposure rate, adjust the total movie recording time to yield a cumulative exposure of ~50–65 e$^-$/ Å^2 (for single particle analysis) or ~80–100 e$^-$/Å^2 (for tomography of thick specimens). Adjust the per-frame exposure time (e.g., 100–250 ms) to yield a per-frame cumulative exposure of ~0.8 e$^-$/Å^2/frame or greater, while maximizing the number of frames to limit the cumulative movement per frame to allow for better correction of per-frame movement.

3.5   K2 Gain Corrections

1. Set the microscope to a lower magnification (e.g., LM-280×) and use the trackball to move the stage to the center of an empty area of the calibration grid.

2. Insert the K2 camera. In Digital Micrograph®, go to Camera and select "Prepare Gain Reference". Follow prompts. Please refer to the user guidelines for additional instructions.

3. It is recommended to collect new hardware dark references at least once every 12–24 h of data collection to limit the deleterious effects of charge accumulation on the K2 sensor chip. The frequency of hardware dark reference update increases with lower acceleration voltages.

4   Notes

1. This protocol used: Ted Pella, Inc. Product # 677.

2. The steps outlined in this chapter are optimized for determining parallel illumination on the Talos Arctica; however, the principle steps should apply to nearly all modern 2 condenser lens electron microscopes with only minor modifications necessary.

3. Alternative camera combinations will require slight modifications to the methods outlined in this chapter (e.g., magnifications used, camera interface) but the principle column alignment steps should still apply.

4. It is recommended that only service engineers change the gun settings on an X-FEG.

5. The finite size of the crossover in the front focal plane of the upper objective lens results in a spread of beam tilts at the single point in which the beam hits the specimen. The size of the source image, and thereby the coherence angle α can be tuned by changing the magnification of the source. A higher gun lens (e.g., 8 vs. 4) or a higher spot size number (e.g., 8 vs. 4) gives a smaller source image diameter and thereby a more coherent electron beam. The downside of these settings is the decreased beam current.

6. The conditions for parallel illumination on a two-condenser lens TEM result in a single C2 intensity value. This strength of C2, in combination with spot size, C2 aperture, objective aperture, gun lens, and magnification, determine the exposure rate. As the C2 intensity value must remain unchanged, in order to increase or decrease the measured exposure rate at the camera a user must change the spot size, the size of the C2 aperture, the gun lens, and/or the magnification (*see* **Note 30**).

7. A centered 100 μm objective aperture on a Talos Arctica (or Glacios) allows for the visualization of the second gold powder diffraction ring (dAu(002) = 2.039 Å) and will thus eliminate information beyond ~1.6 Å resolution (Fig. 5h). Although this setup will not limit the useable high-resolution information available for most biological single particle cryo-EM specimens, if information beyond 1.6 Å resolution is anticipated, the user should insert a larger objective aperture (i.e., 150 μm), or remove the objective aperture entirely after aligning the microscope (Fig. 6). However, as the objective aperture removes high-angle inelastically scattered electrons, imaging with a large objective aperture or no aperture at all will decrease the observed amplitude contrast compared to smaller objective apertures. For specimens supported by gold foil, particular caution should be taken during imaging without an objective aperture as considerable high-angle scattering from the gold foil will be observed, which, can deleteriously affect frame alignment, CTF estimation, and 3D reconstruction.

8. The image of the source should be checked to ensure proper gun alignment and the X-FEG is not currently undergoing a splitting event. At a spot size of 1, 150 μm C2 aperture, 73,000 kx magnification, slightly condense the beam until a bright spot (i.e., the source) can be observed within the central disc of the probe. The bright spot should not touch the outer diameter of the central disc or form a tear drop image. High-resolution imaged will not be attainable if the gun tilt/alignments are off.

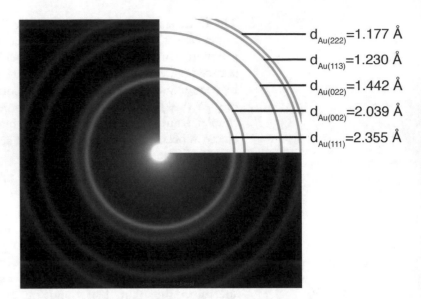

$d_{Au(222)} = 1.177 \text{ Å}$

$d_{Au(113)} = 1.230 \text{ Å}$

$d_{Au(022)} = 1.442 \text{ Å}$

$d_{Au(002)} = 2.039 \text{ Å}$

$d_{Au(111)} = 2.355 \text{ Å}$

Fig. 6 2× zoomed image of the gold powder diffraction rings from a crossed line grating replica calibration grid at parallel illumination on a Talos Arctica. Settings: Nano probe mode, D 850 mm, 70 μm C2 aperture, gun lens 8, spot size 3, 2000 μm C1 aperture. Gold powder diffraction rings (*hkl* indices) are annotated

9. Please refer to the user guidelines provided by FEI/Thermo Fisher Scientific. Clipping grids under liquid nitrogen may result in physical damage to the grid. This will manifest as large cracks within grid squares or microfractures within holes containing vitreous ice that "pop" immediately upon illumination. Therefore, it is recommended that grids are clipped under nitrogen vapor rather than liquid nitrogen.

10. Micro probe mode uses the minicondenser lens located immediately above the upper pole piece of the objective lens to create another crossover above the objective lens (Fig. 4). As a result, parallel illumination in Nano probe mode will yield a much smaller diameter beam (often >35% smaller) than what is obtained using Micro probe mode when using the same C2 aperture, spot size, gun lens, and magnification (Fig. 4). However, alignment errors in Nano probe mode are accentuated more so than would be observed in Micro probe mode.

11. The backlash of the stage will cause a jolting of the image upon initialization of movement, so it is best to make adjustments to Z-height during the smoother section of stage tilting motion.

12. This method can also prove to be very useful to quickly check if the specimen is at eucentric height when diagnosing potential column alignment issues.

13. Care must be taken using this method for determining Z-height since the focused beam will compromise some radiation sensitive specimens (i.e., thin ice, resins, etc.). It is recommended that procedures **steps 2a** or **2b** (Subheading 3.1) be subsequently used to fine-tune eucentric height.

14. Method **step 2c** (Subheading 3.1) for bringing the specimen to eucentric height provides a means to quickly check basic alignments of the objective lens through evaluation of the caustic ring.

    (a) Rotation center: If rotation center is grossly misaligned, then the center of the spot will not be located at the center of the caustic ring. Adjust rotation center until the center of the spot lies at the center of the caustic ring.

    (b) Objective astigmatism: If the objective lens is astigmatic, then the shape of the caustic ring will no longer be a circle. Adjust the currents through the objective lens stigmators until the caustic ring is round.

15. Tecnai Image Acquisition can also be used to view the live FFT of an image acquired using the CCD camera although this method is less sensitive than using a direct electron detector, and higher exposure rates should be used.

16. Misalignment of beam tilt pivot points can result in autofocus errors when using automated software packages. In practice, most autofocus features implemented in automated data collection software packages (i.e., Leginon [14], EPU [16], SerialEM [15]) estimate focus within 200 nm or better of proper focus.

17. The degree of objective astigmatism changes with the amount of defocus introduced [17]. To ensure the amount of astigmatism present in the Contrast Transfer Function (CTF) is minimized during imaging of the specimen, stigmation of the objective lens should be performed at the mean defocus value planned for data collection. For conventional imaging using the Talos Arctica, underfocus values between 400 and 1500 nm are recommended [11, 13].

18. In addition to manual minimization of objective astigmatism using Digital Micrograph® or the program s² stigmator [18], automated measurement and correction of astigmatism are implemented within Leginon [14], EPU [16], and SerialEM [15].

19. Whenever the objective aperture is removed, inserted, or recentered, it is best to measure and correct for any astigmatism as the physical movement of the aperture motor is imperfect.

20. Significant changes in the rotation center alignment will alter the centering of the objective aperture in the back focal plane of the objective lens. Although the aperture has not physically

moved, the electron path through the objective lens has changed and therefore the objective aperture will no longer be properly centered. To ensure that ideal imaging conditions are maintained, and high-spatial frequencies are not inadvertently eliminated, verify the objective aperture is centered after making changes to the rotation center alignment.

21. A misalignment of rotation center will result in an apparent translation of the image upon increasing defocus.

22. Coma-free alignment can also be performed using Tecnai User Interface, SerialEM [15], and/or Auto-CTF within EPU [16]. Please refer to the respective user manual for detailed instructions.

23. Coma-free alignment effectively "fine-tunes" the rotation center alignment performed in **step 9** (Subheading 3.1). If this alignment was performed correctly, then only minor adjustments will be required. Please refer to **Note 20** when making adjustments to ensure coma-free alignment.

24. It is recommended to adjust the defocus (e.g., 300–500 nm underfocus) to observe only 1–2 Thon rings in the tableau FFT images. Large underfocus (i.e., >1000 nm) will result in too many Thon rings and will make it more difficult to assess astigmatism and differences between the negatively and positively tilted images.

25. Making adjustments to focus in diffraction mode does not physically move the objective aperture but rather places the image plane in the back focal plane of the objective lens (i.e., where the objective aperture is physically located).

26. The exact camera length in Diffraction Mode will vary for each microscope depending on camera and imaging accessory combinations.

27. If only a portion of the objective aperture is crisp (i.e., crisp left and right edges versus blurry top and bottom edges), or the width of the gold powder diffraction rings varies around the central spot, then the diffraction lens is not properly stigmated. Follow **steps 6–9** (Subheading 3.3).

28. It is good practice to keep track of C2 intensity values for parallel illumination under varying combinations of spot size, C2 aperture, and gun lens, as significant deviations between data collections indicate a potential misalignment of the column.

29. If the C2 aperture is properly centered and stigmated, and the spot size remains unchanged, then the C2 intensity value for parallel illumination should remain unchanged as well. However, it is still recommended to verify the parallel illumination condition is preserved when changing between C2 apertures (refer to Subheading 3.3).

30. As long as magnification-dependent zoom is disabled, changes in magnification will not alter the size of the electron beam at the specimen level. Changing magnification will alter how the camera samples the electron beam, with an increase in magnification resulting in a decrease in observed exposure rate (e^-/pixel/s). However, the flux, as measured in $e^-/\text{Å}^2$/s, should, in theory, remain unchanged.

31. Exposure rates greater than 10 e−/pixel/s will result in increased coincidence loss and decreased K2 camera recording capability of spatial frequencies lower than half Nyquist [19]. For other direct electron detectors, please refer to the user manual for optimal exposure rates and dose fractionation.

32. The condenser lenses of the Talos Arctica use a zoom system and thus if properly aligned, changing spot size number will change the C2 intensity value without affecting parallel illumination. If the spot size-dependent zoom alignment is off, then changing spot size will alter parallel illumination and **steps 1–10** (Subheading 3.3) should be repeated.

Acknowledgments

I would like thank Gabe Lander, Mengyu Wu, and Mr. Bill Anderson at The Scripps Research Institute (TSRI) and Matthijn Vos for helpful advice and discussion regarding TEM alignments and data acquisition. I would like to thank Mr. Bill Anderson (TSRI), Kevin Corbett (UCSD), Gabe Lander (TSRI), Andres Leschziner (UCSD), Sergey Suslov (UCSD), and Mengyu Wu (TSRI) for critical reading of the chapter and helpful discussions.

References

1. Eyidi D, Hebert C, Schattschneider P (2006) Short note on parallel illumination in the TEM. Ultramicroscopy 106:1144–1149

2. Glaeser RM, Typke D, Tiemeijer PC, Pulokas J, Cheng A (2011) Precise beam-tilt alignment and collimation are required to minimize the phase error associated with coma in high-resolution cryo-EM. J Struct Biol 174:1–10

3. Christenson KK, Eades JA (1988) Skew thoughts on parallelism. Ultramicroscopy 26:113–132

4. Zivanov J et al (2018) New tools for automated high-resolution cryo-EM structure determination in RELION-3. elife 7:e42166

5. Zhang K (2016) Gctf: Real-time CTF determination and correction. J Struct Biol 193:1–12

6. Punjani A, Rubinstein JL, Fleet DJ, Brubaker M (2017) A. cryoSPARC: algorithms for rapid unsupervised cryo-EM structure determination. Nat Methods 14:290–296

7. Noble AJ et al (2018) Routine single particle CryoEM sample and grid characterization by tomography. elife 7:e34257

8. Danev R, Yanagisawa H, Kikkawa M (2019) Cryo-electron microscopy methodology: current aspects and future directions. Trends Biochem Sci 44:837–848. https://doi.org/10.1016/j.tibs.2019.04.008

9. Tan YZ et al (2017) Addressing preferred specimen orientation in single-particle cryo-EM through tilting. Nat Methods 14:793–796

10. Avila-Sakar A, Li X, Zheng SQ, Cheng Y (2013) Recording High-Resolution Images of

Two-Dimensional Crystals of Membrane Proteins. In: Schmidt-Krey I, Cheng Y (eds) Electron Crystallography of Soluble and Membrane Proteins. Methods in Molecular Biology (Methods and Protocols), vol 955. Humana Press, Totowa, NJ. https://doi.org/10.1007/978-1-62703-176-9_8

11. Herzik MA, Wu M, Lander GC (2017) Achieving better-than-3-Å resolution by single-particle cryo-EM at 200 keV. Nat Methods 14:1075–1078

12. Herzik Mark AJ, Wu M, Lander GC (2017) Setting up the Talos Arctica electron microscope and Gatan K2 direct detector for high-resolution cryogenic single-particle data acquisition. https://doi.org/10.1038/protex.2017.108

13. Herzik MA, Wu M, Lander GC (2019) High-resolution structure determination of sub-100-kDa complexes using conventional cryo-EM. Nat Commun 10:1032

14. Suloway C et al (2005) Automated molecular microscopy: the new Leginon system. J Struct Biol 151:41–60

15. Mastronarde DN (2005) Automated electron microscope tomography using robust prediction of specimen movements. J Struct Biol 152:36–51

16. Janus M, Voight A (2016) EPU Software User's Guide. FEI Company, Hillsboro, pp 1–106

17. Yan R, Li K, Jiang W (2018) Defocus and magnification dependent variation of TEM image astigmatism. Sci Rep 8:344

18. Yan R, Li K, Jiang W (2017) Real-time detection and single-pass minimization of TEM objective lens astigmatism. J Struct Biol 197:210–219

19. Li X, Zheng SQ, Egami K, Agard DA, Cheng Y (2013) Influence of electron dose rate on electron counting images recorded with the K2 camera. J Struct Biol 184:251–260

Chapter 7

Multi-body Refinement of Cryo-EM Images in RELION

Takanori Nakane and Sjörs H. W. Scheres

Abstract

Single-particle analysis of electron cryo-microscopy (cryo-EM) images allows structure determination of macromolecular complexes. But when these molecules adopt many different conformations, traditional image processing approaches often lead to blurred reconstructions. By considering complexes to be comprised of multiple, independently moving rigid bodies, multi-body refinement in RELION enables structure determination of highly flexible complexes, while at the same time providing a characterization of the motions in the complex. Here, we describe how to perform multi-body refinement in RELION using a publicly available example. We outline how to prepare the necessary files, how to run the actual multi-body calculation, and how to interpret its output. This method can be applied to any cryo-EM data set of flexible complexes that can be divided into two or more bodies, each with a minimum molecular weight of 100–150 kDa.

Key words Cryo-EM, RELION, Single-particle analysis, Image classification, Molecular motions, Structural biology, Regularized likelihood optimization

1 Introduction

Relative motions between different domains underlie the functional mechanisms of macromolecular complexes that are involved in many processes in the living cell. Structural characterization of these complexes in their multiple states is key to understanding how they work. Electron cryo-microscopy (cryo-EM) allows visualization of individual complexes in a thin layer of vitreous water. 3D classification of the resulting particles into subsets of structurally homogeneous complexes can then be used to obtain atomic structures of these complexes in their different states [1]. However, subdivision of the cryo-EM data into a discrete number of subsets is not well suited for the description of continuous types of molecular motion, for which an infinite number of subsets would, in principle, be needed.

We recently proposed multi-body refinement as an alternative to discrete classification to describe continuous molecular motions in cryo-EM projection images [2]. The underlying model of

Tamir Gonen and Brent L. Nannenga (eds.), *CryoEM: Methods and Protocols*, Methods in Molecular Biology, vol. 2215, https://doi.org/10.1007/978-1-0716-0966-8_7, © Springer Science+Business Media, LLC, part of Springer Nature 2021

multi-body refinement is that the dynamics in a macromolecular complex can be described by a discrete number of independently moving, rigid bodies. This model builds on the observation that the structure of individual protein domains, or tertiary structure, often remains intact upon continuous changes in the overall (quaternary) structure of flexible complexes. During multi-body refinement, the signal from all-but-one body is iteratively removed from the experimental images using partial signal subtraction [3–5], and the signal from the remaining body is aligned with respect to reference projections of only that body. Thereby, alignments of the individual bodies are not compromised by differences in the relative orientations between the bodies among the images in the data set, and reconstructions from the individual bodies are better defined than in a single refinement of the entire data set. Moreover, upon convergence of the multi-body refinement, principal component analysis on the relative orientations of all bodies for every experimental image in the data set is used to characterize the most dominant motions in the complex.

This chapter describes how to perform multi-body refinement of a flexible macromolecular complex in RELION, and how to interpret the results. We use a previously published data set on the pre-catalytic spliceosomal B-complex [6] as an example. These data, including the necessary files to perform multi-body refinement, are publicly available through the EMPIAR data base [7], under accession number 10180. We describe how to design real-space masks that define the individual bodies (Fig. 1); we provide detailed instructions on how to choose the various parameters; and we describe how to analyze the results by post-processing the individual body reconstructions (Figs. 2 and 3) and by characterizing the motions through principal components analysis (Fig. 4). A minimum molecular weight for the individual bodies, which is required for accurate alignment, limits the applicability of this method to relatively large complexes that can be divided into two or more rigid bodies of at least 100–150 kDa. Besides this caveat, multi-body refinement is generally applicable to flexible macromolecular complexes.

2   Materials

2.1   Computer Hardware Requirements

1. A computer with a reasonably modern CPU with at least 8 cores. We used a computer with a 12-core Intel Xeon ES-2620v3 CPU at 2.4 GHz. Hyper-threading resulted in 24 visible cores in the operating system.

2. At least 400 GB of disk space to execute the example case described in this chapter. We used a BeeGFS file system of

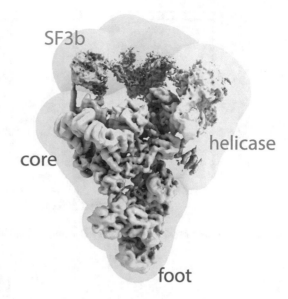

Fig. 1 The four body masks, called core, foot, helicase, and 3F3b, used for the spliceosome multi-body refinement. The masks are shown in semi-transparent colors on top of the consensus map

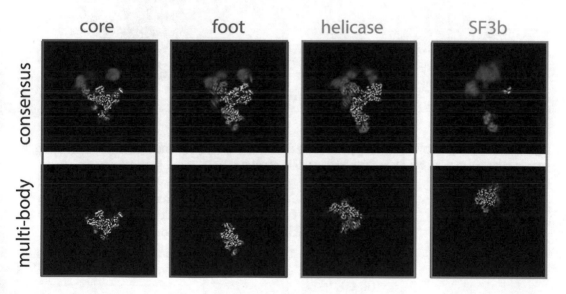

Fig. 2 Slices through the density of the four bodies. The slices are shown after the consensus refinement (top) and after multi-body refinement (down)

600 TB. For processing of your own data, storage solutions of at least several TB are recommended.

3. At least 32 GB random-access memory (RAM). We used a computer with 64 GB.

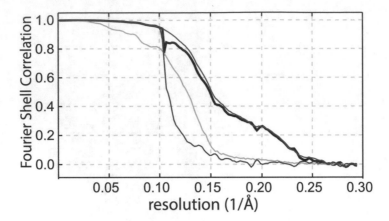

Fig. 3 FSC plot generated by the post-processing job-type of RELION for the helicase body. *See* **Note 16** for details

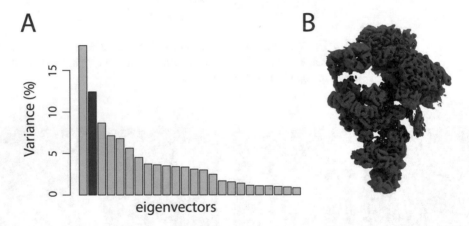

Fig. 4 (**a**) The contributions of all principal motions to the variance. (**b**) The third and eighth map along the second principal motion, which is highlighted in red in panel **a**

4. At least 2 NVIDIA graphics processing units (GPU), each with at least 6 GB of memory. We used a computer with 4 NVIDIA GTX1080s, each with 8 GB of memory (*see* **Note 1**).

5. If your data is stored on a central storage system, a local solid-state scratch disk (SSD) is recommended. We used a 400 GB SSD.

2.2 Software Requirements

1. A Linux-based operating system. We used Scientific Linux 7.

2. CUDA. We used version 9.0. Download from `https://developer.nvidia.com/cuda-zone`.

3. MPI. Any MPI flavor may work. Not only the runtime (mpirun) executable, but also the header files are required to compile RELION. We used OpenMPI version 1.5.4. Download from `https://www.open-mpi.org`.

4. RELION. We used version 3.0 [8]. Download from `https://www3.mrc-lmb.cam.ac.uk/relion`.

5. A molecular volume viewer like UCSF Chimera [9] and/or Pymol (Schrödinger, LLC). We used Chimera version 1.11.2. Download from `https://www.cgl.ucsf.edu/chimera`.

2.3 Test Data

1. Download the test data from the EMPIAR data base, under accession number 10180: `https://www.ebi.ac.uk/pdbe/emdb/empiar/entry/10180`.

3   Methods

3.1 Prepare Input Files

1. Multi-body refinement starts from the so-called consensus refinement, where all particles have been refined with respect to a single 3D reference. Use the consensus map to define a division of the complex into a discrete number of independently moving bodies. To ensure enough signal for accurate alignments of the individual bodies, each body should comprise a molecular weight of at least 100–150 kDa. In the spliceosome example, we chose to divide the complex into four bodies: the central core, the foot domain, the helicase domain, and the SF3b domain (Fig. 1), also *see* **Notes 2** and **3**.

2. Define a mask for each of the bodies. The masks should have a maximum value of 1 inside the body, and a minimum value of 0 outside the body. The masks should all be on the same Cartesian grid as the reference map of the consensus refinement, i.e., they should have the same pixel and box size as the consensus reference. To minimize artifacts in Fourier-space, the edges of the masks should be soft, i.e., they should gradually decrease from 1 to 0 over multiple pixels (*see* **Note 4**). The masks for the spliceosome example are stored as `Mask-and-Ref/*_mask.mrc`.

3. Determine the order of bodies. Because earlier bodies are subtracted first, it is best to place larger and more stable bodies earlier in the list.

4. For each body, define one neighbouring body to express how the two bodies rotate relative to each other. This is necessary because priors on the relative orientations between bodies are determined along axes between the centers of mass of their respective masks. For the spliceosome example, we chose to have the foot, the helicase, and the SF3B domains that rotate relative to the central core domain. The central core domain was chosen to rotate relative to the second largest body, i.e., the foot domain (*see* **Note 5**).

5. Determine the width, i.e., the standard deviation, of the Gaussian priors on the relative rotations and translations between pairs of bodies. Rotational and translational searches of the individual bodies will be performed locally, with a search range of three times these standard deviations. Therefore, these values should express the expectation by how much the bodies move relative to each other in the data set. For the spliceosome example, the fuzziness in the consensus structures indicated that the core and the foot are relatively stable, whereas the helicase, and especially the SF3b domains move more. We thus used widths on the angular priors of 10° for the core and foot, 15° for the helicase domain, and 20° for the SF3b domain. Likewise, the widths on the translational priors were set to two pixels for the core and foot, three pixels for the helicase, and five pixels for the SF3b domain. For each project, these values should be determined empirically, but the values given here provide a useful starting point.

6. Combine the information from **steps 1–5** into a single metadata STAR file. The file for the spliceosome example is stored as" `Example/4-bodies-tight-mask.star`".

3.2 Execute Multi-body Refinement

1. Launch the RELION graphical user interface (GUI) from the project directory. In case of the example data set, this is the directory in which the `Example`, `Mask-and-Ref` and `Micrographs` directories are. If this is the first time you launch the GUI in this directory, the program will ask whether you are sure you want to launch the GUI. Type "`y`" to confirm.

2. On the left-hand, vertical list of job-types click on "`3D multi-body`".

3. On the "`I/O`" tab, set "`Consensus refinement optimiser.star`" to "`Example/consensus_optimiser.star`". This file contains all the information from the previously performed consensus refinement (*see* **Note 6**).

4. On the "`I/O`" tab, set "`Body STAR file`" to "`Example/4-bodies-tight-mask.star`". This is the STAR file that defines the bodies, which was generated in Subheading 3.1, **step 6**.

5. On the "`I/O`" tab, set "`Reconstruct subtracted bodies`" to "`yes`". This will make the program write out reconstructions for each of the bodies where the density of the other bodies is subtracted. Thereby, the density around each body should become ever cleaner and also provide feedback on the subtraction quality, as the alignments of the bodies improve during the subsequent iterations of multi-body refinement. When set to "`No`", fuzzy densities for the other bodies will be present around the reconstructed density for each body.

6. On the "Auto-sampling" tab, leave the default of "Initial angular sampling: 1.8 degrees", "Initial offset range (pix): 3", and "Initial offset step (pix): 0.75". In general, one sets the sampling rates to somewhat coarser values than the estimated accuracy as reported in the "model_classes" table in the model.star file from the consensus refinement. For the spliceosome case, the values are 0.786° for the angular accuracy and 0.535 pixels for the translational accuracy, as given in the file "Example/consensus_half1_model.star". (also *see* **Note 7**).

7. On the "Analyse" tab, leave the defaults. "Run flexibility analyses? Yes" will perform the principal components analysis on the relative orientations of all bodies upon convergence. "Number of eigenvector movies: 3" will select the first three principal motions in the data, and write out ten maps for each of them (*see* **Note 8**). "Select particles based on eigenvalues? No" will switch off the option to select particle subsets based on eigenvalues (*see* **Note 9**).

8. On the "Compute" tab, set "Use parallel disc I/O?" to "Yes" if you have a fast, parallel storage system, or if you are working on a local hard drive. Set this option to "No" if you are using a less parallel, shared storage system like NFS.

9. On the "Compute" tab, set "Number of pooled particles: 30" if you are using GPUs. Otherwise, use a smaller value like 3.

10. On the "Compute" tab, set "Skip padding? Yes", as this will save RAM and GPU memory by up to eight-fold. Only set this option to "No" if your particle occupies such a large volume of the box that artifacts are observed near the edges of the box in runs where this option was set to "Yes".

11. On the "Compute" tab, set "Pre-read all particles into RAM? No", as the example data set occupies more than 120 GB, and each MPI worker would read the full data set into memory. Only set this option to "Yes" if your data set is small enough to fit into your computer's RAM.

12. On the "Compute" tab, set "Copy particles to scratch directory: /ssd", provided your computer has a local SSD scratch drive that is called "/ssd", and reading from this drive is significantly faster than reading from the storage system your data resides on. If your data set is small enough to read into RAM (previous step), then copying the data to a scratch drive is not necessary.

13. On the "Compute" tab, set "Combine iterations through disc? No". Only set this option to "Yes" if the default option of combining iterations through the network gives problems, e.g., MPI-related errors at the end of an iteration.

14. On the "Compute" tab, set "Use GPU acceleration? Yes" since the program will run a lot faster on GPUs than on CPUs.

15. On the "Compute" tab, adjust "Which GPUs to use:" to your computer (*see* **Note 10**). On our computer with four cards, we used "0,1:2,3", where the ":" sign is used to divide tasks among different MPI workers and the "," sign is used to divide tasks within a single MPI worker. We used three MPI processes (*see* **step 16**), i.e., one master and two workers. Therefore, GPUs 0 and 1 jointly do the work of the first worker, and GPUs 2 and 3 divide the work of the second worker.

16. On the "Running" tab, we set "Number of MPI procs: 3" to divide the calculations over one master and two worker processes (*see* **Note 11**). Provided your computer has enough RAM, you could increase the number of MPI processes (*see* **Note 12**).

17. On the "Running" tab, we set "Number of threads: 12" to make optimal use of the 24 (hyper-threaded) cores on our computer (*see* **Note 13**).

18. On the "Running" tab, we set "Submit to queue? No", as we were performing our calculation on a local computer. You can set this option to "Yes" to submit your job to a cluster queuing system instead. *See* the RELION wiki (www3.mrc-lmb.cam.ac.uk/relion) for details on how to set this up for your system.

19. Finally, press the "Run!" button to execute the calculation. On our computer, this job took approximately 14 h.

3.3 Analyze the Body Reconstructions

1. The multi-body refinement program writes out a reconstructed map that is filtered to its estimated resolution for each body as "Multibody/job001/run_body00[1--4].mrc". These maps can be directly visualized in a molecular volume viewer, or in slices from the "Display!" pull-down menu on the GUI (Fig. 2; also *see* **Note 14**).

2. Besides the filtered maps, the program also writes out unfiltered maps for both independently refined half-sets. On the "I/O" tab of the "Post-processing" job-type, input the first half-map of the first body: "One of the 2 unfiltered half-maps: Multibody/job001/run_half1_body001_unfil.mrc". Also set "Solvent mask: Mask-and-Ref/CORE_-mask.mrc" and "Calibrated pixel size (A) 1.699".

3. On the "Sharpen" tab, one can provide a STAR file with the modulation transfer function of the detector (*see* **Note 15**).

4. Leave all other options on the "Sharpen", "Filter", and "Running" tabs to their defaults and execute the calculation by pressing the "Run!" button (*see* **Note 16**).

5. Repeat **steps 2–4** for the remaining three bodies.

6. Inspect the Fourier Shell Correlation (FSC) plot (Fig. 3) in the "logfile.pdf" from the "Display!" pull-down menu for each of the four bodies. Confirm that the red FSC curve is close to zero at the resolution estimate of each body. If this is not the case, their resolution estimate may be unreliable due to your masks being too tight or too sharp (*see* **Note 17**). In that case, repeat multi-body refinement with wider and/or softer-edged masks.

7. The post-processed maps of the four bodies obtained in **steps 2–5** can be opened in molecular volume viewers and used for separate atomic modeling of the individual bodies, for example, in COOT [10] (*see* **Notes 18** and **19**).

3.4 Analyze the Body Motions

1. Select "MultiBody/job001" in the "Finished jobs" window on the lower part of the GUI. Use the "Display:" dropdown menu to visualize the "analyse_logfile.pdf" file. The first plot (Fig. 4a) shows how much variance in the orientations is explained by each of the principal components. The subsequent plots show histograms of the projections of the relative orientations of the bodies of each experimental projection image onto the corresponding component. Unimodal histograms are an indication of continuous motions in the complex, whereas bi- or multi-modal histograms indicate the presence of two or more discrete states. In the spliceosome example, all histograms are unimodal, but in other cases it may be useful to select subsets of particles based on their projections along the principal components (*see* **Note 20**).

2. Make a movie that represents this first principal motion. Ten maps called "analyse_component001_bin0??.mrc" are generated by rotating and translating each body along each principal motion. In Chimera, from the "Tools" menu, select "Volume Data" and then "Volume Series". Click on the "Open..." button and browse to the directory "Multibody/job001". To make a movie for the first principal component, hold down the SHIFT key and use the left-mouse button to select all ten maps called "analyse_component001_bin0??.mrc". Then, change "Play direction" to "oscillate" and press "Play". You can change the sampling rate and threshold of the maps through the "Volume Viewer" tool, or by using the "Command Line" from the "Favourites" menu, and typing: "vol all step 2" and "vol all level 0.015" (*see* **Note 21**).

3. Write a movie to file by selecting from the "Tools" menu "Utilities" and then "Movie recorder".

4. Repeat **steps 2** and **3** for the second and third components. Figure 4b shows the third and eighth maps of the second principal motion.

3.5 Partial Signal Subtraction after Multi-body Refinement

1. To perform focused classifications and/or refinements on part of the complex after multi-body refinement, one can create stacks of particles in which signal from the different bodies is subtracted taking all their orientations into account. For example, to subtract all signal except the SF3b domain, type the following commands from the project directory (also *see* **Note 22**):

```
mkdir Subtract
relion_flex_analyse --data MultiBody/job001/
run_data.star --model MultiBody/job001/run_mo-
del.star --bodies Example/4-bodies-tight-mask.
star --o Subtract/sf3b --subtract --keep_inside
Mask-and-Ref/SF3b_mask.mrc --ctf
```
On our computer, this calculation took 8 h.

2. The resulting STAR file with subtracted particles, "Subtract/subtracted.star", can then be used as input in a conventional masked 3D classification or refinement (*see* **Note 23**).

4   Notes

1. Multi-body refinement can be performed without GPU-acceleration, but to keep computation times within reasonable limits you will need a multi-node CPU cluster, e.g., with more than 200 CPU cores.

2. Displaying the consensus map in a slice-viewer, e.g., for this case by executing "relion_display --i Example/consensus_hall_class001.mrc", may help in determining which parts of the complex are flexible.

3. The individual bodies are allowed to overlap. This may be useful in characterizing the density at the interfaces between different bodies (also *see* **Note 19**). In addition, by including part of an adjacent larger body, overlapping bodies may help in the refinement of relatively small, flexible domains.

4. RELION implements a "Mask creation" job-type that can be used to low-pass filter an input map; binarize it at a specified threshold; grow the binary mask by a specified number of pixels; and add a soft, raised cosine-shaped, edge with a specified width. Besides this functionality, tools for generating 3D masks are not provided in RELION. We find the "Volume Eraser" tool in UCSF Chimera useful to manually edit 3D maps; access it from from the "Tools" menu, under "Volume

`Data`". From the same menu one might also use the "`Segment Map`" option [11]. In addition, if a preliminary atomic model is available for a body, then the "`molmap`" command in UCSF Chimera can be used to convert the fitted PDB into a density map (*see* `https://www.cgl.ucsf.edu/chimera/docs/UsersGuide/midas/molmap.html`); and the volume operation ("`vop`") option "`resample onGrid`" (*see* `https://www.cgl.ucsf.edu/chimera/docs/UsersGuide/framecommand.html`) can be used to generate a map on the same Cartesian grid as the consensus map. This map can then be saved from the "`Volume Viewer`" window, and input into the "`Mask creation`" job-type of RELION. Whatever the method chosen, make sure the final masks do not have high-resolution features and only have values in the range [0,1] by applying low-pass filtering and soft-edge addition in the "`Mask creation`" job-type.

5. If the definition of a body smaller than 150 kDa is necessary, it is possible to fix the relative orientation of this body by defining the widths of its rotational and translational priors to zero.

6. Within a typical RELION project, the `optimiser.star` file is directly visible through the "`Browse`" button, but this is not the case in this simplified example. Therefore, just type its location in the GUI text entry.

7. The value of the offset range specifies the maximum translational offsets searched in every iteration. As the translational searches are centred at the optimal translation from the previous iteration, individual particles can shift more than this search range during multiple iterations. Therefore, this value can be smaller than the total search allowed by the prior on the translations as specified in **step 5**, Subheading 3.1.

8. The ten maps that are generated for each principal motion in the data will have relative orientations of their bodies that correspond to the median orientations of ten equi-populated subsets of particles along the principal motion.

9. By setting "`Select particles based on eigenvalues? Yes`" on the "`Analyse`" tab, one can generate STAR files with subsets of the particles that have eigenvalues above or below the specified minimum or maximum values for any of the principal components, as specified on the GUI. This calculation is run as a continuation of a previously performed multi-body job, by pressing the "`Continue!`" button. Multiple selections can be performed in a single job directory. The output STAR files with the selected subsets can then be used for further 3D classification or refinement.

10. The N available GPUs are typically numbered from 0 to (N-1). The command "nvidia-smi" can be used to see which NVIDIA GPUs are available on your computer.

11. The minimum amount of MPI processes to use in (multi-body) refinement in RELION is 3 because each of the gold-standard half-sets is executed on a different MPI worker, and one process is reserved for the master. Provided you have enough RAM, you may increase the number of MPI processes. It is best to keep the total number of MPI processes an odd number, so that equal amounts of MPI workers work on both halves of the data.

12. The total amount of required RAM is difficult to predict. An important part of the required RAM scales linearly with the number of bodies, but this changes to quadratic scaling if all bodies overlap with each other. For this example, the RAM requirements were up to 13 GB for the workers and approximately 6 GB for the master.

13. Using even more threads than available cores on your computer may lead to further, minor speed-ups, as not all threads will always be occupied at the same time.

14. Multi-body refinement does not write out a single map that somehow combines the individual bodies, as such map would not represent the heterogeneity in the data (also *see* **Notes 19** and **20**).

15. STAR files with the modulation transfer function for commonly used detectors can be downloaded from the RELION Wiki. In the example shown here, the data were collected on a K2 detector at 300 kV.

16. In the "Post-processing" job-type, resolution estimation within a mask region is performed using high-resolution phase randomization [12]. In the FSC plot of the output "logfile. pdf" (Fig. 3), the FSC between the two unmasked half-maps is shown in green. The FSC between the two half-maps to which the solvent mask has been applied is shown in blue. Typically, the blue curve has higher FSC values than the green curve. There are two reasons for this. Firstly, by removing noise from the solvent, and from errors in the subtraction of the other bodies, application of the solvent mask improves the signal-to-noise ratio of the half-maps. Secondly, because application of the solvent mask comprises a multiplication operation in real-space, Fourier-space components from different spatial frequencies will be mixed together through convolution. In particular, mixing of larger and better-correlating Fourier components at lower spatial frequencies with smaller and worse-correlating components at higher spatial frequencies will lead to an inflation of the FSC curve. Application of the

solvent mask to half-maps in which the phases of the high-resolution Fourier components have been randomized provides a measure of the second effect. To this purpose, the post-processing program randomizes the phases of the Fourier components with resolution beyond the one where the green FSC curve drops below 0.8. Then, it applies the solvent mask to both phase-randomized maps and calculates the red FSC curve between the two masked phase-randomized maps. Any non-zero FSC values in the red curve beyond the resolution at which the phases were randomized are due to the convolution effect of the solvent mask. The black FSC curve applies a correction for this effect to the blue curve. The spatial frequency at which this curve drops below 0.143 is the estimated resolution of the combined map. However, the correction can lead to artifacts, most notably under-estimation of resolution, if the red curve does not drop to values close to zero near the final estimated resolution of the map. In particular sharper masks that cut through density in the half-maps lead to larger FSC values in the red curves. Therefore, such masks should be avoided.

17. During multi-body refinement, phase randomization is applied for each body at every iteration to estimate its resolution. It is possible that during the earlier iterations, when some bodies have more spread-out, blurred densities, that the body masks are too tight and the resolution for these bodies is under-estimated due to the artifacts mentioned in **Note 16**. Sometimes these problems disappear by themselves as the body reconstruction improves. In other cases, it may be necessary to use wider and/or softer-edged masks. The "Post-processing" job-type described in Subheading 3.3, can also be performed on the unfiltered half-maps that are written out during the multi-body refinement as: "Multibody/job001/run_it0??_half?_body00?_unfil.mrc". However, during RELION refinements maps are not necessarily calculated to the Nyquist frequency to save computational resources. This may lead to zero values for all higher spatial frequency components in both half-maps, which in turn may result in artificially high FSC values at these frequencies. Therefore, the resulting FSC curves should only be evaluated up to the spatial frequency that was included in that iteration of the refinement, which may be found in the corresponding model.star files.

18. As the quality or the local resolution of the density may vary within each body, it may be useful to calculate multiple post-processed maps for atomic model building. To this end, use the options to provide a user-defined B-factor or a user-defined spatial frequency for a low-pass filter on the "Sharpen" and "Filter" tabs of the "Post-processing" job-type. In

addition, one may want to run the RELION "Local resolution" job-type to generate maps that are filtered according to their local resolution.

19. After interpretation of each of the individual body reconstructions in terms of atomic models, one may want to generate a composite atomic model of the entire complex. One way to do this is to perform rigid-body fitting of the individually refined atomic models for each body in the initial consensus reconstruction. However, it is crucial to understand that a single atomic model can never be a good description for the flexible complex. In particular, the assumption that each body moves as a complete rigid body is most likely flawed at the interfaces between the different bodies. This is because local conformational changes are likely to occur at the interface as one body moves relative to its neighbor. To some extent, overlapping body masks may help in the modeling of the interfaces, but often there may simply be insufficient information in the data set to define the multiple different conformations at the interfaces between bodies. Therefore, one should be particularly careful with refinements of composite models in the consensus map, which may sometimes seem necessary to handle clashes at the interfaces.

20. **Note 19** raises the question of which maps to submit to the Electron Microscopy Data Bank (EMDB) and which models to submit to the Protein Data Base (PDB) [13]. Multiple options are possible, and the field has not yet converged on one. An important consideration is that the maps used for building all parts of the atomic model are preserved. For example, one could make a separate EMDB and PDB entry for each body, together with separate EMDB and PDB entries for the consensus map and the composite atomic model. Alternatively, one could make one EMDB and one PDB entry with the consensus map and the composite model, and then submit reconstructions for the individual bodies that were used for model building to the EMDB as supplementary maps. For each post-processed map used for atomic modeling, it is strongly recommended to submit the unfiltered half-maps and the masks used for body-definition and/or post-processing as supplementary maps to the EMDB.

21. Movies can also be generated in Pymol, where the ten maps should be loaded into different states of a single map object.

22. From RELION release-3.1 onwards, partial signal subtraction after multi-body refinement is different from the one described in Subheading 3.3. Instead of using the "relion_flex_analyse" program, the optimiser.star file from the multi-body refinement can be given directly as

input to the "Particle subtraction" job-type. This new subtraction utility will also allow to change the center and the box size of the subtracted particles.

23. Details about performing conventional and focused classifications are provided in the RELION tutorial, which is available from the RELION Wiki. General advice on classification strategies is also provided in [14].

Acknowledgments

We are grateful to Clemens Plaschka for discussions; to Jake Grimmett and Toby Darling for help with high-performance computing; and to Shabih Shakeel, Rafael Fernandez-Leiro, and Giulia Zanetti for critical reading of the manuscript. This study was supported by the MRC-LMB EM facility. Funding was provided by the UK Medical Research Council to SHWS (MC_UP_A025_1013); and the Japan Society for the Promotion of Science through an Overseas Research Fellowship to TN.

References

1. Fernandez-Leiro R, Scheres SHW (2016) Unravelling biological macromolecules with cryo-electron microscopy. Nature 537:339–346. https://doi.org/10.1038/nature19948

2. Nakane T, Kimanius D, Lindahl E, Scheres SH (2018) Characterisation of molecular motions in cryo-EM single-particle data by multi-body refinement in RELION. elife 7:e36861. https://doi.org/10.7554/eLife.36861

3. Bai X, Rajendra E, Yang G et al (2015) Sampling the conformational space of the catalytic subunit of human gamma-secretase. elife 4:e11182. https://doi.org/10.7554/eLife.11182

4. Zhou Q, Huang X, Sun S et al (2015) Cryo-EM structure of SNAP-SNARE assembly in 20S particle. Cell Res 25:551–560. https://doi.org/10.1038/cr.2015.47

5. Ilca SL, Kotecha A, Sun X et al (2015) Localized reconstruction of subunits from electron cryomicroscopy images of macromolecular complexes. Nat Commun 6:8843. https://doi.org/10.1038/ncomms9843

6. Plaschka C, Lin P-C, Nagai K (2017) Structure of a pre-catalytic spliceosome. Nature 546:617–621. https://doi.org/10.1038/nature22799

7. Iudin A, Korir PK, Salavert-Torres J et al (2016) EMPIAR: a public archive for raw electron microscopy image data. Nat Methods 13:387–388. https://doi.org/10.1038/nmeth.3806

8. Zivanov J, Nakane T, Forsberg BO et al (2018) New tools for automated high-resolution cryo-EM structure determination in RELION-3. elife 7:e42166. https://doi.org/10.7554/eLife.42166

9. Pettersen EF, Goddard TD, Huang CC et al (2004) UCSF chimera--a visualization system for exploratory research and analysis. J Comput Chem 25:1605–1612. https://doi.org/10.1002/jcc.20084

10. Emsley P, Lohkamp B, Scott WG, Cowtan K (2010) Features and development of Coot. Acta Crystallogr D Biol Crystallogr 66:486–501. https://doi.org/10.1107/S0907444910007493

11. Pintilie G, Chiu W (2012) Comparison of Segger and other methods for segmentation and rigid-body docking of molecular components in cryo-EM density maps. Biopolymers 97:742–760. https://doi.org/10.1002/bip.22074

12. Chen S, McMullan G, Faruqi AR et al (2013) High-resolution noise substitution to measure overfitting and validate resolution in 3D structure determination by single particle electron cryomicroscopy. Ultramicroscopy 135:24–35.

https://doi.org/10.1016/j.ultramic.2013.06.
004
13. Patwardhan A, Lawson CL (2016) Databases
and archiving for CryoEM. Methods Enzymol
579:393–412. https://doi.org/10.1016/bs.
mie.2016.04.015

14. Scheres SHW (2016) Processing of structurally
heterogeneous Cryo-EM data in RELION.
Methods Enzymol 579:125–157. https://doi.
org/10.1016/bs.mie.2016.04.012

Chapter 8

Evaluating Local and Directional Resolution of Cryo-EM Density Maps

Sriram Aiyer, Cheng Zhang, Philp R. Baldwin, and Dmitry Lyumkis

Abstract

A systematic and quantitative evaluation of cryo-EM maps is necessary to judge their quality and to capture all possible sources of error. A single value for global resolution is insufficient to accurately describe the quality of a reconstructed density. We describe the estimation and evaluation of two additional resolution measures, local and directional resolution, using methods based on the Fourier shell correlation (FSC). We apply the protocol to samples that encompass different types of pathologies a user is expected to encounter and provide analyses on how to interpret the output files and resulting maps. Implementation of these tools will facilitate density interpretation and can guide the user in adapting their experiments to improve the quality of cryo-EM maps, and by extension atomic models.

Key words Anisotropy, Preferred orientation, Fourier shell correlation, Sampling, Cryo-EM, Single-particle analysis

1 Introduction

In a cryo-EM experiment, images of frozen hydrated macromolecules are recorded using an electron microscope. Individual macromolecular "particles" embedded within the layer of vitreous ice can be distributed in different orientations relative to the electron beam. A computational assignment of their orientation distributions is necessary to perform a 3D reconstruction of the imaged object. This gives rise to a reconstructed 3D density map. A number of recent reviews describe the various procedures in detail [1–3]. The quality of the reconstructed density is then assessed using a global resolution metric.

A single global resolution metric may not be sufficient to ensure that a map can be properly interpreted with an atomic model. Several other resolution measures should be satisfied to better interpret a map. We discuss here two such measures. The first is a windowed, or "local resolution" evaluation approach [4], where sections of final half-maps are compared using the usual

Tamir Gonen and Brent L. Nannenga (eds.), *CryoEM: Methods and Protocols*, Methods in Molecular Biology, vol. 2215, https://doi.org/10.1007/978-1-0716-0966-8_8, © Springer Science+Business Media, LLC, part of Springer Nature 2021

Fourier shell correlation (FSC) criterion [5], defined below, for the windowed volumes. The second is a purely Fourier approach which compares conical sectors of Fourier space of the half-maps for consistency [6]. It may be thought of as an averaged (and therefore smoother) version of 3D SSNR that was introduced earlier [7]. We can therefore assign a direction to each resolution curve (the direction being down the axis of the aforementioned cone); we have used the term "directional resolution" for this reason. The problems in a reconstructed map cannot be accurately captured using a single global resolution criterion and should be quantitatively evaluated using both local and directional resolution measures. High-quality reconstructions will show consistency using the given measures.

The global resolution of the reconstructed map across all spatial frequencies is typically determined using the Fourier shell correlation (FSC) criterion [5]. To obtain an FSC, it is necessary to split the dataset into two equal independent halves and reconstruct individual "half-maps":

$$\mathrm{FSC}(k) \equiv \frac{\sum_{|\vec{k}'| \approx k} F\left(\vec{k}'\right) G^*\left(\vec{k}'\right)}{\mathrm{Norm}_F\left(\vec{k}\right) \mathrm{Norm}_G\left(\vec{k}\right)},$$

$$\mathrm{Norm}_F(k) \equiv \sqrt{\sum_{|\vec{k}'| \approx k} F\left(\vec{k}'\right) F^*\left(\vec{k}'\right) \cdots},$$

and $\mathrm{Norm}_G(k)$ defined similarly. Here, F and G refer to the Fourier transforms of each of two half-map reconstructions, and k is the spatial frequency vector. The FSC is the cross-correlation of the half-maps divided by the self-correlation evaluated over some fixed radius of Fourier space. It is the three-dimensional equivalent of the two-dimensional Fourier ring correlation [8, 9]. The nominal global resolution of the map can be obtained based on some criterion for determining a spatial frequency cutoff [10, 11]. There have been other resolution metrics previously proposed in the literature, but due to the widespread acceptance of the FSC, those will not be addressed here. A detailed background on the FSC and other measures can be found in multiple reviews [12, 13].

The local resolution is obtained by calculating the FSC in small fields (patches) and scanning across the map. Several approaches make use of this "windowed" local resolution evaluation approach [4]. The central voxel within each windowed patch would then be assigned the resolution value computed within this patch. The patches would be scanned across the map, enabling the assignment of local resolution to the entirety of the reconstructed density.

Assignment of local resolution values to individual voxel centroids and scanning the patches across the map produces a 3D array of local resolution values, the same size as the reconstructed object. This local resolution file can be opened and visualized just like any density within a visualization program (e.g., UCSF Chimera) [14]. The reconstructed object is colored by the values of the local resolution file.

Variations in local resolution arise from intrinsic dynamics inherent to macromolecules and macromolecular assemblies, from differences in the (dis)assembly state of macromolecular species, or can also be affected by errors in orientation assignment. Because averaging many particles into a 3D reconstruction is an inherent aspect of single-particle analysis, variations in local resolution, more generally, describe differences in the structural states of individual protein particles within averaged ensemble reconstructions. In practice, most specimens in cryo-EM have at least some amount of heterogeneity. Virtually all macromolecules are better ordered, and therefore better resolved, within their central core regions that are tightly packed and occluded from solvent molecules. In contrast, protein loops and solvent exposed regions often respond dynamically to changing cellular environments and are thus typically resolved to lower resolutions. There can also be multiple different species in the sample (e.g., from transiently or sub-stoichiometrically associated components) and apparent compositional heterogeneity can also result from protein degradation, or denaturation at the air–water interface during plunge freezing [15]. All of these result in lower resolution in specific regions. The inclusion of a local resolution analysis is therefore essential to properly evaluate a reconstructed map. Loss of resolution from some local region does not necessarily correspond to any particular direction in Fourier space, so that the directional resolution, as we have defined it and will be described next, may not be very informative for such types of errors.

To define the directional resolution, it is possible to investigate the agreement of F and G over specific viewing angles. One could alter the calculation and determine the resolution in cones. As adapted from earlier concepts applied in tomography [16, 17], we previously defined the conical measure of FSC, the 3D FSC [6], as:

$$\text{FSC}_{\delta\theta}\left(\vec{k}\right) = \frac{\sum_{\left|\vec{k}'\right|\cong k,\left|\widehat{\vec{k}'\cdot\vec{k}}\right|\geq\cos\delta\theta} F\left(\vec{k}'\right)G^*\left(\vec{k}'\right)}{\left|F_{\delta\theta}\right|^2\left(\vec{k}'\right)}$$

Here, the FSC is taken over all values of k' that are on the same shell as k, and sufficiently close to the direction governed by the parameter $\delta\theta$, which is the half-angle of the cone. The half-angle is defaulted to $20°$, but can be varied by the user. A set of 1D FSC curves is therefore computed over distinct angular directions, and these individual 1D curves are compiled into a 3D array that we term the 3D FSC. This 3D FSC can be visualized as an isosurface density just like any other map within a visualization software (e.g., in Chimera) and at some threshold value. When projections are evenly distributed within the Fourier transform and individual lattice points are approximately equally sampled across any individual shell, the directional resolution should not vary with different viewing angles. In such a scenario, the directional resolution is isotropic. However, when projections are not evenly distributed, there could be large variations in apparent structural features depending on the viewing angle, and the directional resolution is anisotropic. As a result, structural features might be elongated in one direction, which could lead to misinterpretations of the density and problems during derivation or refinement of the atomic model. As we showed recently, this will also affect the global resolution [18]. Careful examination of the 3D FSC, especially under conditions when anisotropy can be significant, can lead to a better understanding of how experimental data is affected.

Due to the geometry of the imaging experiment, the direction that is parallel to the electron beam shows the poorest resolution for preferentially oriented samples. If the electron beam is assumed to be along the Z direction, then the X/Y plane of Fourier space should be best resolved, whereas resolution curves along Z (e.g., the X/Z or Y/Z planes) would be compromised. Depending on the angular coverage of the projection views, the quality of the features evident along distinct viewing directions may vary. In experimental cases where the projections are dominated by more than just a few orientations (or when there is a high molecular symmetry), there may be only minor, if any, variations when viewing the map along any of the three directions. In the more severe cases, an anisotropic distribution of projection orientations may lead to apparent elongation of structural features within the map along the Z-direction. This may affect the interpretation of the map—sometimes severely—due to the appearance of artifactual (low-resolution) density parallel to the dominant view [6]. In some cases, an additional consequence of severe variations in directional resolution is that the refinement of individual projection orientations could be affected, leading to orientation misassignment. Therefore, this may lead to additional overfitting and/or artifacts during the refinement process.

Variations in directional resolution arise from preferred orientation of macromolecules within the vitrified ice layer, resulting in incomplete sampling of orientations in Fourier space [6, 19, 20]. If the macromolecules are in random orientations and approximately uniformly distributed within the ice layer (this means that there is no single dominant view), then the projections are also evenly sampled, and we expect the directional resolution curves, described above, to be very similar. Unfortunately, particles usually have an orientation bias, which arises from their adherence to one of two (top and bottom) interfaces on a grid. These can either be an air–water or support–water interface [19, 20]. Adherence to an interface results in an uneven distribution of particle projections. Consequently, the reconstruction of a 3D density is typically not evenly resolved in all directions, and the directional resolution curves can be very different [6]. Just as flexibility affects primarily local resolution, but not necessarily directional resolution, preferred orientation affects directional resolution, but not necessarily local resolution. There are exceptions, for example, if preferred orientation also causes particle denaturation (in which case local resolution would also be affected) [15], but such special cases are not discussed here.

Here, we will describe protocols for calculating local and directional resolution for reconstructed density maps. Other than commenting on some general features of the curves, we do not describe the calculation of the global resolution, which is standard in all cryo EM processing packages. Different samples will be presented to show distinct aspects of resolution evaluation, including structural homo/heterogeneity and directional (an)isotropy. We will provide general guidelines when evaluating the results from these measures using experimental samples.

2 Materials

2.1 Software

1. Local resolution: SPARX [21] (*see* **Note 1** about the use of alternative softwares, e.g., Relion or CryoSparc). In order to run local resolution estimation, the software described in the current protocol can be downloaded from:
   http://sparx-em.org/

2. Directional resolution: 3DFSC. In order to run directional resolution estimation using the command line, the software described in the current protocol can be downloaded from:
   https://github.com/LyumkisLab/3DFSC

   The software requires an installation of the Anaconda distribution (the scientific platform for Python), which can be downloaded from:
   https://www.anaconda.com

   The web version of 3D FSC is available through a web-server at:
   https://3dfsc.salk.edu

3. Visualization: UCSF Chimera. This program allows visualizing and analyzing the results from local and directional resolution analysis. Chimera can be downloaded from:
   https://www.cgl.ucsf.edu/chimera/

2.2 Cryo-EM Maps

1. Fullmap. This is the final reconstruction derived from the cryo-EM experiment.

2. Half-map #1. First reconstruction from one half of the data.

3. Half-map #2. Second reconstruction from second half of the data.

4. Map derived from PDB model. Density map computed from a fitted molecular model (optional, and only for map-to-model FSC calculations).

5. Mask. Used to remove solvent regions dominated by noise (optional).

---

3 Methods

Subheading 3.1 will describe the protocol for local resolution analysis. Subheading 3.2 will describe the protocol for directional resolution analysis. Subheading 3.3 will provide the results of these steps and analyses.

3.1 Estimating and Evaluating the Local Resolution of a Reconstructed Map

One of the first implementations of a measurement for evaluating local resolution in cryo-EM based on conventional FSC analysis was the blocres program [4] within the Bsoft processing package [22]. There are now several other alternatives that have been described in the literature (*see* **Note 1**). To estimate the local resolution, we use the program sxlocres.py, which is also based on FSC analysis. Because we have used this program for most of our published structures, we adhere to its description, although many programs offer similar implementations (*see* **Note 1**). This algorithm is implemented within the SPARX [21] package and distributed with EMAN2 [23]. In adhering to conventional FSC-based methods, we focus on describing this method here. The resolution is computed in 3D patches windowed across the map, and the nominal resolution value within the windowed region is assigned to the centroid of the patch. The output is an identically sized local resolution map, wherein a value is assigned to each voxel within the density and each voxel corresponds to the local resolution value of the specific region of the input map. The evaluation of the output from local resolution analysis is generally similar across packages.

1. Navigate to the directory where both half-maps are located. Optionally, you may wish to generate a mask to be used for the analysis (*see* **Notes 2** and **3**).

2. Run a command, similar to the following, within the folder that contains both half-maps.

```
>> /path/to/sxlocres.py /path/to/halfmap1.mrc /path/to/half-map2.mrc local_resolution_volume.mrc --cutoff=0.143 --fsc=resolution.txt
```

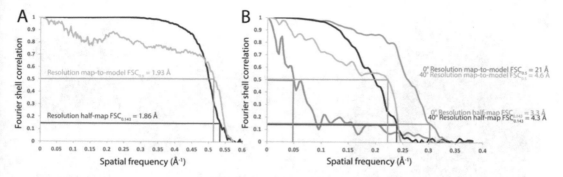

Fig. 1 Global half-map and map-model FSC curves. FSC curves with the corresponding resolutions indicated at the respective cutoffs of half-map and map-model with the axes as indicated in the two panels. (**a**) AAV2 [24] half-map (red) and map-model (orange). (**b**) HA trimer [6] untilted half-map (blue) and map-model (green) and tilted half-map (red) and map-model (orange)

Here, "halfmap1.mrc" and "halfmap2.mrc" are the two half volumes. The resolution cutoff here is set to 0.143 (*see* **Notes 4** and **5**). The output is the "resolution.txt" file, which describes the global resolution (also computed in all refinement software packages, e.g., *see* Fig. 1), and the "local_resolution_- volume.mrc", which is a volume file that contains a resolution value for each voxel across the full volume. This latter file is subsequently used for coloring and evaluating the local resolution of the input volume, and optionally to apply a local resolution filter (*see* **Note 6**).

Once the local resolution volume file has been computed, it is used to evaluate the local resolution of the input volume. The latter steps will be identical among all methods for computing local resolution described in **Note 1**.

3. Open Chimera [14] and load the volume files "local_resolution_volume.mrc" and "fullmap.mrc", where the latter corresponds to the final density map that is used for structural interpretations. Open the "Volume Viewer" panel (Tools → Volume Data → Volume Viewer), go to the "Features" tab and open the "Coordinates" panel. Make sure the Origin index of both maps has been set to 0 and the Voxel size has been set to the correct value.

4. Go to the "Tools" tab in "Volume Viewer" and select the "Surface Color" panel. Use the following settings to color the volume by resolution:

   - Color Surface: fullmap.mrc

   - By: Volume data value

   - Volume file: local_resolution_volume.mrc

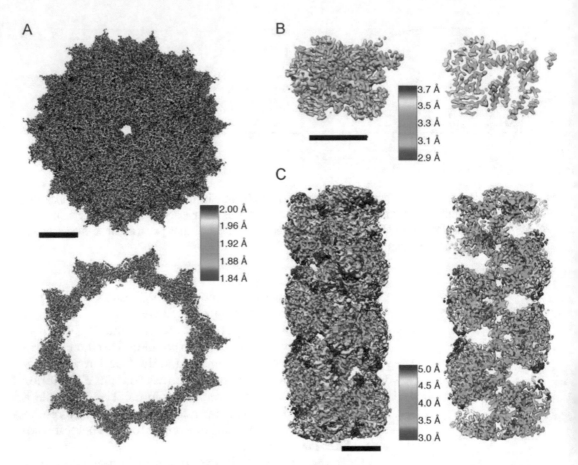

Fig. 2 Local resolution estimates colored onto experimental maps. Isosurface views and central slices of different EM maps color coded according to the local resolution gradient in Ångstroms. (**a**) AAV2 capsid [24]. (**b**) Cas13d ternary complex [26]. (**c**) Helical assembly of the SgrAI oligomer [27]. Scale bars are 50 Å

5. Press the "Set" button. This will set the full range of voxel values within the local resolution volume file. Chimera automatically estimates these based on their range within the volume file (*see* **Note 7**).

6. Press the "Color" button. This will color the input reconstruction by the local resolution file. There are other options available in the program, which are supplied within the Chimera documentation.

7. The map should now be colored by local resolution (*see* Fig. 2). Each voxel in the "local_resolution_volume.mrc" file corresponds to the resolution (in absolute frequency, ranging from 0 to 0.5) at that position. You can use the formula below to convert these voxel values to units of spatial frequency to obtain the nominal resolutions:

$$(\text{Spatial frequency}) = (\text{Absolute frequency})/(\text{Voxel size}(\text{Å}))$$

$$\text{Resolution}(\text{Å}) = 1/(\text{Spatial frequency})$$

$$= (\text{Voxel size}(\text{Å}))/(\text{Absolute frequency})$$

For example, if the voxel size is 1.31, and you want to color the surface with local resolution of >5 Å (red), 4.5–5 Å (yellow), 4.0–4.5 Å (green), 3.5–4.0 Å (cyan), and <3 Å (blue), set the values to 0.262 (1.31/0.262 = 5.0), 0.291 (1.31/0.291 = 4.5), 0.3275 (1.31/0.3275 = 4.0), 0.374 (1.31/0.374 = 3.5), and 0.437 (1.31/0.437 = 3.0), respectively.

8. Inspect the results and compare to the global resolution (*see* **Note 8**). The values displayed in the "Surface color" panel can also be adjusted and used to recolor the reconstructed map by a different range of resolution values (*see* **Notes 7** and **8**).

3.2 Estimating and Evaluating the Directional Resolution of a Reconstructed Map

The directional resolution, which is distinct from local resolution, provides a measure for how the apparent information content within a map changes as a function of viewing angle. The two metrics are independent, provide distinct pieces of information, and are complementary in evaluating the quality of a reconstructed density. While there may, on occasion, be overlap in the regions of a map that are poor by both local and directional resolution measures, these two criteria should not be confused with one another. There have been multiple implementations of directional resolution evaluation, but all of them are similar in that they evaluate the FSC in cones (*see* **Note 9**). Here, we focus on the implementation and analysis of the 3D FSC, proposed and described in detail in [6]. Resolution is computed in conical sectors (the size of the cone can be varied), and an FSC curve is assigned to the central axis of the cone. Varying the cones across all directions of Fourier space accordingly produces a set of directional FSC curves. Therefore, the output of the method is a set of 1D curves computed over distinct angular directions that are combined into a 3D array that we term the 3D FSC. The latter can be used to quantitatively evaluate the map at distinct viewing directions.

1. Navigate to the directory where both half-maps are located.

2. Run a command, similar to the following, within the folder that contains both half-maps

```
>> <path_to_3dfsc_program>/ThreeDFSC_Start.py --halfmap1=/
path/to/halfmap1.mrc --halfmap2=/path/to/halfmap2.mrc --full-
map=/path/to/fullmap.mrc --apix=1.31 --ThreeDFSC=Output
```

The results will appear in the "Results_Output" directory, specified by the --ThreeDFSC option. The command line version has numerous additional options that are described in the software (*see* **Notes 10–12**).

3. Alternatively, the 3DFSC program can be run through a web interface, as described in **steps 3a–3h**. The web address for running directional resolution estimation online is:
   https://3dfsc.salk.edu

   (a) Click "Register" on the navigation bar and follow the instructions to create an account.

   (b) Once an account is created, navigate to the processing form via the "Submit job" link.

   (c) The online job submission requires the user to enter a valid email address and other parameters in the form.

   (d) You must upload a unique job name, two half-maps (.mrc format), a full map (also .mrc format) (*see* **Note 13**), and an appropriate pixel size. Click "Submit job".

   (e) Optional parameters include submission of a mask file (.mrc format), appropriate cone angle, cutoff for FSC, a threshold for sphericity and a high-pass filter (in Ångstroms) (*see* **Note 10**).

   (f) You should receive an email to confirm your processing job. If you do not receive an email, please check your spam folder.

   (g) When your job is complete, you will receive another email with a link to view the results.

   (h) A typical output will include a compressed folder that contains the input files used and a range of outputs. These can be downloaded from the website. There is a password on the output page that needs to be used in order to open and decompress the folder. Once the folder is decompressed, the outputs will be similar to the outputs within the command-line version.

4. Inspect the outputs. These include the 3D FSC volume (*see* Figs. 3 and 4), the global FSC curve overlaid onto a histogram plot computed from all individual angular FSC values (individual angular FSC curves corresponding to ±1 standard deviation of the FSC values within the histogram are also displayed on the plot, *see* Fig. 5), and raw data used for computing these plots are provided as spreadsheets (*see* **Note 14**).

5. Open UCSF Chimera and load the "3DFSCPlot_Chimera. cmd" file from the output directory. This script is designed for convenient visualization using Chimera, but individual mrc files can also be opened independently. Once executed, the Chimera visualization program opens with the full map and the 3D FSC volume along with an FSC plot that compares global resolution with the Z-directional resolution relative to the viewing angle (*see* Fig. 4).

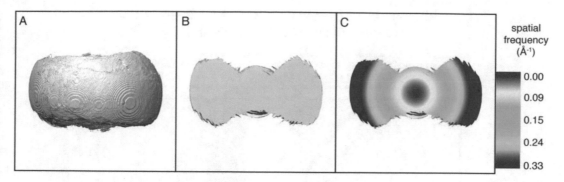

Fig. 3 A 3D FSC volume displayed in three different manners in Chimera. The volumes are oriented such that the Y axis in the figure (top/down) corresponds to the worst directional resolution (this is also the axis of preferred orientation relative to the electron beam). The 3D FSCs are displayed at a threshold of 0.143 and as (**a**) an isosurface, (**b**) an uncolored slice through the central section, or (**c**) the same slice colored by spatial frequency

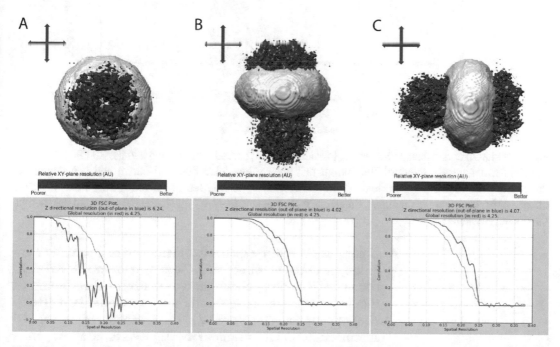

Fig. 4 Distinct angular views of the output from the 3DFSC program visualized in Chimera. When the output "3DFSCPlot_Chimera.cmd" is loaded into Chimera, the resulting 3D FSC volume is aligned with the input experimental map (here, the reconstruction of HA from 40° tilted images, [6]). A color gradient bar is included and provides a qualitative metric for evaluating the in-plane (X/Y) resolution of the map. Below that is a plot of the out-of-plane Z-resolution for a given view, along with the global FSC. Here, three different views of HA are provided: (**a**) View down the symmetry axis (and also along the axis of preferred orientation), showing good X/Y resolution, but bad Z-resolution. (**b**) The volume in panel A is rotated around the X-axis by 90°. Now, the new Y-resolution is the worst (leading to the map being colored red), but the new Z-resolution is improved. (**c**) The volume in panel A is rotated around the Y-axis by 90°. Now, the new X-resolution is the worst (also leading to the map being colored red), but the new Z-resolution is likewise improved

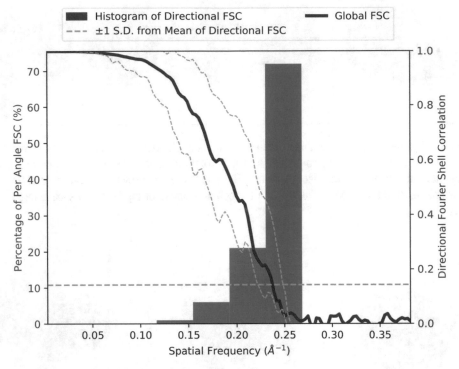

Histogram and Directional FSC Plot for T40HalfmapHA
Sphericity = 0.874 out of 1. Global resolution = 4.25 Å.

Fig. 5 Histogram of directional FSC curves. Individual 1D FSC curves are compiled into the 3D FSC and can be represented within a histogram. The current plot shows the global FSC (red curve) and a histogram of directional FSC values (ranging from ~1/8.0 to 1/3.7 Å^{-1}) for the reconstruction arising from 40°-tilted images of HA (*see* also Figs. 4 and 7). Green curves correspond to +/− 1 standard deviation from the average directional resolution curve

6. Color the 3D FSC by spatial frequency and calculate the resolution in different directions. Once you have the 3D FSC opened in Chimera, check the voxel size of the volume. It should be expressed in Å^{-1} and therefore should be 1/(voxel size) of your data (*see* **Note 17**). The Chimera script in **step 5** automatically sets the origin of the 3D FSC to be at the center of the box. If you open it by hand, you may have to set the origin manually (e.g., volume #0 origin Index 256, 256, 256 for a 3D FSC with a box size of 512). Go to the "Tools" tab in "Volume Viewer" and select the "Surface Color" panel. Use the following settings to color the 3D FSC volume by resolution:

   • Color Surface: your_3dfsc_map.mrc

   • By: Radius

   The "Surface Color" panel also lets you set the origin, if you did not do this in the previous step. Press the "set" button to set the full range of surface values and make sure the "report value at mouse position" is checked. Now, whenever you drag your mouse on a point in the 3D FSC, the "radius" value should be displayed in the dialog box in the bottom left-hand corner of Chimera. To get the resolution at that particular point, you simply have to divide the box size by that value (*see* **Note 17**). It can also be convenient to display radial slices of the 3D FSC. To do this, go to Tools → Viewing Controls → Side View and drag the clipping panels to constrain to the center of the volume. The results of this procedure, for various reconstructions, are displayed (*see* Figs. 3c and 7).

7. Rotate the map. You should see variations in the Z-resolution relative to the viewing angle. If the map is anisotropic, the Z-resolution will change, as will the color of the map (*see* Fig. 4 and **Notes 15** and **16**).

3.3 Analyses of the Individual Results

Here, we will elaborate on the results of the operations above and comment on analyzing individual case studies. The different case studies show variations in both local and directional resolution.

3.3.1 Half-Map and Map-Model FSCs

The first step in evaluating map quality is computing global half-map, and optionally (if a model is previously available or has been derived from the data), map-model FSCs. A half-map FSC describes the internal consistency between two reconstructions computed from half-sets of the data, while a map-model FSC describes the consistency between an experimental map and an atomic model (*see* **Note 11**). These steps are standard in every refinement program and have not been described in the current protocol, but we provide the results of these operations for reference and as examples distinguished by different types of anisotropies. A suitable mask was applied to the experimental maps and to the maps generated from the atomic models (*see* **Note 12**). The AAV2 capsid has icosahedral symmetry. The 60-fold averaging not only boosts the signal-to-noise ratio, but also results in uniform distribution of projection views (it also improves the sampling compensation factor, as described recently [18]. As a result, the reconstruction is isotropic, and a smooth half-map FSC curve indicates a healthy reconstruction (*see* Fig. 1a). The half-map and map-model nominal resolution values are also similar to one another. In contrast, the HA trimer is anisotropic, and the true "top" projection views predominate over other projection views. A comparison of two HA trimer datasets was performed to demonstrate the effects of directional resolution anisotropy: one dataset was collected without tilting (0°) and one dataset was collected with a 40° tilt angle of the specimen stage [6]. The half-map FSC curve from the 0°-dataset shows an apparent resolution of 3.3 Å (*see* Fig. 1b, blue). However, the map-model FSC curve (using an externally derived and previously available atomic model derived from X-ray crystallography, PDB 3WHE) shows a much poorer global resolution of ~21.0 Å (*see* Fig. 1b, green). This discrepancy is problematic, and, assuming that the model is properly docked and fitted into the map, there could be several possible explanations: either the model is severely underfit into the density (too much weight has been placed onto idealizing model parameters, rather than the experimental density) or the map contains errors. In this particular case, due to the large amount of anisotropy, coupled to reinforcement of misassigned orientations during refinement, the main reason for the discrepancy is the worsening of the map. The fact that the half-map resolution belies the actual resolution of the reconstructed density also highlights some of the limitations of a simple global self-consistency metric. On the other hand, the reconstruction from 40°-tilted images shows a slightly decreased half-map (*see* Fig. 1b, red) but much improved map-model (*see* Fig. 1b, orange) global resolution when compared with the reconstruction from the 0° images. Now, there is less discrepancy between the half-map and map-model FSC curves, and the resulting values are consistent with the structural features (*see* Fig. 9). The origin underlying these issues is partially described in [6] and is also subject to ongoing investigation.

3.3.2 Interpreting the Local Resolution of the Map

We selected three experimental cases with which to demonstrate the result of local resolution analysis using the FSC-based method. In the case of the icosahedrally symmetric reconstruction of AAV2 capsid (EMDB: EMD-9012) [24], the resolution varies only slightly about the global resolution (*see* Fig. 2a). Most areas of the capsid proteins are resolved approximately evenly, although some loops on the periphery show slightly lower resolutions. This is not surprising because many of these loops are known to be dynamic and functionally relevant, as they interact with cell surface receptors and are also most prone to variation between viral subtypes [25]. The second experimental data analyzed is the ternary complex of Cas13d with CRISPR RNA and target RNA (EMDB: EMD-9014) [26]. In this structure, Cas13d associates with RNAs to form a stable ribonucleoprotein assembly with a global resolution value of 3.3 Å and a local resolution that ranges from 2.9 to 3.7 Å (*see* Fig. 2b). Here, the variation in local resolution is more significant. The regions that form extensive electrostatic interactions with RNA have the highest resolution of 3.1 Å, while most of the remaining areas range from 3.3 to 3.5 Å. The worst resolution of 3.7 Å is only found on a solvent exposed RNA loop, which is expected to be dynamic. The third experimental case is a helical reconstruction of the SgrAI oligomer with associated double-stranded DNA (dsDNA) (EMDB: EMD-20015) [27]. The global resolution within an asymmetric unit is 3.5 Å. The best resolved regions are again located within a tightly packed core, as is typical for most macromolecular assemblies. However, the local resolution varies more substantially, ranging from 3.0 to 5.0 Å (*see* Fig. 2c). The enzyme is dynamic and there are conformational changes apparent in multiple regions, with the unprotected ends of dsDNA resolved to ~5 Å or worse. There is also compositional heterogeneity, as the filaments are symmetrized from multiple species containing a distinct number of underlying asymmetric units. In summary, the global resolution of a 3D volume does not ensure that all regions can be interpreted equally, and local resolution analysis is a fundamental measure that informs the experimentalist about variations in resolution, which helps with structural interpretations.

3.3.3 Interpreting the Shape of the 3D FSC Volume

The 3D FSC can be displayed and visualized as an isosurface at a fixed threshold (e.g., 0.143 for half-map and 0.5 for map-model). Another way to represent the information content in 3D FSC volumes is to display the central slice and color the volume according to the Fourier radius, which is directly related and can be converted to the spatial frequency and resolution (*see* **Note 17**). These different displays are shown in Fig. 3 for a severely anisotropic reconstruction containing a large missing cone. Within the 3D FSC volumes, the uncorrelated regions showing missing density correspond to the direction of preferred orientation, and thus the Z-direction relative to the electron beam (*see* Fig. 4). Rotating the maps in Chimera will change the color, as well as the out-of-plane Z resolution that is displayed in the second window (*see* **Notes 15** and **16**). Displaying 3D FSC volumes is complementary to displaying the Euler angle distribution profiles, but provides additional, quantitative insight pertaining to directional resolution. The Euler angle profiles giving rise to the AAV2 and HA reconstructions discussed below are shown in Fig. 6 (*see* **Note 18**). For a perfectly isotropic sample, the shape of the 3D FSC volume should be a sphere. Expectedly, this is the case for the icosahedrally symmetric reconstruction of AAV2 (*see* Fig. 7a). In contrast, for an anisotropic sample like HA, the 3D FSCs will deviate from spherical to varying extents. In the half-map 3D FSCs, both the reconstructions from untilted and tilted images show a volume with some evidence of the missing cone. (Fig. 7b top untilted and bottom tilted). The most anisotropic reconstructions will have 3D FSCs that will have a clear indication of a missing cone (in practice, they may appear "pancake-shaped"). For the particular case of HA, there is likely some amount of overfitting evident in the half-map 3D FSCs, especially for the reconstruction from untilted images. This explains the presence of some correlations within the 3D FSC in regions that should normally correspond to the missing cone. For this reason, it is often useful to display the 3D FSC volumes at different thresholds, as indicated in Fig. 8 (*see* **Note 19**). For the

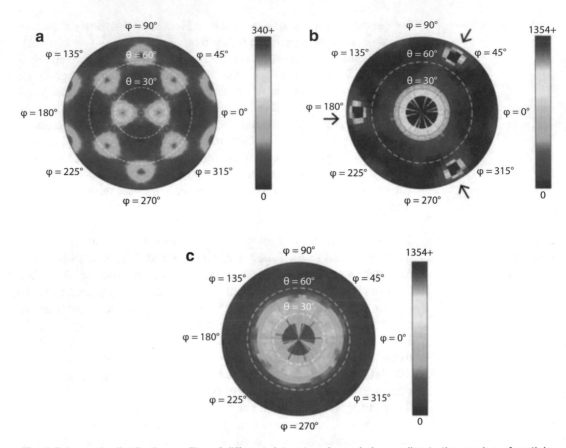

Fig. 6 Euler angle distribution profiles of different datasets color coded according to the number of particles occupying a particular orientation. Plots refer to: (**a**) AAV2, (**b**) HA trimer reconstructed from 0° images, and (**c**) HA trimer reconstructed from 40° images. These profiles inform the user of the degree of preferred orientation. For preferentially oriented samples, tilting the stage will ameliorate the effects of preferred orientation by altering the distribution profile (compare panels **b** and **c**). In panel **b**, the arrows point to false-positive orientation assignments, which disappear in **c** when the map is more isotropic, as described in [6]

map-model analysis (the model, here, is an external reference standard that is expected to be isotropic, *see* **Note 20**), the 3D FSC describing the reconstruction from untilted images is severely flattened relative to the 3D FSC describing the reconstruction from tilted images. Because we are using an external standard for the analysis, this is likely a more realistic representation of the true directional resolution (*see* **Notes 20** and **21**).

3.3.4 Interpreting the Experimental Map

The presence of icosahedral symmetry fills in all directional views and expectedly results in a high-quality map for AAV2 capsid, where the interpretation of the map and derivation of the subsequent model is not hampered by anisotropy artifacts (*see* Fig. 9a). Viewing the map along three different planes shows uniform map features. In contrast, an anisotropic map will show artifactual densities in the direction parallel to preferred

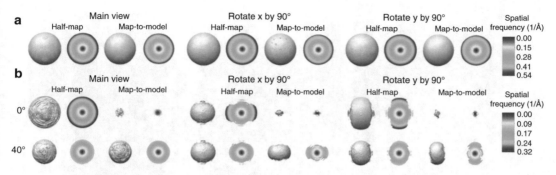

Fig. 7 3D FSC volumes of three different orientations with left panels showing volumes computed from two sets of experimental "half-maps" at 0.143 threshold and right panels showing volumes computed from a synthetic map derived from a model and the experimental "full map" at 0.5 threshold. The three different orientations displayed correspond to the XY plane (left, "main view"), the XZ plane (middle, "rotate x by 90°"), or the YZ plane (right, "rotate y by 90°"). For each 3D FSC result, the entire volume is depicted alongside a central slice color coded according by spatial frequency, as indicated in the figure (*see* **Note 17**). (a) AAV2 (b) HA trimer (top row untilted, bottom row tilted). *See* **Note 19** and Fig. 8 with regard to displaying the 3D FSCs at different thresholds

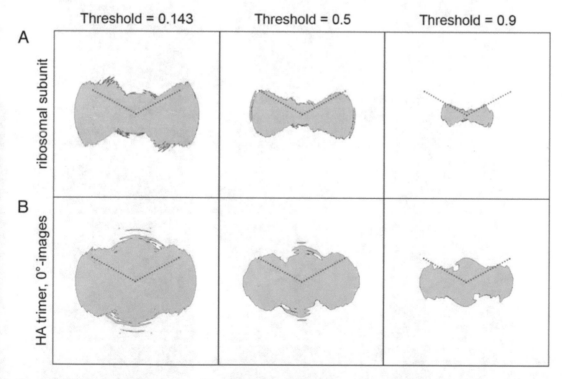

Fig. 8 3D FSC volumes displayed at different thresholds. For anisotropic reconstructions, it may be useful to display the 3D FSCs at different thresholds, because sometimes the pathologies can be better identified. (a) For a clean reconstruction that does not show signs of overfitting, there is clear evidence of a missing cone in the 3D FSC at all correlation thresholds. (b) For a reconstruction such as the HA trimer from untilted, 0° images, which shows signs of overfitting (described in [6]), displaying the 3D FSC at high correlation thresholds (e.g., 0.143) can obscure some of the problems with anisotropy. In contrast, at lower thresholds (e.g., 0.5, 0.9), the 3D FSC often appears flatter, and the missing cone is more pronounced

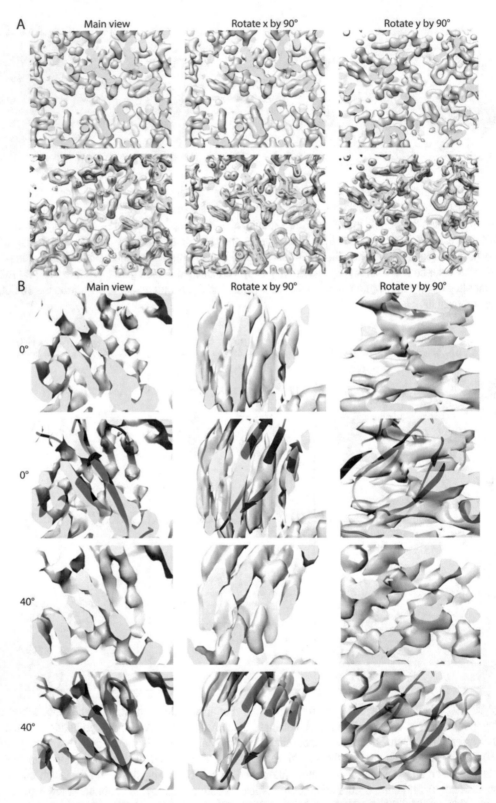

Fig. 9 Panels showing three different projection views of different maps. (**a**) AAV2 (map color grey, first row) and AAV2 (map color gray at 50% transparency, second row) overlaid with the model (carbon in yellow,

orientation. Such artifacts make *de novo* modeling more challenging and could lead to incorrect map interpretation. Although tilting ameliorates the effects of artifactual features, the problem of preferred orientation is not necessarily completely overcome because there is still a missing cone that accounts for residual elongation and loss of resolution in the corresponding direction within the resulting map (*see* Fig. 9b). Importantly, it is not necessary to have a perfectly isotropic map in order to accurately interpret the structural features. In fact, even with moderate levels of anisotropy, the map may appear normal to the eye, and can be used for model building and refinement. However, care must be taken during the interpretation of the map and the atomic model, in order to make sure that structural features are arising from the true features in the imaged macromolecular object, and not due to artifacts caused by loss of directional resolution.

A combination of the analyses described above can guide the user in quantitatively characterizing the directional resolution of an EM map and thus altering strategies for improving resolution and overall map quality.

4 Notes

1. FSC-based methods, similar to the one described in the current protocol, are also implemented in most of the major processing packages, including Relion [28] and cryoSparc [29]. A popular alternative approach to the FSC-based analysis is the ResMap approach, which assigns significance to features within an experimental density map if and only if a 3D local sinusoid of a nominal wavelength is statistically detectable above the noise level [30]. The ResMap approach only requires a single map as input. Another recent alternative, MonoRes [31], is implemented within the Xmipp/Scipion processing packages [32, 33]. MonoRes is based on the identification of a monogenic amplitude across different spatial frequencies in a map and determining whether, at the specific resolution and location in the map, this amplitude is significantly higher than what is expected from pure noise. MonoRes also only requires a

Fig. 9 (continued) nitrogen in blue, oxygen in red and sulfur in green). All three views show uniform map density allowing for unambiguous interpretation during model building. (**b**). HA trimer, first row map from 0° untilted images (map color grey) and second row the same map overlaid with an atomic model (blue). Map features appear stretched, especially when rotated across the X and Y axes, resulting in artifactual density. The third (map only) and fourth rows (model overlayed into map at 50% transparency) show a reconstruction from 40° tilted images. Now, the map features appear visibly better with less stretching of map density. HA reconstructions depict a β sheet region. Map and model for both AAV2 and HA were aligned using the *fit* function in UCSF Chimera [14]

single map as input. The resulting files from all these programs can generally be analyzed using a similar procedure as that described in the current protocol.

2. The mask constrains the analysis to specific regions of a map and, by means of excluding regions of solvent outside of the density, provides a more accurate measure of global resolution. In Sparx, this mask will also define the region to which an optional local filter will be applied (*see* **Note 6**). The mask can be generated using most processing packages.

3. Sharp or hard edges from the mask can skew and/or artificially inflate the local resolution within specific regions of the map. The amount of inflation can be measured using the high-resolution noise substitution test [34]. Care must be taken when generating the mask to make sure that it maintains soft edges and does not cut into the density. Each package has its own protocol to generate a mask, but generally, the procedures are: binarize the experimental map at a defined threshold → extend the mask → soften the edges.

4. A different threshold (e.g., 0.5) can be selected for computing the nominal resolution value at each voxel centroid.

5. For more usages of sxlocres.py, you can use the command sxlocres.py -help or check the website:
    http://sparx-em.org/sparxwiki/sxlocres

6. The output local resolution map can also be used to locally filter the input volume. This step is not covered in the current protocol, but can be optionally performed.

7. In practice, identifying the best resolution range by which to color the input map involves some trial and error and depends on what the user wants to highlight. The user can usually assume that the distribution of local resolution values is centered around the global resolution of the map (although this need not be the case, e.g., for dynamic regions) and select a set of resolution values that color the volume with, e.g., ~0.5 Å intervals across the resolution range. Chimera provides a five-color palette as default; further adjustments of the resolution intervals can be performed, with the general goal of displaying the full range of colors. The range can also be estimated by looking at a histogram of the local resolution voxel values, which is displayed in Chimera. Sometimes, the range of local resolution values is very narrow (e.g., this is the case with the AAV2 example that is described in the current protocol, *see* Fig. 2a). In this case, the user should decrease the resolution interval range. Other times, especially when there is substantial heterogeneity (conformational or compositional) in the complex, the range could be larger (e.g., this is the case with the SgrAI helical filament described in the current protocol, *see*

Fig. 2c). We have also observed that severe preferential orientation can also apparently skew the local resolution values, as is the case with the reconstruction of HA from untilted images described in the current protocol. However, we note that local and directional resolution measurements provide fundamentally distinct assessments of the map, and a detailed analysis as to the origin of this phenomenon is beyond the scope of the current protocol. It is always worthwhile to carefully inspect the map and the structural features within every local resolution patch.

8. In general, the values for local resolution should vary slightly around the global resolution of the volume, as estimated by the global FSC (the extent can vary). If this is not the case, then something in the protocol may be wrong, and the user should inspect the inputs. A common mistake is to not set the proper values for the voxel size or the origin.

9. There have been several implementations of the conical resolution estimates applied to tomographic reconstructions [16, 17] and several implementations applied to single-particle reconstructions [6, 35]. These measures are all similar conceptually, with minor variations in how the data is weighted and/or how the cone is defined.

10. For more options, the user is directed to the help documentation of the standalone 3DFSC program.

11. Computing the 3D FSC requires two half-maps to calculate the half-map FSC. In order to calculate map-to-model FSC, a map has to be generated from a model derived from the density. In the absence of such a model, the map-to-model FSC cannot be computed. In order to generate a map from a model for the methods described here, the e2pdb2mrc command from the EMAN [36] software package was used, and the map was computed at a resolution corresponding to Nyquist frequency. For example, the command used to obtain a synthetic map from the AAV2 model (PDB ID 6E9D) was:t

```
>> e2pdb2mrc.py 6E9D.pdb 6E9D.mrc --apix 0.79 --res 1.58 --box 512
```

There are other methods to calculate a molecular model, for example, in Chimera [14], or in other software packages. Once the map is generated, it must be aligned to match the dimension and origin of the experimental map to be used for analysis. This can be achieved in Chimera by resampling the map generated from the model onto the coordinates of the experimental map.

12. For accurate global FSC calculations, the experimental map should be masked to remove background noise from the solvent. Many packages can create masks (*see* also **Notes 2** and **3**); we performed this step using the relion_mask_create command implemented within the Relion software package. For example, the command used for the AAV2 dataset was:

```
>> relion_mask_create --i experimental_map.mrc --o output.mrc --
ini_threshold 0.001 --width_soft_edge 6 --extend_inimask 6
```

13. A common mistake is to upload volume files in a format that is not mrc (e.g., hdf, spi, vol). The current version of the program only supports mrc format.

14. Other outputs associated with the run are provided for convenience, such as a thresholded and binarized 3D FSC volume, as well as supplementary curves corresponding to the X, Y, and Z resolutions relative to the internal coordinate system of the map (*see* **Note 15** for a word of caution on the lack of a relation between the internal coordinate system of the map and the direction relative to the electron beam). The 3D FSC is used to evaluate the directional resolution of the reconstructed map.

15. In Chimera, the X/Y plane corresponds to the image that is displayed on the screen, whereas the Z direction is out-of-plane and is not evident to the eye. The experimental map and the 3D FSC volume are loaded in the first window. The second window shows the dynamic plot that reflects the FSC for a given Z-direction (which changes as the map is rotated), as well as a fixed global FSC curve (*see* Fig. 4). Due to the geometry of the imaging experiment, and the fact that particles adhere along a plane perpendicular to the electron beam, the Z-resolution *relative to the electron beam* is typically the worst resolved direction. However, a word of caution here: the Z-direction relative to the electron beam does not necessarily correspond to the Z-direction of the internal coordinate system of the map. The former can be precisely identified, e.g., using the tilt-pair test and with knowledge of the tilt angle and tilt axis of the microscope [10]. The latter inherently depends on how the orientation angles are assigned to the map. In practice, the Z-direction relative to the electron beam can correspond to any direction when the map is initially opened in Chimera. Due to the geometry of the imaging experiment, and the fact that particles adhere to the air–water interface perpendicular to the electron beam, the Z-direction relative to the electron beam can often be identified by locating the direction of worst Z-resolution within the second window (*see* **Note 16**).

16. The map is colored according to the XY plane resolution. Rotating the map automatically changes its color to reflect the relative resolutions in the plane of view; it will also change the relative Z-resolution curve in the second window. The user can actively monitor these changes to determine the worst resolution direction. An intuitive way to analyze these outputs is to rotate the map to a point where the Z-resolution curve in the second window reflects the lowest resolution. Once this is accomplished, then the out-of-plane direction is the worst resolved view; this direction is usually (but not necessarily) also the "true" Z-direction relative to the electron beam for this particular orientation (*see* **Note 15**). There could also be multiple preferred orientations relative to the electron beam. The out-of-plane view is not visible to the user, and therefore the user cannot visually judge the detrimental effects on the map in this orientation. Conversely, the view that is displayed on the screen (the X/Y plane) should be the best resolved view. Accordingly, the map should now be colored blue (the "best" resolution in the X/Y plane). If the user now rotates the map 90° about the local X-axis, this would place the old XZ plane in the field of view and the old Y direction out-of-plane. Now the map should be colored red, because the "worst" resolution (the Z-resolution relative to the electron beam) is now in the field of view. Likewise, if the user rotates the map back, and then 90° about the local Y-axis, this would place the old YZ plane in the field of view, and the old X direction out of plane. Again, the map should be colored red because the "worst" resolution is still in the field of view. These procedures, and how the Z-resolution and map coloring change with rotations, are summarized in Fig. 4. Rotating the map around its different axes, carefully monitoring the best and worst Z-resolution, and inspecting the density will help to show how features are elongated in the Z-direction relative to the electron beam.

17. The radial points of the 3D FSC correspond to the Fourier lattice points in the 3D transform, multiplied by the box size, with units of inverse Angstroms (Å^{-1}). Therefore:

Spatial frequency in Å^{-1} = (Radial_point_on_3DFSC)/(Box size).

The reason for the incorporation of the box size factor is to make the numbers more user friendly. For example, if a map has a voxel size of 0.79 Å, then the 3D FSC displays a value for the voxel size as $1/0.79 = 1.266$, and it is easy to read off the voxel values of the 3D FSC from the chimera plots. The appropriate voxel size value is defined in the "Coordinates" tab under "Volume Viewer" in Chimera. To convert the radial lattice points of the 3D FSC into resolution (as displayed in

Figs. 3c and 7), one can use the usual relationship between resolution and spatial frequency:

$$\text{Resolution} = 1/(\text{Spatial frequency}).$$

Combining the last two relationships gives:

$$\text{Resolution in } \text{Å} = (\text{Integer box size})/(\text{Radial\_point\_on\_3DFSC})$$

18. Euler angle distribution profiles are standard outputs in all refinement packages and are not described in the current protocol. To display the profiles for the datasets described here, the parameter file and associated stack were imported into cisTEM [37] and Euler angle distribution profile was visualized in the results of 3D refinement tab.

19. It is useful to view the 3D FSCs at high correlation thresholds. Although the nominal values for global resolution should be obtained at fixed thresholds (e.g., 0.143, 0.5) or variable thresholds (as described in [38]), we have found that regions of the 3D FSC corresponding to the directional resolution along the axis of preferred orientation (i.e., those that should be the worst resolved) often have correlations at high spatial frequencies. This makes the 3D FSC look more "healthy" and isotropic at low thresholds (e.g., 0.143). However, when viewed at higher thresholds, the 3D FSC is less spherical, and the missing cone is more evident. Because this issue is exacerbated during the course of orientation refinement, and as a function of increasing the number of particles (including false positives), we believe this has to do with the reinforcement of improperly assigned orientations and/or noise (i.e., overfitting). For this reason, *viewing* the 3D FSC at high correlation thresholds (e.g., 0.75–0.95), which typically correspond to lower spatial frequencies, may actually provide a better overall impression of the true level of anisotropy. This is summarized in Fig. 8. For the time being, this appears to be one way of identifying pathologies in the map, although in the future, it would be of interest to address the root cause of the problem.

20. Modeling into slightly anisotropic density maps is typically not a problem when the refinement is properly restrained. However, problems may occur when the level of anisotropy is severe, and the density is substantially elongated. In this case, even properly restrained refinements may be insufficient to limit overfitting into density features resulting from artifactual elongation due to limited directional resolution, as opposed to true structural features of the object. Care must

be taken with such samples, and the maps, along with 3D FSCs, should be carefully inspected. It would be better to alter the orientation distribution (e.g., through tilting) and obtain a less anisotropic map.

21. For new structures, an external reference standard does not exist, and the model must be derived from the density. The model would be refined into the density, which would accordingly improve map-model FSC. In such a case, an analysis comparing differences between half-map and map-model FSCs is less likely to yield insight into any potential overfitting. The user is advised to take care when interpreting anisotropic maps, and preferably, improve the orientation distribution and obtain a less anisotropic map.

Acknowledgments

The authors acknowledge the input from Yong Zi Tan. Molecular graphics and analyses were performed with the USCF Chimera package (supported by NIH P41 GM103311). This work was supported by the Pioneer Fund to C.Z. and by grants DP5 OD021396, U54 GM103368, and R01 AI136680 of the National Institutes of Health, grant MCB-1933864 from the National Science Foundation, and the Margaret T. Morris Foundation to D.L.

Author Contributions: *S.A. performed analyses and wrote sections for directional resolution. C.Z. performed analyses and wrote sections for local resolution. P.R.B. develops and maintains the 3D FSC code and helped with writing and analyses. D.L. guided and supervised the analyses and helped with writing. All authors contributed to manuscript preparation and review.*

References

1. Vinothkumar KR, Henderson R (2016) Single particle electron cryomicroscopy: trends, issues and future perspective. Q Rev Biophys 49:e13. https://doi.org/10.1017/S0033583516000068

2. Cheng Y, Grigorieff N, Penczek PA, Walz T (2015) A primer to single-particle cryo-electron microscopy. Cell 161(3):438–449. https://doi.org/10.1016/j.cell.2015.03.050

3. Lyumkis D (2019) Challenges and opportunities in cryo-EM single-particle analysis. J Biol Chem 294(13):5181–5197. https://doi.org/10.1074/jbc.REV118.005602

4. Cardone G, Heymann JB, Steven AC (2013) One number does not fit all: mapping local variations in resolution in cryo-EM reconstructions. J Struct Biol 184(2):226–236. https://doi.org/10.1016/j.jsb.2013.08.002

5. Harauz GavH M (1986) Exact filters for general geometry three dimensional reconstruction. Optik 73:146–156

6. Tan YZ, Baldwin PR, Davis JH, Williamson JR, Potter CS, Carragher B, Lyumkis D (2017) Addressing preferred specimen orientation in single-particle cryo-EM through tilting. Nat Methods 14(8):793–796. https://doi.org/10.1038/nmeth.4347

7. Penczek PA (2002) Three-dimensional spectral signal-to-noise ratio for a class of reconstruction algorithms. J Struct Biol 138(1–2):34–46

8. van Heel M (1982) Detection of objects in quantum-noise-limited images. Ultramicroscopy 7(4):331–341

9. Saxton WO, Baumeister W (1982) The correlation averaging of a regularly arranged bacterial cell envelope protein. J Microsc 127 (Pt 2):127–138. https://doi.org/10.1111/j.1365-2818.1982.tb00405.x

10. Rosenthal PB, Henderson R (2003) Optimal determination of particle orientation, absolute hand, and contrast loss in single-particle electron cryomicroscopy. J Mol Biol 333 (4):721–745

11. van Heel M, Schatz M (2005) Fourier shell correlation threshold criteria. J Struct Biol 151(3):250–262. https://doi.org/10.1016/j.jsb.2005.05.009

12. Sorzano CO, Vargas J, Oton J, Abrishami V, de la Rosa-Trevin JM, Gomez-Blanco J, Vilas JL, Marabini R, Carazo JM (2017) A review of resolution measures and related aspects in 3D Electron Microscopy. Prog Biophys Mol Biol 124:1–30. https://doi.org/10.1016/j.pbiomolbio.2016.09.005

13. Penczek PA (2010) Resolution measures in molecular electron microscopy. Methods Enzymol 482:73–100. https://doi.org/10.1016/S0076-6879(10)82003-8

14. Pettersen EF, Goddard TD, Huang CC, Couch GS, Greenblatt DM, Meng EC, Ferrin TE (2004) UCSF Chimera--a visualization system for exploratory research and analysis. J Comput Chem 25(13):1605–1612. https://doi.org/10.1002/jcc.20084

15. D'Imprima E, Floris D, Joppe M, Sanchez R, Grininger M, Kuhlbrandt W (2019) Protein denaturation at the air-water interface and how to prevent it. elife 8:e42747. https://doi.org/10.7554/eLife.42747

16. Dudkina NV, Kudryashev M, Stahlberg H, Boekema EJ (2011) Interaction of complexes I, III, and IV within the bovine respirasome by single particle cryoelectron tomography. Proc Natl Acad Sci U S A 108 (37):15196–15200. https://doi.org/10.1073/pnas.1107819108

17. Diebolder CA, Faas FG, Koster AJ, Koning RI (2015) Conical Fourier shell correlation applied to electron tomograms. J Struct Biol 190(2):215–223. https://doi.org/10.1016/j.jsb.2015.03.010

18. Baldwin PR, Lyumkis D (2020) Non-uniformity of projection distributions attenuates resolution in cryo-EM. Prog Biophys Mol Biol 150:160–183. https://doi.org/10.1016/j.pbiomolbio.2019.09.002

19. Noble AJ, Dandey VP, Wei H, Brasch J, Chase J, Acharya P, Tan YZ, Zhang Z, Kim LY, Scapin G, Rapp M, Eng ET, Rice WJ, Cheng A, Negro CJ, Shapiro L, Kwong PD, Jeruzalmi D, des Georges A, Potter CS, Carragher B (2018) Routine single particle CryoEM sample and grid characterization by tomography. elife 7:e34257. https://doi.org/10.7554/eLife.34257

20. Taylor KA, Glaeser RM (2008) Retrospective on the early development of cryoelectron microscopy of macromolecules and a prospective on opportunities for the future. J Struct Biol 163(3):214–223. https://doi.org/10.1016/j.jsb.2008.06.004

21. Hohn M, Tang G, Goodyear G, Baldwin PR, Huang Z, Penczek PA, Yang C, Glaeser RM, Adams PD, Ludtke SJ (2007) SPARX, a new environment for Cryo-EM image processing. J Struct Biol 157(1):47–55. https://doi.org/10.1016/j.jsb.2006.07.003

22. Heymann JB, Belnap DM (2007) Bsoft: image processing and molecular modeling for electron microscopy. J Struct Biol 157(1):3–18. https://doi.org/10.1016/j.jsb.2006.06.006

23. Tang G, Peng L, Baldwin PR, Mann DS, Jiang W, Rees I, Ludtke SJ (2007) EMAN2: an extensible image processing suite for electron microscopy. J Struct Biol 157(1):38–46. https://doi.org/10.1016/j.jsb.2006.05.009

24. Tan YZ, Aiyer S, Mietzsch M, Hull JA, McKenna R, Grieger J, Samulski RJ, Baker TS, Agbandje-McKenna M, Lyumkis D (2018) Sub-2 A Ewald curvature corrected structure of an AAV2 capsid variant. Nat Commun 9(1):3628. https://doi.org/10.1038/s41467-018-06076-6

25. DiPrimio N, Asokan A, Govindasamy L, Agbandje-McKenna M, Samulski RJ (2008) Surface loop dynamics in adeno-associated virus capsid assembly. J Virol 82 (11):5178–5189. https://doi.org/10.1128/jvi.02721-07

26. Zhang C, Konermann S, Brideau NJ, Lotfy P, Wu X, Novick SJ, Strutzenberg T, Griffin PR, Hsu PD, Lyumkis D (2018) Structural basis for the RNA-guided ribonuclease activity of CRISPR-Cas13d. Cell 175(1):212–223.e217. https://doi.org/10.1016/j.cell.2018.09.001

27. Polley S, Lyumkis D, Horton NC (2019) Indirect readout of DNA controls filamentation and activation of a sequence-specific endonuclease. bioRxiv:585943. doi:https://doi.org/10.1101/585943

28. Scheres SH (2012) RELION: implementation of a Bayesian approach to cryo-EM structure

determination. J Struct Biol 180(3):519–530. https://doi.org/10.1016/j.jsb.2012.09.006

29. Punjani A, Rubinstein JL, Fleet DJ, Brubaker MA (2017) cryoSPARC: algorithms for rapid unsupervised cryo-EM structure determination. Nat Methods 14(3):290–296. https://doi.org/10.1038/nmeth.4169

30. Kucukelbir A, Sigworth FJ, Tagare HD (2014) Quantifying the local resolution of cryo-EM density maps. Nat Methods 11(1):63–65. https://doi.org/10.1038/nmeth.2727

31. Vilas JL, Gomez-Blanco J, Conesa P, Melero R, Miguel de la Rosa-Trevin J, Oton J, Cuenca J, Marabini R, Carazo JM, Vargas J, Sorzano COS (2018) MonoRes: automatic and accurate estimation of local resolution for electron microscopy maps. Structure 26(2):337–344. e334. https://doi.org/10.1016/j.str.2017.12.018

32. de la Rosa-Trevin JM, Oton J, Marabini R, Zaldivar A, Vargas J, Carazo JM, Sorzano CO (2013) Xmipp 3.0: an improved software suite for image processing in electron microscopy. J Struct Biol 184(2):321–328. https://doi.org/10.1016/j.jsb.2013.09.015

33. de la Rosa-Trevin JM, Quintana A, Del Cano L, Zaldivar A, Foche I, Gutierrez J, Gomez-Blanco J, Burguet-Castell J, Cuenca-Alba J, Abrishami V, Vargas J, Oton J, Sharov G, Vilas JL, Navas J, Conesa P, Kazemi M, Marabini R, Sorzano CO, Carazo JM (2016) Scipion: a software framework toward integration, reproducibility and validation in 3D electron microscopy. J Struct Biol 195(1):93–99. https://doi.org/10.1016/j.jsb.2016.04.010

34. Chen S, McMullan G, Faruqi AR, Murshudov GN, Short JM, Scheres SH, Henderson R (2013) High-resolution noise substitution to measure overfitting and validate resolution in 3D structure determination by single particle electron cryomicroscopy. Ultramicroscopy 135:24–35. https://doi.org/10.1016/j.ultramic.2013.06.004

35. Dang S, Feng S, Tien J, Peters CJ, Bulkley D, Lolicato M, Zhao J, Zuberbuhler K, Ye W, Qi L, Chen T, Craik CS, Jan YN, Minor DL Jr, Cheng Y, Jan LY (2017) Cryo-EM structures of the TMEM16A calcium-activated chloride channel. Nature 552 (7685):426–429. https://doi.org/10.1038/nature25024

36. Ludtke SJ, Baldwin PR, Chiu W (1999) EMAN: semiautomated software for high-resolution single-particle reconstructions. J Struct Biol 128(1):82–97. https://doi.org/10.1006/jsbi.1999.4174

37. Grant T, Rohou A, Grigorieff N (2018) cis-TEM, user-friendly software for single-particle image processing. elife 7:e35383. https://doi.org/10.7554/eLife.35383

38. Rohou A (2020) Fourier shell correlation criteria for local resolution estimation. bioRxiv, It20, 2020.03.01.972067. https://doi.org/10.1101/2020.03.01.972067

Chapter 9

Automated Modeling and Validation of Protein Complexes in Cryo-EM Maps

Tristan Cragnolini, Aaron Sweeney, and Maya Topf

Abstract

The resolving power of cryo-EM experiments has dramatically improved in recent years. However, many cryo-EM maps may still not achieve a resolution that is sufficiently high to allow model building directly from the map. Instead, it is common practice to fit an initial atomic model to the map and refine this model. Depending on the resolution and whether the structure suffers from inherent flexibility or experimental limitations, different methods can be applied, to obtain high-quality, well-fitted atomic model of the macromolecular assembly represented by the map, and to assess its properties. In this review, we describe some of these methods, with the main focus on those that have been developed in our group over the last decade.

Key words Cryo-EM, Assembly, Density fitting, Validation, Structure modeling

1 Introduction

The resolving power of cryo-EM experiments has dramatically improved in recent years, now rivalling crystallography, with near-atomic resolutions for many systems. Technical advances, in particular, direct electron detectors and better image processing methods have made it possible to obtain resolutions better than 3.5 Å in some cases, making cryo-EM a popular tool to probe the structure of macromolecular assemblies, even for low purity samples. Other related methods, such as cryo-ET and subtomogram averaging can elucidate the structure of complex samples, entire cell components, tissues, or full viral particles.

However, the relatively low resolution in still most cryo-EM maps limits their applicability to model building based on the information from the map only. Instead, it is common practice to fit an initial atomic model into the density map and refine it. These initial models can either be experimentally derived (e.g., from X-ray crystallography, NMR spectroscopy, or high-resolution cryo-EM), or comparative models (homology models build from sequence

Tamir Gonen and Brent L. Nannenga (eds.), *CryoEM: Methods and Protocols*, Methods in Molecular Biology, vol. 2215,
https://doi.org/10.1007/978-1-0716-0966-8_9, © Springer Science+Business Media, LLC, part of Springer Nature 2021

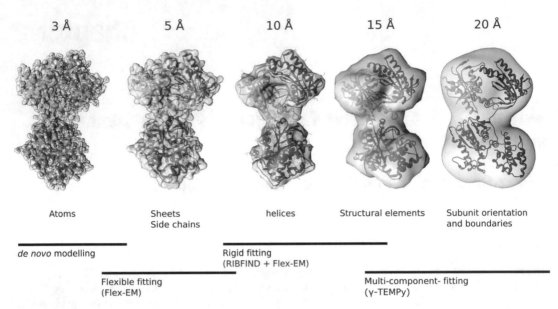

Fig. 1 Structure of a protein complex, with maps at different resolutions. The smallest elements usually visible at a given resolution are listed, as well as the computational methods that are used. The range over which they are applied is represented by black bars of various lengths

and structural homology templates) [1, 2], ab initio models (models built from sequence based on physical principles), or a combinatory approach.

While near-atomic maps allow the use of tools similar to those developed for X-ray crystallography (and for very high resolutions may not require an initial model at all), in others, depending on the resolution and whether the structure suffers from inherent flexibility or experimental limitations, different methods can be applied, to obtain high-quality, well-fitted atomic model of the macromolecule represented by the map. At 20 Å, there may be significant ambiguity regarding the position of specific protein components in the map, while a higher resolution map will provide this information much more clearly (Fig. 1). However, secondary structure elements (SSE) such as alpha helices will not be resolved. At 10 Å, long alpha helices may be visible, but with significant ambiguity regarding their turn or register, for example. Thus, the nature of the method used often correlates with the protein structural organization: as the map resolution improves, large structural elements are easier to identify, and the method will focus on describing smaller structural features. Below, we describe some of these methods, focusing on those that have been developed in our group over the last decade.

2   Fit Assessment

2.1   Comparing Structure and Maps

Given a structural model of a protein (*candidate model*), a common task is then to define how well its conformation "fits" in the map, that is, how well does the structure describe the observed density of the map.

2.2   Map Blurring

Direct comparison from a map, with density values in each voxel, and no exact assignment of this density to atoms, to a structure, with highly accurate position for each atom nucleus, is complicated. It is possible to blur a structure, that is, to obtain a map with voxel densities computed based on the proximity of atoms in the structure. However, many tools and scoring methods have been developed to describe how well two maps matches. These blurred maps will be used in the following to evaluate the quality of proposed structures, and to find regions of the maps that are missing or ill-represented by the modeled structures.

In the TEMPy software, first, a grid is constructed around the protein, with voxels of 1 $Å^3$. Then, the density value of voxels containing heavy atom nuclei (i.e., the position of the atom) is increased by the sum of the corresponding atomic number. The blurring itself is done by convoluting that map with a gaussian function. The gaussian sigma factor (controlling the width of the gaussian function, and the level of blurring) is usually expressed as a constant times the resolution, with different possible formulation for the constant, with 0.225 and 0.187 the most common (also *see* **Note 1**). These two choices make the width of the Fourier Transform of the gaussian distribution fall to $1/e$ and ½ at a wavenumber of 1/resolution. Finally, if the blurred map is to be compared against an existing map, it is resampled to ensure their grid matches.

2.3   Scoring Functions

Finding an atomic structure with the best fit to a density map is an issue central to cryo-EM. This can be assessed with a "scoring function," that provides a numerical value for any possible pair of structure and map. The "best" structure, then, is dependent on the definition of this scoring function. Global scoring functions are useful to provide a general picture of the fit between candidate structures and map but may miss smaller scale modeling errors. Local scoring functions are needed to detect and potentially correct those errors [7, 8].

In the following, we will describe scoring functions to estimate the match between two maps, X and Y, and assume there is a one-to-one correspondence between their voxels. Although this condition is not guaranteed, it is usually possible to estimate the density values for one map at voxel positions matching the other, thus aligning them.

The map can also be transformed or filtered before the scoring takes place, *see* **Note 2**.

2.3.1 Global Scoring

Cross-Correlation

Cross-correlation coefficient (CCC) is a popular metric to compare maps [9]. The most common metric is the Manders' coefficient:

$$\mathrm{CCC}(X, \Upsilon) = \sum_i \frac{X_i \Upsilon_i}{|X_i||\Upsilon_i|},$$

where $X_i = \sqrt{\sum_i X_i^2}$ is the standard norm or Euclidean distance, this normalizes the root sum of squares of the densities to be normalized to 1. A discussion regarding the alternative definitions of the cross-correlation is provided in **Note 3**.

Mutual Information

The *mutual information* (MI) is an information-theoretic measure, representing the amount of information shared by two probability distributions. It has been used in many fields to quantify the difference and similarity between various quantities. The mutual information between two maps is calculated as:

$$\mathrm{MI}(X, \Upsilon) = \sum_{x \in X} \sum_{y \in \Upsilon} p(x, y) \, \log \left(\frac{p(x, y)}{p(x) p(y)} \right)$$

with $p(x)$ representing the probability distribution of the density values in map X, $p(y)$ representing the same for map Υ, and $p(x,y)$ is the joint probability distribution, computed over the aligned voxels. In TEMPy, this is implemented by binning the values in a histogram for both maps although other density estimates are possible (e.g., a gaussian kernel density estimate). Although those results confirm that the cross-correlation function (CCC) remains an excellent scoring method, the mutual information (MI) shows similar or higher performance [10, 11].

It can be advantageous for this type of measure to be normalized, and several such variations have been used in the context of map comparison [12, 13]. For example,

$$\mathrm{NMI}(X, \Upsilon) = \frac{\mathrm{MI}(X, \Upsilon)}{\min (H(X), H(\Upsilon))} \text{ with } \mathrm{H}(X)$$

$$= -\sum_{x \in X} p(x) \, \log \, p(x).$$

the entropy of the probability distribution $p(x)$, as before. Other normalizations are possible, by dividing the MI by $H(X) + H(\Upsilon)$ [14].

Overlap Score

To compute the *overlap score* (OVR), we first define a *contour*, that is a surface within which all voxels are above a certain intensity. This is done for both maps that we want to compare. Then, the score is the number of overlapping voxels divided by the number of voxels in the smallest of the two maps.

Several of those scores as well as others such as the *normal vector score*, *Chamfer distance*, and *envelope score* have been implemented in TEMPy, and benchmarked in term of speed and performance [10].

2.3.2 Local Scoring

Although it is sometimes convenient to provide a single number summarizing the quality of a fit, for example during an optimization method such as a rigid-body fit, it is also useful to have a more local information regarding the quality of a fit: a loop or some other structural elements may not fit well to the map, if for example there is some conformational difference between the proposed structure and the one represented by the map. This can be assessed by using a local score that will provide several numerical values, usually along voxels or groups of voxels and this can be guided by the sequence or known structural elements (such as secondary structure or domains)

Local Cross-Correlation: SCCC

The *segment-based cross-correlation* score (SCCC) is computed on a subset of the map, using the cross-correlation (*see* **Note 2**). The part of the map that is chosen for each SCCC calculation is based on a segmentation of the protein into rigid bodies, or along its sequence. The initial identification of segment (rigid bodies) in the structure to be assessed can be done using RIBFIND (*see* Subheading 3.2). After normalization of the data, and alignment of the fragments, the score is calculated.

SMOC

The SMOC score is computed in the same way, using the Manders' coefficient presented above. The score can be calculated over overlapping windows of neighboring residues along the sequence (SMOCf), or by selecting voxels in the map close to the residues' atoms (SMOCd, *see* [15]), and computing the Manders' overlap coefficient over these voxels. By repeating this procedure over windows of nine residues at a time, moving the window by one residue each time, it is possible to obtain an SMOCf profile over the entire protein sequence (and structure), giving an indication as to which regions are well fitted in the map, and which are not [16].

Several of those functions have been developed and tested in our group [10, 11]. The *overlap score* displayed excellent properties in terms of discriminating good and bad fits. A combination of OVR and SMOC had better statistical properties for low-resolution maps, with all scores displaying similar performances at high resolution.

2.3.3 Consensus Scoring

The different scoring methods are not always consistent with one another, with different models getting ranked lower or higher. This is problematic when limitations in the processing require the choice of a limited subsets, or even a single model to use in further refinement. By computing the score and ranks with multiple methods, and combining them using the *Borda score*, it has been shown that most of the ambiguity in the rankings can be resolved [17].

All the scores described above are available within TEMPy [17]. TEMPy is a python library developed for the assessment and manipulation of macromolecular structures specifically designed to integrate with cryo-EM density maps. TEMPy implements functions for the reading, writing, assessment, and fitting of cryo-EM maps and structures within them.

3 Structure-Density Fitting

3.1 Protein Structure Prediction

The specific approach used for generating an atomic model from a cryo-EM map is dependent upon the information available. When the data contained within the cryo-EM map is at a sufficiently high resolution that the positions of amino acid side-chain atoms can be determined, it is possible to build a model de novo into the map [18, 19]. Although this is becoming more commonplace, in most cases so far initial candidate models are derived from an a priori structural model. This can be a known structure of the candidate protein complex or any of its components from other experiments (e.g., X-ray crystallography). One issue that can arise in this scenario is incomplete structures (e.g., flexible regions not observed during experiments). Another issue is differences in conformation as a result of structure determination at different experimental conditions. However, it is often the case that a suitable experimental structure is not available at all. In this instance, protein structure prediction approaches (such as comparative [20, 21] or ab initio modeling can be used, either to generate models for the missing components/regions or obtain a complete model of a protein complex.

Such methodology has been used to generate initial structures for a kinesin-6 family motor MKLP2 [22]. Briefly, for each nucleotide state of MKLP2 100 homology models were generated using MODELLERv9.51. Multiple templates were selected using sequence identity and comparison of the MKLP2 predicted secondary structure with the DSSP-assigned [23] secondary structure elements of candidate templates. The best models were chosen using the QMEAN (Qualitative Model Energy ANalysis) score, a measure of structure quality [24].

3.2 Rigid-Body Fitting

Once a candidate model is identified, it has to be fitted (docked) into the EM map. A common approach to this task involves the systematic maximization of the CCC between the candidate model and the map. This however is a nontrivial task, especially if the density corresponding to individual proteins cannot be easily distinguished. Therefore, rigid fitting can be thought of in two parts, namely global and local rigid fitting. If a reasonable approximation regarding the initial placement of individual proteins into the map is possible, a local search to refine the models about their approximate positions may yield a sensible candidate model of the complex. However, if individual components cannot be placed into the map with any confidence, a more thorough search is needed to rigidly place individual subunits.

One program that can be employed for both local and global rigid fitting is Mod-EM [25] implemented in MODELLER [2]. Briefly, given a density map and candidate model, the model is converted into a candidate density by the blurring procedure described in Subheading 2.2 above. The best fit is obtained by altering the position of the map to maximize the CCC between candidate and cryo-EM density.

When the cryo-EM map corresponds one-to-one with the density from the candidate model, a local rigid fitting protocol is performed (similar to the fit-in-map approach in UCSF-Chimera [3]). This can be the case when the map represents the candidate model only, or when a protein complex map has been segmented around an individual protein (e.g., using tools such as Segger [26] implemented in Chimera [3]). In this case, the candidate model center-of-mass is translated to the map center-of-mass, and a local search is performed either exhaustively, or with a Monte Carlo method. A single step consists of a translation of the probe for a single grid unit in the map and a search of three Euler angles.

When segmentation of the density corresponding to the candidate model from the map is not possible, a global rigid fit protocol is employed. There are two available protocols for this in Mod-EM and their use is dependent on the relative size of the candidate model to the map. Typically, if the candidate model represents a significant portion of the map (>35% volume) a Monte Carlo method is recommended. This begins with an initial arbitrary placement of the model into the map before executing a number of Monte Carlo steps. When the candidate model represents a small volume of the map (≤35% volume), a scanning Monte Carlo protocol is used, where the map is divided into "cells" approximately the volume of the candidate model and a local search is performed in each cell for which the probe initial CCC is positive.

Global rigid fitting methods such as that implemented in Mod-EM generally work best at resolutions where the boundaries of proteins and domains can be identified in the map. However, at lower resolutions (>15 Å) this is not always possible and the use of CCC can give misleading results regarding the fit of individual components in the map as their densities overlap upon complex formation, and without taking this into account global optimization has been shown to worsen the fit [27]. In fact, with advances in cryo-EM and cryo-ET, it is now realistic to obtain maps for very large complexes, like the nuclear pore complex [28]. One way around this problem is to simultaneously optimize the fit of all the components in the map (namely, *assembly/simultaneous/multi-component fitting*).

3.3   Assembly Fitting

Even with the help of high-resolution candidate structures of the individual components, predicting the overall architecture of a complex remains highly challenging. The multi-component search for the best-fitting assembly model is complicated by the existence of many local maxima, corresponding to models that produce densities similar to the experimental map, but with incorrectly swapped components, or wrongly oriented ones. This is particularly salient at lower resolutions. Additional experimental sources may help resolve this ambiguity, but may not fully lift it. If the initial fitting of a candidate models in the map can be obtained, for example by homology [29], the search space could be reduced significantly, avoiding unnecessary calculations. Such an approach was used more than a decade ago to generate models for the first mammalian ribosome structure at 8.7 Å resolution (PDB ID: 2ZKR, 2ZKQ) [30] and the 80S ribosome from *T. lanuginosus* at 8.9 Å resolution (PDB ID: 3JYX, 3JYW, 3JYV) [31], where the fit of each modeled protein was optimized locally, using Mod-EM local exhaustive optimization. However, if there is no initial knowledge about the approximate position of a candidate individual or multiple models in the map, efficient methods to conduct that search, such as gmfit [32], MultiFit [33], SITUS [4], and γ-TEMPy [34], are necessary.

3.3.1   Initial Placement

To obtain initial positions for the components in the map, an analysis of the map is conducted, using vector quantization (VQ) (Fig. 2) [35]. This method reduces the complex density data to a limited set of points (also known as codebook vectors), and iteratively updates the set of points to reduce the distortion, a measure of the distance between each point and the part of the data it represents, until convergence. This method is similar to the k-means procedure. Other methods can produce the components' placement, for example by using a gaussian mixture model to obtain a reduced representation formed by a set of gaussian functions [32]. These methods provide powerful ways to obtain initial placement for components in the map, but there is no guarantee that the codebook vectors or gaussian centers will correspond to the centroid of components. Therefore, the placement of components with respect to those vectors need to be refined, and the initial centroid placements strongly affect the refinement [34].

3.3.2   IQP and wICP

After generating code vectors with VQ, it is possible to follow the same procedure on individual map components and obtain a putative match with the IQP algorithm. This match can then be iteratively refined using the wICP procedure, to reassign the data points to a center, recomputing the centers, and iterating until convergence. This is similar in spirit to the expectation-maximization procedure. The combination of IQP and wICP showed results qualitatively similar to gmfit [35].

Fig. 2 Example of a set of three vectors generated by VQ (with TEMPy) from a map of the 1C34 PDB structure

3.3.3 γ-TEMPy (Genetic Algorithm for Modeling Macromolecular Assemblies with Template and EM Comparison Using Python)

In γ-Tempy, a genetic algorithm is used to iteratively refine candidate assemblies and improve their match to the map, judged by a global score (*see* Subheading 2.3 above) [34]. After an initial VQ run, positions and rotations are generated for each component, each around a separate codebook vector. A number of those sets are generated, producing the initial "population" of assembly fits. A new "generation" of assemblies are created by mutating (i.e., randomly changing) the position and rotation values for the components and by combining values from two "parents." The set of "parents" and "children" are scored and ranked, and the best candidates are kept for the next "generation." The fitness (scoring) function that is used combines a map score (MI) and a penalty clash score to keep the components apart from each other. The result from multiple runs can be collected, refined with Flex-EM (or some other flexible refinement method, *see* Subheading 3.3 on flexible fitting), and the best results brought forward for analysis. This method was shown capable of producing reasonable fit for assemblies up to eight components (with no symmetry), in particular at 10 Å resolution (Fig. 3) but also at as low as 20 Å [34].

When a large number of models are generated, it can prove useful to cluster them: many models may represent closely related alternative fits, and clustering may reveal common features of the fit even in cases where different assemblies are scored similarly [17, 36].

3.4 Flexible Fitting (Refinement)

Once an atomic candidate model (or multiple models) have been fitted rigidly in a region of the cryo-EM map, it is often noticeable that its conformation is different from the conformation

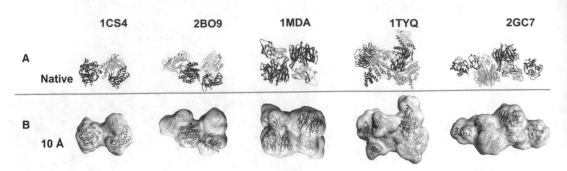

Fig. 3 The best-predicted assemblies found in the 20 GA runs of assembly fitting using γ-TEMPy with simulated maps at 10 Å resolution shown (**B**) below their corresponding native assemblies (**A**). The method achieves high-quality reconstructions. Individual components of the assemblies are shown in cartoon representation with unique colors. The same coloring schemes are used for the individual components of the native and the predicted assemblies. The PDB id of each structure is indicated above the native structure. Adapted from [34]

represented by the map. Differences can arise from variable experimental conditions used from structural determination, distortions from crystal packing in X-ray crystallography and experimental noise. Additionally, in single particle cryo-EM experiments, sample heterogeneity (i.e., the presence of multiple sample conformational states) can yield multiple maps, for which a single candidate model cannot usually account for. When the candidate model is generated by protein structure prediction approaches inaccuracies in the model, for example, the misassignment of SSEs can also lead to differences between map and model. Finally, proteins usually change their conformation due to biomolecular interactions, for example, to stabilize a protein complex or upon ligand binding. These conformational changes are commonly facilitated by the movement of domains and SSEs within proteins via flexible regions. Thus, if the candidate models are the individual components of a complex then they are likely to differ in conformation from corresponding component represented by the map of the intact complex. Interfaces can be optimized with programs such as HAD-DOCK or Rosetta. This approach was used to improve the MKLP2 and tubulin interface (PDB ID: 5ND7, 5ND4, 5ND3, EMDB: 3623, 3622, 3621) [22].

To further improve the agreement between the model and the map *flexible fitting* (or refinement) methods can be applied. One approach could be to generate multiple conformations (e.g., by comparative modeling, ab initio modeling, or molecular dynamics) and select the conformation that fits best in the density [25, 37, 38] based on one or more scores (described above). This approach, which is sometimes necessary at intermediate-to-low resolutions, was used in the case of the mammalian (PDB ID: 4V5Z, EMDB: 1480) and *T. lanuginosus* ribosomes (PDB ID: 4V7H, EMDB: 1345) (mentioned above [30, 31]) where for each modeled

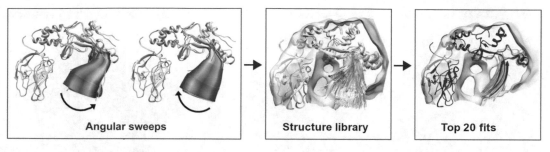

Fig. 4 Angular sweeps used to generate an ensemble of structures and assess them in the density using TEMPy. The top 20 fits were submitted to the PDB (PDB ID: 4V3M; EMDB: 2795) (Adapted from [39])

protein, hundreds of candidate models were assessed (following local optimization of the fit) and the top-scoring model was selected. It was also used to describe conformational changes during pore formation by the perforin-related protein pleurotolysin using cryo-EM maps at ~11 Å resolution [39]. To describe a prepore state, the top 20 conformations out of thousands generated by angular sweeps using TEMPy were selected based on SCCC (Fig. 4). Since the resolution was not sufficiently high, multiple models were reported (PDB ID: 4V3A, EMDB: 2794).

Yet, especially at higher resolutions, it is more efficient to guide conformational changes by iteratively optimizing the fit into the density [40], and this can be done while maintaining geometric and mechanistic properties, as in Flex-EM [5].

3.4.1 Flex-EM

The MODELLER/Flex-EM method integrates rigid and flexible fitting of a component structure into the cryo-EM map of their assembly [5]. Atomic positions are optimized by utilizing a scoring function that incorporates terms for cross-correlation with density, stereochemical and non-bonded interaction terms. The latter two terms are included to yield an optimized structure which makes sense from a physico-chemical point of view. The heuristic optimization uses conjugate-gradients minimization and simulated annealing rigid-body molecular dynamics. The method relies on identifying elements of the structure that are not expected to change their internal conformation. Rigid-body elements could range from SSEs to domains or whole proteins. The success of such an approach is sensitive to the definition of rigid bodies. If the number of rigid bodies is small, local structural rearrangements may be missed during fitting. Alternatively, refining all atoms in the map greatly increases the level of computation needed and the danger of overfitting. This will lead to a reduction in the efficiency of refinement and the model may become stuck in a local energy minimum. Thus, finding appropriate rigid bodies within a structure render this computation feasible even for large structures.

Human TRPA1 ion channel   S. cerevisiae Polymerase epsilon   ATP-bound N-ethylmaleimide sensitive factor

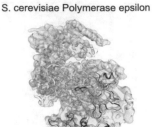

Fig. 5 Example of RIBFIND rigid-body clusters (each cluster is colored uniquely). In the bottom row, the complexes are shown within corresponding cryo-EM maps (from left to right: PDB 3J9P, EMD-6267; PDB 6HV8, EMD-0287; PDB 3J94, EMD-6204)

3.4.2   RIBFIND

The RIBFIND method employs a spatial clustering algorithm to identify clusters of rigid bodies with a protein structure/complex (Fig. 5). The use of RIBFIND has been shown to improve the fitting of structures in simulated and experimental maps [41, 42]. Briefly, the algorithm takes an input of atomic coordinates and assigns SSEs using DSSP [23]. An SSE is randomly assigned as a cluster seed, and SSEs above the threshold defined for spatial proximity are added to the cluster, based on the cluster cutoff parameter. This process is repeated until no more SSEs can be assigned to that cluster. If the number of SSEs in the cluster is greater than one, the SSEs can no longer be assigned to another cluster. If the number of SSEs is one, the SSE is defined as unassigned. This process is repeated until all SSEs are defined within a cluster or unassigned. SSEs that are defined within a cluster along with flexible regions connecting them are grouped and treated as a single rigid body. The output contains all of the rigid-body clusters along with unassigned SSEs and includes results for all cluster cutoff values (0–100%) although the recommended value is the one that corresponds to the maximal unique number of clusters. The output can be used as an input for Flex-EM refinement (but could also be useful for other flexible fitting methods).

3.4.3   Resolution-
Dependent Refinement

As described above, here too the appropriate fitting method to use is dependent on the resolution of the density map. For high-resolution maps (better than 4 Å), which contain data such as the position of side chains, alpha helices, and individual beta strands, approaches utilizing restraints that hold correct model geometry, SSEs, and hydrogen bonding are used to refine models into the map. PHENIX [18] is one such program that has been adapted for the refinement of high-resolution cryo-EM structures in both real and reciprocal space. Recent examples of PHENIX refinement include the structure of γ-secretase in complex with the amyloid precursor protein (PDB ID: 6IYC; EMDB: 9751) [43] and notch fragments (PDB ID: 6IDF; EMDB: 9648) [44].

Since the reliable information contained within a map at intermediate resolutions (~4–15 Å) can be variable (from domains at ~15 Å to long alpha helices at resolutions closer to 10 Å, to short helices and even separated beta strands at resolutions close to 4 Å), at these resolutions, maintaining some rigidity of the amino acids and larger elements during the refinement is necessary. Using Flex-EM in conjunction with RIBFIND, it was shown that a hierarchical refinement approach (*see* also Subheading 3.5), whereby first rigid bodies of subdomains are restraints and refined followed by a further refinement of restrained SSEs can help to reduce the search space and reduce overfitting, thereby achieve a more accurate model [42] (Fig. 6).

Such an approach was used to flexibly fit the atomic structures of *apo* GroEL to multiple intermediate resolution maps (7–9 Å) (EMDB: 1997, 1998, 1999, 2000, 2001, 2002, 2001) representing different GroEL-ATP conformations (PDB ID: 4AAQ, 4AAR, 4AAS, 4AAU, 4AB2, 4AB3) [45].

Along with Flex-EM, various flexible fitting methods have been introduced. For example, MDFF [46, 47] works by complementing a standard force field with a potential based on the density profile of the map, such that atoms in the model are subject to forces in a way proportional to the gradient of the density map. Rosetta uses the agreement of a simulated candidate model density with the map along with a weighted energy score [48] to refine candidate models into the map [49]. NMFF [50] and iMODFIT [51] use normal mode analysis for the fitting.

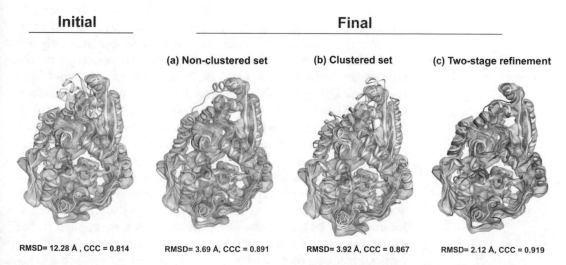

Initial   **Final**

(a) Non-clustered set   (b) Clustered set   (c) Two-stage refinement

RMSD= 12.28 Å , CCC = 0.814   RMSD= 3.69 Å, CCC = 0.891   RMSD= 3.92 Å, CCC = 0.867   RMSD= 2.12 Å, CCC = 0.919

Fig. 6 Improvements in reconstruction with no clustering (**a**), RIBFIND clustering (**b**) and a two-stage hierarchical refinement (**c**). RMSD is the Cα RMSD of each of the models from the X-ray structure (white) (PDB ID: 1DPE); CCC is calculated between each model and the 5 Å resolution corresponding simulated map (grey). Adapted from [42]

In the case of the lower end of intermediate range and even lower (worse than 15 Å), the use of flexible fitting techniques is limited but may include, for example, refinement of the orientation between domains.

3.5 Consensus Refinement Approach

The importance of consensus flexible fitting methods was highlighted in one study that showed that even with diverse methodologies, flexible fitting methods generally converge to the same solution [52]. This solution was seen to better represent the conformations of the density map when compared to rigid-body fitting and formed the basis of the idea of using consensus flexible fitting methodologies to improve the quality of fitted models. The idea is to compare the results of the different fits produced by different methods using a local goodness-of-fit score (such as the SCCC described in Subheading 2.3) [53]. Additionally, the results of all the fitting programs are compared between themselves, for example, using Cβ RMSD. By this method, consensus local regions (regions with a low RMSD between programs) can be identified. Furthermore, non-consensus regions (regions with a high RMSD between programs) can identify areas where the fit is incorrect and requires further refinement, and these usually correlate with worse fit to the density.

This approach has been shown to be able to detect errors propagated from incorrect comparative models. An actin subunit comparative model was flexibly fit into a 9 Å resolution simulated density map from a high-resolution X-ray structure (PDB ID: 2A40, chain A) in a distinct conformation, using both Flex-EM and iMODFIT [53]. The SCCC scores were calculated from the SSEs and the average SCCC values were very similar for Flex-EM and iMODFIT, respectively. The RMSD for the fitted model compared to the ground truth model was seen to be approximately 4 Å for both programs. This apparent inability to converge to the correct conformation was hypothesized to be due to errors from the model. Using the QMEAN model assessment score [24], unreliable residues in the initial comparative model were identified. Identified unreliable residues were seen to be within loop regions connecting SSEs with low consensus fits. A hybrid approach was then used by running Flex-EM with the iMODFIT output, and relaxing constraints on non-consensus SSEs. In this way, the final model was more representative of the ground truth model (with averaged Cβ RMSD over all SSEs of 3.6 Å), with the SCCC improving in 84% of cases. The protocol was then applied to the case of the mature and empty capsids of Coxsackievirus A7 (CAV7) by flexibly fitting comparative models into the corresponding cryo-EM density maps at 8.2 and 6.1 Å resolution.

A similar methodology was used more recently to fit comparative models of the mouse MKLP2 (kinesin 6) in complex with ADP. ALFx, ADP, AMPPNP and in an *apo* state were determined at 5.5,

7, 8, and 7 Å resolution, respectively (PDB ID: 5ND4, 5ND2, 5ND7, 5ND3; EMDB: 3622, 3620, 3623, 3621) [22]. Rigid-body restraints were initially defined as SSEs, and flexible fitting was run with Flex-EM and iMODFIT. To produce a hybrid fit, the best-fitting SSEs from both programs were combined, and an all-atom refinement with Flex-EM was run. For each model, an increase in global CCC was reported and a QMEAN analysis showed no degradation in model quality.

3.6 Hierarchical Refinement Approach

Recently, a hierarchical refinement protocol, in the spirit of the approach mentioned in Subheading 3.3, has been introduced [16] (Fig. 7). The method involves using progressively smaller rigid bodies at successive iterations and provides a way to significantly reduce the search space needed for optimization of the fit. Initial rigid bodies were defined at the domain level using RIB-FIND. A second round of flexible fitting was conducted with rigid bodies defined as SSEs, and a final stage of fitting was conducted using an all atom refinement. The local fit assessment was determined using the TEMPy SMOC score (*see* Subheading 2.3). The applicability of the method is dependent upon the resolution of the map since at lower resolutions data regarding subtle local conformational changes may not be present in the map. However, the protocol can be adjusted according to the resolution of the map by stopping the optimization at the appropriate level of rigid-body representation.

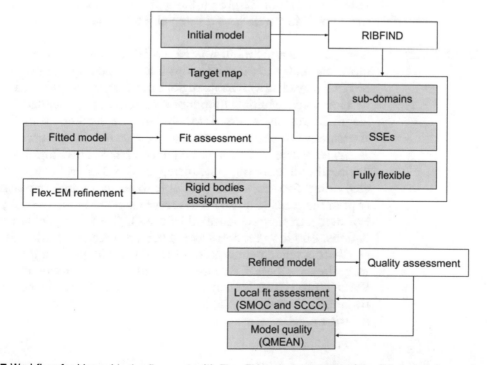

Fig. 7 Workflow for hierarchical refinement with Flex-EM and assessment of the fitted model

To demonstrate the utility of the protocol for refinement in high-resolution maps, two maps from a 2.5 Å resolution X-ray structure of *E.coli* Adenylate Kinase (PDB ID: 1AKE) were simulated at resolutions of 2.5 and 3.5 Å (Fig. 8a). A comparative model based on 46% sequence identity to a mutant adenylate kinase from *S. cerevisiae* at a different conformation (distinct from the experimentally determined structure) was used as the input. Following refinement into the maps, the Cα- and all-atom RMSD between the model and X-ray structures were 0.41 Å and 0.96 Å, respectively, for the 2.5 Å resolution map, and 0.64 Å and 1.19 Å, respectively, for the 3.5 Å resolution map [35].

Additionally, this protocol was successfully used to model the structure of *D. discoidium* initiation factor 6 (eIF6) into a 3.3 Å experimental cryo-EM map of its complex with 60S ribosomal subunit (EMD-3145). An initial comparative model was used based on the structure of a *M. jannaschii* eIF6. At each stage of refinement, the SMOCf score, which was used to assess the local fit of the model to the density, was improved (Fig. 8). The final model was assessed using QMEAN score [24] and seen to satisfy typical spatial constraints for proteins of similar size.

The hierarchical methodology has also been shown to be applicable to the fitting of atomic models into intermediate resolution EM maps, for example, in the refinement of nicotinic acetylcholine receptor structures [54]. Recently, it was applied to model a Kinesin-8/inhibitor complex bound to microtubules, using an initial model derived from X-ray crystallography in an MT-free state (PDB ID: 3LRE) [55]. SSEs of the model were refined into the cryo-EM map, followed by an all-atom refinement step.

3.6.1 *Loop Optimization*   A major challenge for automated model refinement is the accurate modeling and placement of loops into the density. Scores such as SMOCf may identify flexible regions in the structure as represented by a low value. Multiple loop models can then be generated for the identified region, assessed locally, and be further refined. The assessment can also be done using a Z-score. In the example of *E. coli* adenylate kinase described above (*see* Subheading 3.5), following the all-atom refinement step, an SMOCf profile was generated using Z-scores, which were able to better capture two regions of poor fit quality (Fig. 8b). These regions corresponded to loops of low model quality as identified by the QMEAN analysis. For both regions, 200 loop models were generated with MODELLER and an all-atom refinement was conducted with Flex-EM on the model with the best SMOCf score. A significant improvement in the SMOCf profile was seen at these regions, indicating the comparative model errors were the limiting factor in generating a good fit for the model.

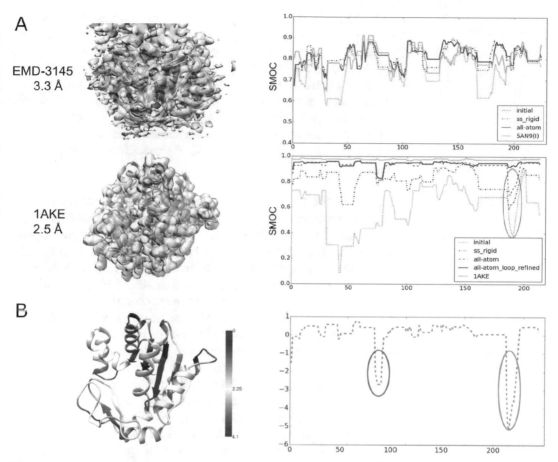

Fig. 8 (**a**) Flex-EM refinement at high-resolution. (**a**) The top row corresponds to refinement of a homology model of *E. coli* adenylate kinase in a 2.5 Å resolution density map representing an inhibitor-bound form (PDB: 1AKE). The bottom row shows the refinement of a homology model of eIF6 into a 3.3 Å resolution map (EMD-3145). Left: The starting model is shown in blue and the X-ray or deposited model in light brown. Associated density map is shown in transparent grey; Right: SMOCf profile showing scores in each stage of Flex-EM run. (**b**) Left: Starting homology model for fitting (using a homolog from S. cerevisiae) colored by local residue errors based on QMEAN scores. Right: Residue segments with significantly low *Z*-scores are circled in pink and green and are also indicated in the corresponding SMOCf refinement plot above (**a**, top right). The profile of SMOC *Z*-scores was calculated using $Z = (Sr-\mu)/\sigma$, where Sr is the SMOCf score for an individual residue, μ is the mean, and σ the standard deviation of the SMOCf scores of all residues. Figure adapted from [16]

In the case of Kinesin-8/BTB-1 complex [55] to improve poorly fit loop regions (identified with SMOCf) the loop optimization was used to generate 250 models per loop and the top-scoring conformations were selected based on the SCCC of the loop. Regions of scores at the lower 50% percentile and residues within 5 Å of them were subjected to further all-atom refinement and this process was repeated iteratively until more than 80% of the regions fitted with SCCC score values higher than 0.85 (the scores of the best-fitted regions were in the range of 0.92–0.93).

An alternative approach is to build loop models ab initio. The program Rosetta offers multiple algorithms for this purpose. Including a Monte Carlo Cyclic coordinate descent method [56, 57], and Kinematic closure methods [58] that have been implemented for both ab initio loop modeling and loop refinement. In spite of these advances, if the resolution is not sufficiently high, it may be "safer" to generate multiple loop conformations and cluster them to provide a better representation of the fit to the density, as done in the case of MKLP2 described above [22] (*see* Subheading 3.4).

3.6.2 *Ligand Fitting*

At near-atomic resolutions, where the density allows for a more accurate placement of atoms, it is sometimes possible to refine ligand directly in the density. This was the case with the ligand cadazolid an inhibitor of the bacterial 50S ribosomal subunit. The authors used real-space refinement in PHENIX [18] to model the ligand directly into the density in a 3 Å map (PDB ID: 6QUL, EMDB: 4638) [59]. However, for lower resolution maps the correct placement of ligand atoms into the density can be ambiguous. One common approach that can aid this process is molecular (or ligand) docking. This involves the prediction of the predominant binding mode(s) of a ligand with a protein, given its three-dimensional structure. Multiple conformations of the complex are generated, scored, and ranked. Scoring functions for docking programs aim to estimate the free energy of the complex, with more stable conformations ranked higher than others, which has the advantage of considering the effects of physical interactions between the ligand and protein. However, the results can be heavily biased by the accuracy of the positions of the side chains that line the binding sites, which for intermediate resolution cryo-EM maps is not adequate. This problem can be somewhat attenuated by programs such as GOLD [60] or HADDOCK [61] that allow for side-chain flexibility during docking calculations. Additionally, the ability of docking software to generate a correct ligand conformation is not ideal [62]. Multiple studies have shown that the confidence in identifying correct conformations can be increased by consensus docking [63]. Where the ligand is docked into the receptor using multiple docking software and conformations predicted by more than one program are taken to be correct.

A difference map is the subtraction of one density map from another and represents a way of identifying changes in conformations and compositions of the complex. Depending on the resolution, it is sometimes possible to identify density corresponding to a bound ligand if one map contains the ligand while the other represents an unbound state. This operation is implemented in many software packages that allow for the analysis of 3D cryo-EM maps, such as Chimera [25] or TEMPy [17]. However, for a difference to contain meaningful information certain criteria must be met, including accurate map-map alignment, structural blurring of one map to the lower of the two resolutions, and scaling of the amplitudes between maps.

One case where volume data and difference mapping was exploited to identify the density corresponding to a ligand was for the Kinesin-8 (Kif18A) specific inhibitor BTB-1 [55, 64]. The authors report three density maps for tubulin-bound kinesin, representing three states of kinesin in either the ANP-PNP (Phosphoaminophosphonic acid-adenylate ester; A non-hydrolyzable ATP analog) bound state, the no-nucleotide state or the BTB-1-bound state (EMD:3780, 3778, 3803; PDB: 5ocu, 5oam, 5ogc, respectively). A difference map was created between the BTB-1-bound kinesin map and the no-nucleotide state, utilizing the difference mapping methodology in TEMPy [17]. The difference density corresponded to areas of conformational change in the vicinity of the nucleotide-binding pocket but it also included a prominent peak between helix-α2 and helix-α3 (Fig. 9), which was unoccupied by the fitted atomic model. Remarkably, this region corresponds to one of the well-characterized allosteric inhibitor-binding sites in Kinesin-5 (Kif11). To fit the atomic model of BTB-1 in the difference map at this region, a two-stage docking protocol was used. First, a global search for ligand-binding site was conducted using HADDOCK, where the top scoring conformations were contained within the α2-,α3-binding pocket. A second stage focused on this binding pocket, and BTB-1 was docked by consensus docking using HADDOCK and AutoDock Vina. From the top scoring conformations, two were chosen to be equally likely based on the CCC of the conformations with both the difference and original

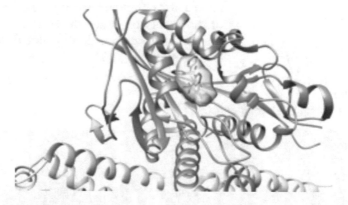

Fig. 9 Density corresponding to BTB-1 (purple) derived from difference mapping between the no-nucleotide state and BTB-1-bound state (EMD:3778 and EMD:3803, respectively) (density around the ligand-binding site has been masked for clarity). The solutions with the best CCC to the difference map, derived from small molecule docking with HADDOCK (orange) and AutoDock Vina (light blue), can be seen occupying the density. Also shown is the kinesin-8 refined model (blue) bound to microtubules (grey)

map. The position of the ligand corresponded well with the difference map density as well as biochemical mutation data of residues surrounding the binding site [55].

3.7 Model Validation

After a model has been obtained, for example using the tools described above, it is necessary to assess its quality, by identifying if its overall fit to the data is reasonable, and if the geometry of the model is chemically possible. Global assessment is useful to provide an overall picture of the model quality and allows for automated sorting, ranking, and filtering of models. Local assessment allows to pinpoint potential modeling errors, which may lead to further refinement. Both type of tools will be presented in this section.

While some tools and methods can be used either at the validation or modeling stage, it is good practice not to use the same for both: scoring and assessment methods may have blind spots in their ability to assess model quality, and a model refined to have a high quality in one may yet be low scoring in another. This would not be easily uncovered unless different methods are used.

3.7.1 Assessing Global Quality

Map Assessment

The cross-correlation remains the most common assessment tool for map-to-map quality and can easily be extended to model-to-map assessment by blurring the structure (*see* Subheading 2.2). This is widely implemented in EM packages and is the de facto standard (often in conjunction with masking or filtering, *see* **Note 2**). The FSC (Fourier Shell Correlation) is the other common measure for map quality. It quantifies the degree to which the real part of the Fourier components of two maps are correlated and is used to estimate the resolution of a density map, by finding the radial frequency at which the FSC falls below a given cutoff (*see* Table 1 for potential cutoffs). The FSC provides a single number for the resolution, and although the curve contains more information, that information is still global, although local variants have been proposed [7, 65].

Table 1
List of common blurring constants

Formula	Value	Explanation	Refs.
$1/(\pi * 2^{1/2})$	0.225	FT falls to $1/e$ its maximum at wavenumber 1/resolution	[3]
$1/(\pi * (2/\log 2)^{1/2})$	0.187	FT falls to 1/2 at wavenumber 1/resolution	[4]
$1/(2 * 2^{1/2})$	0.356	Gaussian falls at $1/e$ its maximum at resolution	[3]
$1/(2 * (2 \log 2)^{1/2})$	0.425	Gaussian falls at 1/2 its maximum at resolution	[5]
	0.5	The distance between the two inflection points being the same length as the resolution	[3]
	1	The gaussian sigma factor equals the resolution	[6]

Structure Assessment

QMEAN is a common method to estimate the quality of a structure. Based on statistical potentials derived from high-quality experimental data, it provides an estimate of the quality of the structure based on the similarity of its geometry to other structures. [24] Other similar potentials, such as DOPE [66] and ProQ3D [67] are in common use [68].

Model Validation by Additional Experiments

To resolve the structures of macromolecular assemblies, multiple experimental methods may be used, and often provide complementary information [69]. By integrating these different sources, it is possible to obtain higher accuracy models. However, some of these experiments can also be used for the purpose of validation, rather than in the actual model generation itself. In TEMPy, one can utilize information from cross-linking mass spectrometry for the validation of the models generated by density fitting.

Jwalk and MNXL: Model validation Using Cross-Linking Mass Spectrometry

Cross-linking mass spectrometry recover information regarding the distance between residues in a complex. This can be used to filter out incorrect conformers, or to validate a proposed model. A compound with two reactivate sites bind with the protein, forming covalent bonds at two distinct sites. MS analysis then recovers the position in the sequence of the amino acids involved.

Using a proposed atomic structure, the Jwalk method computes potential paths along the surface of the molecule between those amino acids using a Breadth-First Search (BFS) algorithm. By comparing the path length to the cross-linker's end-to-end distance, one can determine if the model fits with the available cross-linking information [70].

The Matched and Nonaccessible Crosslink (MNXL) score [71] also includes information regarding the surface accessibility of the lysines putatively involved in a cross-link: matching pairs where either end is buried below the protein's surface are penalized since they are not solvent-accessible and should not be able to form cross-links. For each cross-link, the Solvent-Accessible Surface Distance (SASD), computed as the length of the Jwalk path, is used to compute a score: a penalty of -0.1 is applied if the path length is above a certain cutoff (33 Å), or if the end residues are buried. The cross-link is then scored using a normal distribution calculated from all SASDs <33 Å from a cross-linking database.

This score was then extended to maximize its performance when modeling protein complexes, by treating intra- and inter-unit cross-links differently; the new score, cMNXL, is described in [72] (Fig. 10). Both Jwalk and MNXL are available as web servers [71].

F-Score (15 Å)
meanRMSD = 0.90 Å fnat = 0.95
precision = 0.40

cMNXL
meanRMSD = 7.43Å fnat = 0.11
precision = 0.48

Combined (15 Å)
meanRMSD = 0.34Å fnat = 1
precision = 0.87

Fig. 10 Using a combined score integrating cross-links and cryo-EM-simulated data (from PDB: 1XD3) to select the best model results in a better model selection, in terms of mean RMSD and fraction of native contacts (fnat). Adapted from [72]

3.7.2 Assessing the Local Quality

Map Assessment

By using SMOC or SCCC, it is possible to detect specific regions with low fit to the data (Fig. 11). This can be useful either in terms of refinement, for example, if a putative model was fit rigidly, and loop optimization (*see* Subheading 3.5) is necessary; or to identify a low-resolution or high-flexibility region of the map or a distortion of SSEs [16]. As mentioned above (*see* Subheading 2.3.2), SMOC and SCCC are useful both at high and intermediate range of resolutions. For models where the residues are placed accurately at near-atomic resolution EMRinger [73] can be useful to assess the quality of the backbone; by testing different position of the C_γ atom, by altering the χ_1 dihedral angle, EMRinger produces a profile of the correlation coefficient as a function of the dihedral angle. The score obtained for the current conformation of each side chain can be aggregated in a single global score [74].

Local Structure

Knowledge-based scoring methods often used for global scoring (such as QMEAN, DOPE) also provide a residue-level profile that can be used for local assessment of the structure (*see* Fig. 8). The MolProbity package [75] is commonly used and reported for structural models generated from cryo-EM, X-ray, and other experimental structural data sources. By analyzing the geometric properties of the proposed model, such as distances, angles, and dihedrals between atoms, MolProbity identifies regions that present clashes or have significant outliers compared to other models, highlighting potential issues in the modeling. MolProbity has been used to check and improve interfaces in protein complexes solved by cryo-EM [54, 76]. Although originally developed for high-resolution structure checking, the method has been extended with the Cα-based low-resolution annotation method (CaBLAM) [77], that checks dihedral over multiple residues and provide a quality measure for intermediate-resolution models where the backbone trace can be more reliably fitted to the EM data than a full-atom model.

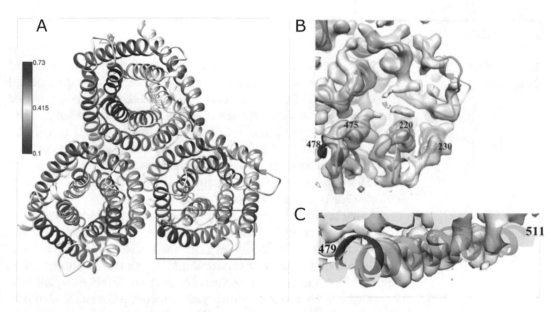

Fig. 11 Illustration of a normalized SMOCf score profile across a protein sequence, projected on the structure of the best CASP13 (http:/predictioncenter.org/casp13/) model for target T1020o (TS004_2o) fitted in the target density map. (**a**) The overall fit is reasonable, but the SMOCf score detects two modeled loops (**b**) and a helix (**c**) that fit poorly to the map. The best model is colored according to the SMOCf score (red to blue), the reference structure is shown in green, and the density map in grey. Adapted from [15]

4   Summary

Despite great advances in cryo-EM techniques, there is still a need to integrate data for solving the structure of macromolecular assemblies. This brings with it a set of problems of modeling atomic structure, density fitting, refinement, and validation. In this review, we described some of these methods and strategies that we have developed in this context. The data used in such an integrative modeling approach is often sparse, noisy, and ambiguous. Therefore, it is important to choose an adequate representation of the structure (depending on the resolution of the map, the accuracy of the candidate model, and the amount of missing information in both the structure and the map) so that the data to fit can be represented in the most accurate way possible. Structure flexibility is an important issue to explore due to the dynamic nature of macromolecular assemblies revealed from EM data. So is the assessment of final models (both in terms of standard geometries and local and global fit to the data), due to variable local resolution and missing information. It is important that the final model(s) reflect uncertainty and completeness of the input information. Future developments will focus on further automation of our protocols, exploration of inherent protein flexibility, incorporation of additional experimental restraints into the model fitting process, and new methods for map and model validation.

5   Methods

5.1   Candidate Model Generation

Having a model close to the final solution increases the speed and efficacy of automated fitting. For Identification of experimental candidate models, the Protein Data Bank (PDB) is a useful resource that contains atomic models derived from experimental data (mainly X-ray crystallography, NMR spectroscopy, and cryo-EM). It is possible to search for structures by sequence (*see* **Note 4**), keyword, experimental technique, and author information among other features. A search by sequence can yield experimental models suitable for use as initial models.

Once an initial model has been identified it is important to consider how complete the model is, as it is common for experimental models to lack flexible regions such as loops. In such cases, it is possible to use further experimental structures to model these regions. To identify further structures from the PDB, a sequence or keyword search can sometimes yield meaningful results. Alternatively, the PDBeFold server (*see* **Note 5**), which compares the 3D fold of input structures with all models in the PDB, can be used to identify structural/sequence homologs. This server takes a PDB coordinate file as input. It is also possible to use a PDB identifier (e.g., 5ogc) to search.

If no suitable experimental structure is available, comparative/ab initio models can be used. Many methods have been proposed to generate models from the protein sequence, and blind competitions have shown that modern methods can achieve very high accuracy in some cases [51]. MODELLER [2, 78] and SWISS-MODEL [21] software suites are among the most popular for comparative modeling. Rosetta provides both ab initio and comparative modeling tools. Other popular tools are I-Tasser [79] and RaptorX [80]. State of the art ab initio (as showcased in recent CASP competitions) often use co-evolution to more accurately predict contacts and help in modeling [81].

5.2   Rigid-Body Fitting

There are many programs that rigidly fit a candidate model into a map by optimizing the CCC. Possibly one of the most commonly used is the *fit-in-map* function implemented in UCSF-Chimera. The method requires an initial manual alignment (*see* **Note 6**) and the cryo-EM map. User input should be defined as *use map simulated from atoms*: Average map resolution. Optimize: correlation. Allow: rotation, shift, and move whole molecules. Generally, this method works well when the initial approximate placement of the candidate model is obvious (*see* **Note 6**).

The Mod-EM method is implemented in the latest version of MODELLER along with the python scripts for executing Mod-EM (*see* **Note 7**). It provides options for both local and global search for a single candidate model. Mod-EM requires the following input: (a) candidate model pdb file, (b) EM density map file, (c) map resolution. The type of fitting (local/global) can be specified by modifying the *grid_search* method (*see* **Note 7**). More details can be found here.

5.2.1 Finding Rigid Bodies with RIBFIND

To use RIBFIND, a web server has been set up (*see* **Note 8**) (although the standalone program can also be downloaded). RIB-FIND accepts the following inputs: (a) the protein coordinates in PDB file format (a single chain or multiple chains with the residues numbered continuously); (b) a description of the protein SSEs in DSSP [20] format (optional, if not provided, the server automatically runs DSSP); (c) the value of the contact distance parameter; (d) the email address of the user; and (e) a density map in MRC format (optional). The RIBFIND client interface reads the input, validates it, and submits it to the server for execution. After completing the job, the server sends an automatic email to the user with a link to the results page. This page contains a NGL Viewer applet showing a cartoon of the user PDB file with each rigid-body uniquely colored. All SSEs and loops that are not part of a cluster are colored white. The rigid-body set that is displayed by default is the one that contains the maximal number of clusters identified by RIBFIND (we previously showed that using this set in cryo-EM flexible fitting produced the best results in most cases [42]). Using the slider control on the results page, the user can view different sets of rigid bodies generated for each cluster cutoff and save the corresponding rigid-body file in a text format which can be used for refinement, e.g., by Flex-EM [5].

5.3 TEMPy

TEMPy is a python library that allows the user to design bespoke pipelines for the manipulation and analysis of candidate models and EM maps [10, 11, 16, 17]. The software package and documentation can be found here: http://tempy.ismb.lon.ac.uk/. The protocols for some commonly used functions are outlined here.

5.3.1 Structure Blurring

Blurred maps derived from candidate models are needed for the analysis of the goodness-of-fit of proposed candidate models into density maps. To blur a map, TEMPy requires (a) the PDB or mmCIF file representing the coordinates and atom types of the candidate model; (b) the resolution (Å) to blur the protein to; and (c) the sigma value (multiplied by the resolution) that controls the width of the Gaussian (with default value of 0.356). An additional option "DensMap" controls whether the blurred map dimensions are based on the candidate structure or the EM map.

5.3.2 Scoring Fits

TEMPy offers a wide range of global and local scoring functions to analyze the goodness-of-fit, both globally and locally. Depending on the map resolution, extent of overlap or shape features, one scoring function may be more useful than others. Both CCC and

MI require two maps or a map and a model to calculate scores. Additionally, map threshold levels can be given for both scores and for MI the number of layers used to bin the maps can be defined (default value of 20).

Local scores can be invaluable for assessing the output of flexible fitting methods such as Flex-EM. Scores such as SMOC or SCCC can be used to measure the local correlation of the model with the map. Both SMOC and SCCC are available in the TEMPy along with example scripts of how to use them. The *"score_smoc.py"* example script included in TEMPy takes as input the map file, a coordinate file and the resolution (*see* **Note 9**). The output is given as a text file of the individual SMOC scores assigned to each residue (e.g., 336 0.88, where 336 is the residue number and 0.88 the SMOC score) and can be plotted to a curve using the matplotlib python library. Additionally, the *"get_sccc.py"* example script included in TEMPy can be used to score the SCCC for individual rigid bodies (*see* **Note 10**). The input requires the EM map, PDB file, resolution, and rigid-body file (that could be produced for example by RIBFIND). The output is given as a Chimera attribute file that can be uploaded into Chimera for a 3-D visualization of the local correlation of rigid bodies with the map.

5.3.3 Ensemble Generation and Scoring

Different scores in TEMPy can also be used to evaluate an ensemble of models and identify one or a few best scoring models (generated by RMSD clustering). A variety of example scripts show workflows from generating to scoring and clustering of ensembles.

Ensembles of structures can be generated either stochastically or using angular sweeps. Both methods implemented require an initial candidate model. The stochastic method requires additional input of: (a) number of structures to generate; (b) maximum translation and rotation permitted; and (c) graining level for the generation of random vectors (default value of 30). For generating ensembles using angular sweeps, the user must define (a) axis and vector for translation; (b) number of structures to generate, (c) and rotation angle for local rotation. The produced output for both methods contains a list of generated structures and their corresponding coordinates.

For scoring of ensemble structures, either a single score or multiple scores can be used. For the latter, the "Consensus" module can be used (based on Borda counting). The required input is: (a) a list of the ensemble candidate models (b) an input list of the scoring functions to use; (c) the target map resolution and sigma coefficient; and (d) the target map. The output contains the list of the structures representing different fits, scored and clustered. The analysis is accompanied by plots and output files that are readable in Chimera, which can help the user to interpret the consensus among the scoring metrics chosen [17].

5.4 γ-TEMPy

γ-TEMPy uses a genetic algorithm to assemble component models simultaneously in a cryo-EM map. The fitness function combines the mutual information score (MI) (*see* Subheading 2.3) to quantify the goodness-of-fit with a penalty score that helps to avoid clashes between components. The script for running γ-TEMPy, which can be run from the command line, can be found in the example scripts from the TEMPy download. The necessary input to run γ-TEMPy are: (a) coordinate file in PDB or mmCIF format. The file should contain all the components (with chain IDs) separated with the TER keyword. (b) Input density map in MRC file format, and (c) the resolution of the map (Å). Optional inputs include the number of solutions to generate, the number of generations for the genetic algorithm to use, and the population size (number of assembly fits). Since GA is a heuristic technique, multiple runs may or may not produce similar results. The resulting output folder will include the initial VQ points in a pdb file, a log file containing details regarding the models generated, and the best scores achieved, with the best model being output as separate pdb file.

5.5 Flexible Fitting with Flex-EM

Flex-EM is available for download as a collection of python scripts to be run from the command line (*see* **Note 9**) together with the MODELLER software package [2]. The input parameters need to be defined within the script "*flex-em.py*" to run Flex-EM. The following parameters are essential: (a) The optimization method must be set to "MD" (for simulated annealing molecular dynamics) or "CG" (conjugate gradient minimization), (b) the input PDB coordinate file that has been initially fit into the map (*see* **Note 11**); (c) the cryo-EM density map file in ".mrc" or ".xplor" format; (d) the voxel size (*see* **Note 10**), (e) the average resolution of the input map, (f) the *X, Y, Z* origin of the map (*see* **Note 12**), the number of CG iterations or MD runs, (g) The cap shift (maximal atom movement allowed) must also be set (*see* **Note 10**). Additional input regarding the rigid bodies must also be supplied (*see* **Note 13**). The output of Flex-EM contains PDB files for models produced at each iteration of MD or CG, along with a final model from the run, that can be used as input for another flex-EM run.

5.6 Density Difference Mapping

A density difference mapping workflow is included with TEMPy ("*difference_map.py*") that can be run from the command line. The input takes: (a) either two maps (for map-map difference) or a map and atomic model (for map-model differences) (*see* **Note 14**), (b) The resolution of both maps (in the case of map-map) or the resolution of the map (in the case of map-model). The output is given as two maps (map 1–map 2 and vice versa). Maps can be visualized in Chimera and used as input for TEMPy.

5.7 Fitting Small Molecules

For fitting small molecules, it is first necessary to obtain a collection of physically plausible ligand conformations. This can be achieved using ligand docking software such as GOLD [60] or AutoDock

Vina [82] (*see* **Note 15**). The input for the docking software is generally unique for that program. However, there are common steps necessary for most software, which include the preparation of receptor files, ligand file, choice of scoring function (some programs such as GOLD contain multiple scoring functions), and receptor flexibility, if allowed. If there is experimental evidence which residues are involved in ligand binding, it may be beneficial to set these residues as flexible, else rigid docking is advised.

The output of docking software is usually given as a set of predicted ligand-binding conformations. To interpret this, conformations predicted by an individual program and scoring function should be clustered. Once clustered, conformations that were predicted by multiple programs should be considered first (consensus docking, *see* [54]).

To identify ligand conformations that can describe the density in the map, the goodness-of-fit between the ligand and the map can be calculated. This can be done with both the ligand-bound conformation map and the difference density map (calculated between the ligand-bound conformation map and an unbound conformation, either from a map or a model) (*see* **Note 16**). Conformations with a high correlation for both difference and density maps should also be visually inspected for factors such as electrostatic complementarity between protein and ligand, the fit of the ligand into the density and ligand-residue interactions.

5.8 Jwalk and MNXL

Jwalk calculates distance between cross-linked residues that can be used to validate candidate models. Two distance metrics are used, Euclidean distance and the Solvent-accessible surface distance (SASD). The SASD has the advantage that it computes the distance across the protein surface, as opposed to the Euclidean distance (which does not take into account the fact that cross-linkers cannot travel through the protein).

Jwalk is available as a web server (*see* **Note 17**), the web server takes a PDB file as input. Two options are available for calculating SASD: (a) calculation of all SASD between two amino acid types (either Lysine, Cystine, Glutamate, or Aspartate). (b) calculation of SASD from an uploaded list of residues (*see* **Note 18**). Additionally, the maximum SASD length and grid spacing (Å) can be specified. The output gives the SASD and Euclidean distance for each calculated link and the cross-links are available for download in PDB format.

The Jwalk output can then be used along with experimental cross-link data to validate multiple models using the MNXL or cMNXL scoring functions. The MNXL web server (*see* **Note 19**) takes an input of: (a) experimental cross-link data. (b) precalculated Jwalk output, or PDB models for validation. The output contains the MNXL score, along with the number of matched, violating, and nonaccessible cross-links.

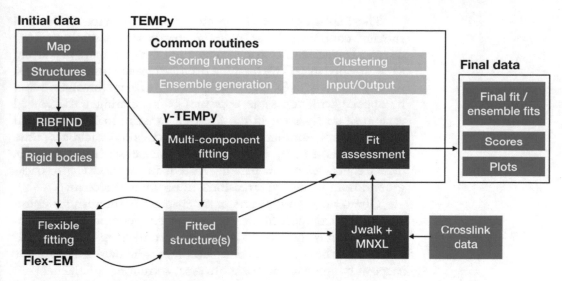

Fig. 12 Overview of our cryo-EM fitting and assessment tools and workflow

5.9 CCP-EM

The CCP-EM software suit comprises a large number of tools to fit, assess, and refine cryo-EM map and structural models. Some of TEMPy tools and scores, Flex-EM, and RIBFIND are available as part of it [83]. An overview of these tools for cryo-EM fitting and assessment can be found in Fig. 12.

6 Notes

6.1 Resolution

Note 1: Multiple definitions of resolution exists and are commonly used: the Rayleigh resolution, the crystallographic resolution, or the Full-Width at Half-Maximum (FWHM); more variants exist, and do not generally agree [84] (also, *see* Table above). In addition, the exact definition of resolution for a cryo-EM map is not simply an experimental parameter, but also a function of the reconstruction quality [85]. The use of a mask or other preprocessing on the map will also alter its estimated resolution.

These variety of approaches cannot be easily reconciled, and the reported numerical value of the resolution should be considered in the context of the definition and criterion of resolution used, and the method used to compute it.

6.2 Density Map
Preprocessing

Note 2: Methods from 2D image processing can be applied to 3D density maps generated from cryo-EM, in order to enhance their contrast, trace contours, segment the map, or remove background noise. This can be used in conjunction with the scoring methods described above to obtain better estimates when comparing maps and models.

The Laplacian filter is commonly used to detect edges and enhance contrast in noisy maps. A matrix representation of the discrete Laplacian operator is convoluted with the map, producing a transformed map. This filter has been successfully used to improve scoring results with CCC in cryo-EM [86]. Other convolutions can be applied with the same approach of generating a discretized version of an operator in the form of a matrix to be convoluted with the map. Another common approach is to obtain a Fourier transform of the map, remove part of the power spectrum (e.g., the high frequencies to low pass filter the map to smooth over background noise), and back transform to obtain a filtered map.

Often, only specific parts of the map are of interest, for example, protein components vs. the solvent or membrane regions. In those cases, it is common to apply a mask to the map, a list of voxels that will not be considered during scoring. The mask can be of any shape although circular masks are most common.

6.3 Cross-Correlation and Manders' Coefficients

Note 3: The Manders' coefficient described above is often called the cross-correlation coefficient.

(*See* for example: Chimera's documentation). Note that its formulation differs from the statistical cross-correlation:

$$\text{Stat CC}(X, \Upsilon) = \frac{\sum\limits_{i}(X_i - \bar{X})(\Upsilon_i - \bar{\Upsilon})}{|X_i - \bar{X}| \, || \, \Upsilon_i - \bar{\Upsilon}|}$$

with $\bar{X} = \sum\limits_{i} X_i / N_X$ the mean value of X. The former Is defined on the interval $[0, 1]$ since all density values are superior or equal to 0, while the latter is defined on $[-1, 1]$.

Another common modification to the CCC involves preprocessing the map before computing the score, for example, using a Laplace filter, or a low-pass filter, or by removing all intensities below a certain threshold or in a given region (masking).

6.4 Model Generation

Note 4: PDB sequence search (http://www.rcsb.org/pdb/search/searchSequence.do)

Note 5: PDBeFold server (http://www.ebi.ac.uk/msd-srv/ssm/) With both the model PDB and EM density map open in Chimera, use the "*fit in map*" tool with options: use map simulated from atoms, *Resolution* (average resolution of map), optimize: correlation, and allow: rotation, shift, and move whole molecules. If the model is outside the map, it must be roughly placed into the map first.

Note 6: With both the model PDB and EM density map open in Chimera, use the "fit in map" tool with options: use map simulated from atoms, Resolution (average resolution of map), optimize: correlation, and allow: rotation, shift, and move whole molecules. If the model is outside the map, it must be roughly placed into the map first.

6.5 Rigid-Body Fitting

Note 7: Mod-EM scripts are available from https://salilab.org/modeller/tutorial/cryoem/fit.html. Documentation on modeler density and grid_search classes are available here: https://salilab.org/modeller/9.21/manual/

6.6 RIBFIND

Note 8: RIBFIND web server available at http://ribfind.ismb.lon.ac.uk/

6.7 Flex-EM

Note 9: Flex-EM scripts are available at http://topf-group.ismb.lon.ac.uk/flex-em/

Note 10: If the rigid bodies are defined at the domain level, and a relatively large conformational change is expected (e.g., an ion channel moving from open to closed state) the recommended cap shift is 0.39 Å. If rigid bodies are defined at the SSE level, the recommended cap shift is 0.15 Å and 0.1 Å for all atom level.

Note 11: Initial fits can be achieved in Chimera using the "fit in map" tool and saving the PDB with new coordinates relative to the input density map.

Note 12: The X, Y, Z origins can be found in Chimera using the "volume viewer" command. If the origin is given in pixels (e.g., from Chimera), the values are calculated as -1 * voxel size.

Note 13: The output of RIBFIND can be used directly as input for Flex-EM. Additionally, custom rigid-body files can be created. The format of these files are: lines starting with # are comments (e.g., #Helix), subsequent lines define a single rigid body. Residues contributing to a rigid bodies are defined by the start and stop residue (e.g., 24 33) multiple segments can be assigned to a single rigid body (e.g., "2 6 28 30" means that residues 2–6 and 28–30 will be included in the same rigid body). For multi-chain models residues can be specified within chains (e.g., "2:A 6:A 28:B 30:B" means that residues 2–6 from chain A and 28–30 from chain B will be included in the same rigid body). For including ligands and other hetero-atoms (HETATM), the molecule number and chain ID can be added to the rigid-body line in the same format (but it should not form a separate rigid body, e.g., "100:A 100:A" means that the HETATM 100 from chain A will be considered as a rigid body).

6.8 Difference Maps

Note 14: If the input is given as two maps, both maps must first be aligned. This can be done automatically in Chimera using the "*fit in map*" tool. However, if it is expected that the two maps represent vastly different conformations (e.g., a large conformational change upon ligand binding) the optimum fit may not be correct" in this case, a manual alignment can be used. However, it is not advisable, and a map-model difference map may be best.

6.9 Docking

Note 15: A graphical interface for AutoDock Vina is implemented in Chimera.

Note 16: This can be done using the "*scoring functions*" module in TEMPy; however, a simulated density map must first be calculated at a suitable resolution using the TEMPy "*structure blurrer*" module.

6.10 Jwalk

Note 17: The Jwalk web server: http://jwalk.ismb.lon.ac.uk/jwalk/

Note 18: The format for uploading a list is each link on a single line of a text file. Each link has the format "residue A|chain|residue B|chain|" (e.g., 52|A|129|A| is a link between residue 52 of chain |A to residue 129 of chain A).

Note 19: The MNXL web server: http://mnxl.ismb.lon.ac.uk/mnxl/.

Acknowledgments

We are grateful for funding from the Wellcome Trust (209250/Z/17/Z and 208398/Z/17/Z) and the Medical Research Council Doctoral Training Programme (UCL). We thank the Topf group and CCP-EM team for their help with software development.

References

1. Moult J, Fidelis K, Kryshtafovych A et al (2018) Critical assessment of methods of protein structure prediction (CASP)-round XII. Proteins Struct Funct Bioinforma 86:7–15

2. Webb B, Sali A (2014) Comparative protein structure modeling using MODELLER. Curr Protoc Bioinformatics 47:5.6.1–5.6.32

3. Pettersen EF, Goddard TD, Huang CC et al (2004) UCSF chimera?A visualization system for exploratory research and analysis. J Comput Chem 25:1605–1612

4. Wriggers W, Milligan RA, McCammon JA (1999) Situs: a package for docking crystal structures into low-resolution maps from Electron microscopy. J Struct Biol 125:185–195

5. Topf M, Lasker K, Webb B et al (2008) Protein structure fitting and refinement guided by Cryo-EM density. Structure 16:295–307

6. Tama F, Miyashita O, Brooks CL (2004) Flexible multi-scale fitting of atomic structures into low-resolution Electron density maps with elastic network Normal mode analysis. J Mol Biol 337:985–999

7. Cardone G, Heymann JB, Steven AC (2013) One number does not fit all: mapping local variations in resolution in cryo-EM reconstructions. J Struct Biol 184:226–236

8. Herzik MA, Fraser JS, Lander GC (2019) A multi-model approach to assessing local and global Cryo-EM map quality. Structure 27:344–358.e3

9. Roseman AM (2000) Docking structures of domains into maps from cryo-electron microscopy using local correlation. Acta Crystallogr D Biol Crystallogr 56:1332–1340

10. Vasishtan D, Topf M (2011) Scoring functions for cryoEM density fitting. J Struct Biol 174:333–343

11. Joseph AP, Lagerstedt I, Patwardhan A et al (2017) Improved metrics for comparing structures of macromolecular assemblies determined by 3D electron-microscopy. J Struct Biol 199:12–26

12. Hryc CF, Jeong H-H, Fei X et al (2017) Subunit conformational variation within individual

GroEL oligomers resolved by Cryo-EM. Proc Natl Acad Sci 114:8259–8264

13. Estevez PA, Tesmer M, Perez CA et al (2009) Normalized mutual information feature selection. IEEE Trans Neural Netw 20:189–201

14. Studholme C, Hill DLG, Hawkes DJ (1999) An overlap invariant entropy measure of 3D medical image alignment. Pattern Recogn 32:71–86

15. Kryshtafovych A, Malhotra S, Monastyrskyy B et al (2019) Cryo-EM targets in CASP13: overview and evaluation of results. Proteins 87:1128–1140

16. Joseph AP, Malhotra S, Burnley T et al (2016) Refinement of atomic models in high resolution EM reconstructions using flex-EM and local assessment. Methods 100:42–49

17. Farabella I, Vasishtan D, Joseph AP et al (2015) TEMPy: a Python library for assessment of three-dimensional electron microscopy density fits. J Appl Crystallogr 48:1314–1323

18. Afonine PV, Poon BK, Read RJ et al (2018) Real-space refinement in *PHENIX* for cryo-EM and crystallography. Acta Crystallogr Sect Struct Biol 74:531–544

19. Emsley P, Lohkamp B, Scott WG et al (2010) Features and development of *Coot*. Acta Crystallogr D Biol Crystallogr 66:486–501

20. Song Y, DiMaio F, Wang RY-R et al (2013) High-resolution comparative modeling with RosettaCM. Structure 21:1735–1742

21. Waterhouse A, Bertoni M, Bienert S et al (2018) SWISS-MODEL: homology modelling of protein structures and complexes. Nucleic Acids Res 46:W296–W303

22. Atherton J, Yu I-MI-M, Cook A et al (2017) The divergent mitotic kinesin MKLP2 exhibits atypical structure and mechanochemistry. elife 6:e27793

23. Kabsch W, Sander C (1983) Dictionary of protein secondary structure: pattern recognition of hydrogen-bonded and geometrical features. Biopolymers 22:2577–2637

24. Benkert P, Biasini M, Schwede T (2011) Toward the estimation of the absolute quality of individual protein structure models. Bioinformatics 27:343–350

25. Topf M, Baker ML, John B et al (2005) Structural characterization of components of protein assemblies by comparative modeling and electron cryo-microscopy. J Struct Biol 149:191–203

26. Pintilie GD, Zhang J, Goddard TD et al (2010) Quantitative analysis of cryo-EM density map segmentation by watershed and scale-space filtering, and fitting of structures by alignment to regions. J Struct Biol 170:427–438

27. Wu X, Milne JLS, Borgnia MJ et al (2003) A core-weighted fitting method for docking atomic structures into low-resolution maps: application to cryo-electron microscopy. J Struct Biol 141:63–76

28. Kim SJ, Fernandez-Martinez J, Nudelman I et al (2018) Integrative structure and functional anatomy of a nuclear pore complex. Nature 555:475–482

29. Kuzu G, Keskin O, Nussinov R et al (2016) PRISM-EM: template interface-based modelling of multi-protein complexes guided by cryo-electron microscopy density maps. Acta Crystallogr Sect Struct Biol 72:1137–1148

30. Chandramouli P, Topf M, Ménétret J-F et al (2008) Structure of the mammalian 80S ribosome at 8.7 Å resolution. Structure 16:535–548

31. Taylor DJ, Devkota B, Huang AD et al (2009) Comprehensive molecular structure of the eukaryotic ribosome. Structure 17:1591–1604

32. Kawabata T (2008) Multiple subunit fitting into a low-resolution density map of a macromolecular complex using a Gaussian mixture model. Biophys J 95:4643–4658

33. Tjioe E, Lasker K, Webb B et al (2011) MultiFit: a web server for fitting multiple protein structures into their electron microscopy density map. Nucleic Acids Res 39:W167–W170

34. Pandurangan AP, Vasishtan D, Alber F et al (2015) γ-TEMPy: simultaneous fitting of components in 3D-EM maps of their assembly using a genetic algorithm. Structure 23:2365–2376

35. Zhang S, Vasishtan D, Xu M et al (2010) A fast mathematical programming procedure for simultaneous fitting of assembly components into cryoEM density maps. Bioinformatics 26:261–268

36. Zeev-Ben-Mordehai T, Vasishtan D, Hernández Durán A et al (2016) Two distinct trimeric conformations of natively membrane-anchored full-length herpes simplex virus 1 glycoprotein B. Proc Natl Acad Sci 113:4176–4181

37. Baker ML, Jiang W, Wedemeyer WJ et al (2006) Ab initio modeling of the Herpesvirus VP26 Core domain assessed by CryoEM density. PLoS Comput Biol 2:12

38. Sachse C, Chen JZ, Coureux P-D et al (2007) High-resolution Electron microscopy of helical specimens: a fresh look at tobacco mosaic virus. J Mol Biol 371:812–835

39. Lukoyanova N, Kondos SC, Farabella I et al (2015) Conformational changes during pore formation by the Perforin-related protein Pleurotolysin. PLoS Biol 13:e1002049

40. Chen Z, Chapman MS (2001) Conformational disorder of proteins assessed by real-space molecular dynamics refinement. Biophys J 80:1466–1472

41. Pandurangan AP, Topf M (2012) RIBFIND: a web server for identifying rigid bodies in protein structures and to aid flexible fitting into cryo EM maps. Bioinformatics 28:2391–2393

42. Pandurangan AP, Topf M (2012) Finding rigid bodies in protein structures: application to flexible fitting into cryoEM maps. J Struct Biol 177:520–531

43. Zhou R, Yang G, Guo X et al (2019) Recognition of the amyloid precursor protein by human γ-secretase. Science 363:eaaw0930

44. Yang G, Zhou R, Zhou Q et al (2019) Structural basis of notch recognition by human γ-secretase. Nature 565:192

45. Clare DK, Vasishtan D, Stagg S et al (2012) ATP-triggered conformational changes delineate substrate-binding and -folding mechanics of the GroEL Chaperonin. Cell 149:113–123

46. Trabuco LG, Villa E, Mitra K et al (2008) Flexible fitting of atomic structures into Electron microscopy maps using molecular dynamics. Structure 16:673–683

47. Trabuco LG, Villa E, Schreiner E et al (2009) Molecular dynamics flexible fitting: a practical guide to combine cryo-electron microscopy and X-ray crystallography. Methods 49:174–180

48. Alford RF, Leaver-Fay A, Jeliazkov JR et al (2017) The Rosetta all-atom energy function for macromolecular modeling and design. J Chem Theory Comput 13:3031–3048

49. DiMaio F, Tyka MD, Baker ML et al (2009) Refinement of protein structures into low-resolution density maps using Rosetta. J Mol Biol 392:181–190

50. Tama F, Miyashita O, Brooks CL III (2004) Normal mode based flexible fitting of high-resolution structure into low-resolution experimental data from cryo-EM. J Struct Biol 147:315–326

51. Lopéz-Blanco JR, Chacón P (2013) iMOD-FIT: efficient and robust flexible fitting based on vibrational analysis in internal coordinates. J Struct Biol 184:261–270

52. Ahmed A, Whitford PC, Sanbonmatsu KY et al (2012) Consensus among flexible fitting approaches improves the interpretation of cryo-EM data. J Struct Biol 177:561–570

53. Pandurangan AP, Shakeel S, Butcher SJ et al (2014) Combined approaches to flexible fitting and assessment in virus capsids undergoing conformational change. J Struct Biol 185:427–439

54. Newcombe J, Chatzidaki A, Sheppard TD et al (2018) Diversity of nicotinic acetylcholine receptor positive allosteric modulators revealed by mutagenesis and a revised structural model. Mol Pharmacol 93:128–140

55. Locke J, Joseph AP, Peña A et al (2017) Structural basis of human kinesin-8 function and inhibition. Proc Natl Acad Sci 114: E9539–E9548

56. Canutescu AA, Dunbrack RL (2003) Cyclic coordinate descent: a robotics algorithm for protein loop closure. Protein Sci 12:963–972

57. Wang C, Bradley P, Baker D (2007) Protein–protein docking with backbone flexibility. J Mol Biol 373:503–519

58. Mandell DJ, Coutsias EA, Kortemme T (2009) Sub-angstrom accuracy in protein loop reconstruction by robotics-inspired conformational sampling. Nat Methods 6:551–552

59. Scaiola A, Leibundgut M, Boehringer D et al (2019) Structural basis of translation inhibition by cadazolid, a novel quinoxolidinone antibiotic. Sci Rep 9:5634

60. Verdonk ML, Cole JC, Hartshorn MJ et al (2003) Improved protein–ligand docking using GOLD. Proteins Struct Funct Bioinforma 52:609–623

61. Dominguez C, Boelens R, Bonvin AMJJ (2003) HADDOCK: a protein–protein docking approach based on biochemical or biophysical information. J Am Chem Soc 125:1731–1737

62. Wang Z, Sun H, Yao X et al (2016) Comprehensive evaluation of ten docking programs on a diverse set of protein–ligand complexes: the prediction accuracy of sampling power and scoring power. Phys Chem Chem Phys 18:12964–12975

63. Houston DR, Walkinshaw MD (2013) Consensus docking: improving the reliability of docking in a virtual screening context. J Chem Inf Model 53:384–390

64. Catarinella M, Grüner T, Strittmatter T et al (2009) BTB-1: a small molecule inhibitor of the mitotic motor protein Kif18A. Angew Chem Int Ed 48:9072–9076

65. Kucukelbir A, Sigworth FJ, Tagare HD (2014) Quantifying the local resolution of cryo-EM density maps. Nat Methods 11:63–65

66. Shen M, Sali A (2006) Statistical potential for assessment and prediction of protein structures. Protein Sci 15:2507–2524

67. Uziela K, Menéndez Hurtado D, Shu N et al (2017) ProQ3D: improved model quality assessments using deep learning. Bioinformatics 33:1578–1580

68. Elofsson A, Joo K, Keasar C et al (2018) Methods for estimation of model accuracy in CASP12. Proteins Struct Funct Bioinforma 86:361–373

69. Joseph AP, Polles G, Alber F et al (2017) Integrative modelling of cellular assemblies. Curr Opin Struct Biol 46:102–109

70. Bullock JMA, Schwab J, Thalassinos K et al (2016) The importance of non-accessible crosslinks and solvent accessible surface distance in modeling proteins with restraints from crosslinking mass spectrometry. Mol Cell Proteomics 15:2491–2500

71. Bullock JMA, Thalassinos K, Topf M (2018) Jwalk and MNXL web server: model validation using restraints from crosslinking mass spectrometry. Bioinformatics 34:3584–3585

72. Bullock JMA, Sen N, Thalassinos K et al (2018) Modeling protein complexes using restraints from crosslinking mass spectrometry. Structure 26:1015–1024.e2

73. Barad BA, Echols N, Wang RY-R et al (2015) EMRinger: side chain–directed model and map validation for 3D cryo-electron microscopy. Nat Methods 12:943–946

74. Afonine PV, Klaholz BP, Moriarty NW et al (2018) New tools for the analysis and validation of cryo-EM maps and atomic models. Acta Crystallogr Sect Struct Biol 74:814–840

75. Chen VB, Arendall WB, Headd JJ et al (2010) MolProbity: all-atom structure validation for macromolecular crystallography. Acta Crystallogr D Biol Crystallogr 66:12–21

76. Atherton J, Jiang K, Stangier MM et al (2017) A structural model for microtubule minus-end recognition and protection by CAMSAP proteins. Nat Struct Mol Biol 24:931–943

77. Richardson JS, Williams CJ, Hintze BJ et al (2018) Model validation: local diagnosis, correction and when to quit. Acta Crystallogr Sect Struct Biol 74:132–142

78. Webb B, Sali A (2014) Protein structure modeling with MODELLER. In: Kihara D (ed) Protein structure prediction. Springer, New York, NY, pp 1–15

79. Zhang Y (2008) I-TASSER server for protein 3D structure prediction. BMC Bioinformatics 9:40

80. Peng J, Xu J (2011) Raptorx: exploiting structure information for protein alignment by statistical inference. Proteins Struct Funct Bioinforma 79:161–171

81. Schaarschmidt J, Monastyrskyy B, Kryshtafovych A et al (2018) Assessment of contact predictions in CASP12: co-evolution and deep learning coming of age. Proteins Struct Funct Bioinforma 86:51–66

82. Trott O, Olson AJ (2010) AutoDock Vina: improving the speed and accuracy of docking with a new scoring function, efficient optimization, and multithreading. J Comput Chem 31:455–461

83. Burnley T, Palmer CM, Winn M (2017) Recent developments in the *CCP-EM* software suite. Acta Crystallogr Sect Struct Biol 73:469–477

84. Jones A, Bland-Hawthorn J, Shopbell P (1995) Towards a general definition for spectroscopic resolution, In: Astronomical data analysis software and systems IV. ASP Conf Ser 77:503

85. Liao HY, Frank J (2010) Definition and estimation of resolution in single-particle reconstructions. Structure 18:768–775

86. Chacón P, Wriggers W (2002) Multiresolution contour-based fitting of macromolecular structures. J Mol Biol 317:375–384

Part III

Electron Crystallography of 2D Crystals

Chapter 10

2D Electron Crystallography of Membrane Protein Single-, Double-, and Multi-Layered Ordered Arrays

Matthew C. Johnson, Yusuf M. Uddin, Kasahun Neselu, and Ingeborg Schmidt-Krey

Abstract

The electron cryo-microscopy (cryo-EM) approach of 2D electron crystallography allows for structure determination of two-dimensional (2D) crystals of soluble and membrane proteins, employing identical principles and methods once 2D crystals are obtained. Two-dimensional crystallization trials of membrane proteins can result in multiple outcomes of ordered arrays, which may be suited for either 2D electron crystallography, helical analysis, or MicroED.

The membrane protein 2D crystals used for 2D electron crystallography are either single- or double-layered ordered proteoliposome vesicles or sheet-like membranes. We have developed a cryo-EM grid preparation approach, which allows for the analysis of stacked 2D crystals that are neither suitable for MicroED nor for directly applying 2D electron crystallography. This new grid preparation approach, the peel-blot, uses the capillary force generated by submicron filter paper and mechanical means for the separation of stacked 2D crystals into single-layered 2D crystals, for which standard 2D electron crystallography can then be employed. The preparation of 2D crystals, the peel-blot grid preparation, and the structure determination by 2D electron crystallography are described here.

Key words Membrane protein, Structure, Reconstitution, 2D Crystallization, Peel-blot, Cryo-EM, 2D Electron crystallography, 2D Crystallography, Electron crystallography

1 Introduction

The cryo-EM method of 2D electron crystallography provided the first understanding of membrane protein secondary structure [1] and resulted in some of the first atomic models of both membrane proteins [2, 3] and soluble α/β tubulin protein [4, 5]. Just as importantly, 2D electron crystallography allows for structure–function studies and has provided unprecedented resolution of membrane protein structure with detailed structural insights on the entire surrounding phospholipid bilayer [6].

Following the "resolution revolution" [7] of cryo-EM, subsequent ongoing development in single particle cryo-EM in

Tamir Gonen and Brent L. Nannenga (eds.), *CryoEM: Methods and Protocols*, Methods in Molecular Biology, vol. 2215, https://doi.org/10.1007/978-1-0716-0966-8_10, © Springer Science+Business Media, LLC, part of Springer Nature 2021

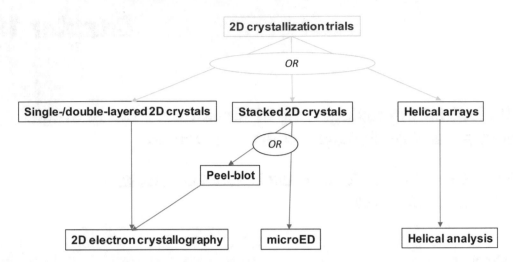

Fig. 1 Schematic representation of 2D crystallization and choice of approach for the structure determination

recent years has made it feasible to study decreasingly smaller membrane proteins, which could previously only be studied via crystallization. Nanodiscs [8] and styrene-maleic acid (SMA) nanodiscs [9] address the difficulties associated with studying purified membrane proteins within a lipid bilayer rather than detergent, which may cause issues for both the protein stability and cryo-EM sample preparation. Yet many small proteins will remain outside the reach of single particle cryo-EM, while neither 2D nor 3D electron crystallography suffer from such a size limitation.

The required 2D crystallization for 2D electron crystallography can have multiple outcomes: not only are multiple 2D crystal types possible, but also helical arrays or 2D crystals stacking out of register (Fig. 1). Attempts at 2D crystallization may also produce small/"micro" 3D crystals, which are suitable samples for MicroED [10, 11]. When 2D crystals stack out of register, or when 3D crystals are too small even for MicroED, samples can be "unstacked" for 2D crystallography via the "peel-blot" technique described here. While in principle unstacking could be applied to 3D crystals used with MicroED, MicroED does not require additional manipulation and most importantly, is fast and results in high-resolution data of suitable samples [10].

The method of 2D electron crystallography involves the following steps: purification, 2D crystallization, cryo-EM grid preparation, and cryo-EM data collection/processing. Like all methods in structural biology, purification is a key foundation of 2D electron crystallography. After purification, the purified protein is then subjected to 2D crystallization, which constitutes the critical, and often most time-consuming step, including substantial time for screening of samples by negative stain TEM to identify conditions to induce or improve 2D crystallization. Once 2D crystallization is successful and quality 2D crystals have been formed, cryo-EM grid preparation is tested with a screening cryo-EM. Cryo-EM grids under the optimal conditions are then used for high-resolution data collection with a high-end cryo-EM and image processing.

Detailed reviews of 2D crystallization of membrane proteins include [12–21]. Cryo-EM grid preparation of 2D crystals is based on the back-injection method developed by Kühlbrandt and Downing [22] and variations of this method with different sugars have also described in detail, including electron diffraction data not discussed here [6, 23–26]. The method was further modified to increase sample and carbon film flatness [27] and to address charging effects of data collection at high tilt angles [28], which can severely limit the resolution. Image processing approaches for high-resolution data from 2D crystals build largely on [1, 2, 29, 30] and references therein. Stahlberg and colleagues have automated data collection of 2D crystals and developed the Focus software package [31] to streamline the process of data collection and processing.

This chapter initially focuses on 2D crystallization of membrane proteins, as these are the protein samples the method has most widely been applied to. However, the methods for screening by negative stain TEM, cryo-EM data collection, and image processing are equally applicable to 2D crystals of soluble and membrane proteins.

2   Materials

Gloves are used for 2D crystallization trials and negative stain as well as cryo-EM grid preparation. Caution and sufficient training is required for work with chloroform and liquid nitrogen. Uranyl acetate may require adherence to specific university regulations for use and disposal.

2.1
Two-Dimensional
Crystallization Trials

1. 1,2-Dimyristoyl-*sn*-glycero-3-phosphocholine (DMPC, Avanti Polar Lipids), in chloroform or powdered *or* 1,2-dioleoyl-*sn*-glycero-3-phosphatidylcholine (DOPC, Avanti Polar Lipids), in chloroform (10 or 25 mg/mL) or powdered (25 mg).

2. Round-bottom flask (5 or 10 mL).

3. Nitrogen gas.

4. Detergent (based on the detergent used for purification).

5. Water-bath sonicator.

6. 4–8 aliquots of 75–100 µL purified, detergent-solubilized membrane protein.

7. 1 L detergent-free dialysis buffer (2 L for 8 crystallization conditions).

8. Ultrapure water.

9. 10–14 K MWCO dialysis tubing, 1 cm flat width.

10. Dialysis clips (3 cm in length or larger).

11. Incubators (heating/cooling).

12. Fumehood for work with phospholipids in chloroform.

2.2 Negative Staining of 2D Crystallization Trials

1. 400-mesh, carbon-coated copper grids.

2. 1–2% uranyl acetate solution.

3. Forceps (Dumont #5) and/or anticapillary forceps (Dumont #N5AC).

4. Whatman #4 filter paper.

5. A glow-discharge unit.

6. Grid storage boxes.

7. Desiccator cabinet.

2.3 TEM Screening of Crystallization Trials

1. 80–120 kV transmission electron microscope (TEM) equipped with suitable detector for FFT assessment.

2.4 Cryo-EM Grid Preparation, Screening and Data Collection

1. 400-mesh copper grids or 400-mesh molybdenum grids.

2. Carbon-coated mica.

3. Anticapillary forceps.

4. 4% trehalose (alternatively, 2% buffered tannin solution or 2% glucose).

5. Whatman #4 filter paper.

6. Liquid nitrogen.

7. Liquid nitrogen dewar.

8. Protective gloves and face shield or protective glasses for work with liquid nitrogen.

9. MilliporeSigma DAWP 0.65 µm filter membrane filter paper (for peel-blot).

10. Parafilm M laboratory film (for peel-blot).

11. 600-mesh copper grids or 600-mesh molybdenum grids (for peel-blot).

12. 120 or 200 kV screening cryo-EM (once crystals are available and ready for cryo-EM).

13. 300 kV cryo-EM for data collection.

2.5 Image Processing

1. PC, Mac, or workstation.

2. Focus software package: https://focus.c-cina.unibas.ch/

3   Methods

The approach of 2D electron crystallography can be applied to both soluble and membrane proteins. For 2D crystals of soluble proteins, the steps from the negative stain grid preparation and beyond are applied. For 2D crystallization of membrane proteins, all of the steps below are applied, unless 2D crystals already exist.

3.1   Crystallization of Membrane Proteins Through Detergent Removal by Dialysis

Membrane protein 2D crystallization requires samples that are purified and contain no or a minimal amount of co-purified lipids. An absolutely minimal amount of co-purified lipids is critical since the 2D crystallization of most membrane proteins to date has been dependent on a carefully controlled lipid-to-protein ratio. At this point, the lipid content of the purified, detergent-solubilized membrane protein sample has likely been characterized by thin layer chromatography, mass spectrometry, and/or initial TEM analysis.

3.1.1   Preparation of Solubilized Lipid Stock

1. If working with lipids in powder form, add buffer-detergent solution for a final lipid concentration of 10 mg/mL (*see* **Note 1**), which will result in the required lipid stock solution. If working with lipids in chloroform (**steps 2–4** below apply to these lipids), pipette 1 mL into a small round-bottom flask.

2. Evaporate the chloroform with a stream of nitrogen gas (*see* **Note 2**).

3. The buffer-detergent solution is pipetted into the round-bottom flask for a final lipid concentration of 10 mg/mL.

4. The round-bottom flask is placed into a sonicator bath and sonicated for several minutes until the solution becomes clear.

5. Aliquots of 10 μL of the solubilized lipid are pipetted into Eppendorf tubes for storage at either −20 °C or −80 °C (*see* **Note 3**).

3.1.2   Setup of 2D Crystallization Trials

1. Soak dialysis tubing in 500–1000 mL of ultrapure water (*see* **Note 4**).

2. Thaw membrane protein, if previously frozen and not freshly purified, and lipid aliquots on ice.

3. Prepare 1 L of dialysis buffer based on the purification buffer, omitting the detergent. In later rounds of crystallization trials, up to 20% glycerol may be added, and the salt may be increased, decreased, or altered.

4. Mix 75–100 μL of the membrane protein with stock solution of lipid at molar lipid-to-protein ratios (LPRs) of 0, 3, 15, and 30 in Eppendorf tubes on ice (this membrane protein-detergent-lipid mixture will be referred to as dialysate in later steps) (*see* **Note 5**).

5. Vortex the sample for 10–20 s.

6. Incubate the dialysate for 10 min on ice. During the incubation, rinse the dialysis tubing with ultrapure water and divide the dialysis buffer into four beakers (400 mL volume).

7. Vortex the dialysate for 10–20 s.

8. Attach a dialysis clip to one end of the dialysis tubing, leaving at least several mm to 1 cm on the bottom end of the clip.

9. Remove excess water from the inside of the tubing by holding the clip and rapidly flicking the long end of the tubing.

10. Hold the clip between the thumb and ring finger of one hand and stabilize the long, open end of the tubing between the index and middle finger in order to hold the tubing upright.

11. Pipette the dialysate in the open end of the tubing, ensuring that the pipette tip is placed as closely as possible to the bottom clip before pipetting the dialysate (*see* **Note 6**).

12. With a second dialysis clip placed at an angle of 90°, clip the open end of the tubing closed, while keeping a space of ~6–7 cm between the clips (*see* **Note 7**).

13. Place the assembled tubing with one dialysate condition per beaker in a beaker covered with aluminum foil (*see* **Note 8**).

14. Dialyze the sample for 2–14 days in an incubator at a temperature just above the phase transition temperature of the lipid (*see* **Note 9**).

15. Remove the sample from the incubator and cut the dialysis tubing below the top clip.

16. Remove the dialysate with a micropipette and place it in an Eppendorf tube. Immediately proceed to Subheading 3.2 unless the dialysate contains air bubbles (*see* **Note 10**).

3.2 Negative Stain TEM Screening for 2D Crystallization Conditions

Screening for 2D crystallization conditions by negative stain TEM will provide important feedback for the protein purification as well. Negative stain preparation of the purified protein before dialysis can provide critical information on, e.g., aggregation or sample heterogeneity.

3.2.1 Negative Staining

1. A carbon-coated 400-mesh grid is picked up with a pair of anticapillary forceps with the carbon-side facing up (*see* **Note 11**).

2. A volume of 2 μL of dialysate is pipetted onto the carbon-side of the grid, incubated for 60 s, and blotted with the torn side of the filter paper (*see* **Note 12**).

3. A same volume of 2 μL 1% uranyl acetate is immediately pipetted on the grid, the stain is incubated for 30 s on the

grid, and blotted with the torn side of the filter paper (*see* **Note 13**).

4. The grid is dried and then either screened by TEM immediately or kept in a grid box in a desiccator cabinet until the scheduled TEM session (*see* **Note 14**).

3.2.2 Screening by TEM

1. Grids are screened with a TEM at 120 kV (*see* **Note 15**).

2. The grid is viewed at a low magnification of approximately 2000–10,000× for a first assessment of sample concentration. For efficient and thorough screening, most grid squares should contain a minimum of 5–10 membranes/proteoliposomes.

3. For samples with low concentrations on the grid, **steps 1–4** in Subheading 3.2.1 are repeated after the sample has been concentrated (*see* **Note 16**).

4. Sample membrane morphology (vesicle, planar-tubular, tubular/helical, membrane sheet, stacked sheets/vesicles), average membrane size and range of sizes, as well as aggregation of either membranes and/or protein are noted.

5. Suitable membranes of a diameter of at least 50–100 nm are selected for identification of ordered arrays at magnifications of ~30,000–50,000× (*see* **Note 17**).

6. Images of membranes are collected for immediate evaluation by Fast Fourier Transform (FFT) (*see* **Note 18**).

7. Each image is assessed by FFT for identification of ordered arrays, size of ordered arrays, initial evaluation of symmetry, and potential crystal mosaicity (*see* **Note 19**).

3.2.3 Refinement of Crystallization Conditions

1. Crystallization conditions are considered in terms of crystal order, size, and morphology.

2. Order is of highest priority and will require consideration of changes in LPR.

3. Once 2D crystals are obtained and the entire membrane is ordered and does not contain significant mosaicity, the 2D crystal size is considered. Dialysis buffer salt concentration, salt mixtures, glycerol, and/or temperature are modified, while carefully considering the activity of the protein.

4. The steps in Subheadings 3.1.2–3.2.2 are repeated with modified conditions until the largest possible 2D crystals are obtained. The 2D crystals will be at least 200 nm and ideally 1 μm or larger in size.

5. Crystal morphology is considered for the cryo-EM approach in terms of methods selection below.

*3.2.4 Decision
on Cryo-EM Approach*

Once crystals are identified, the crystal type will decide on the approach for data collection and image processing. Single- and double-layered 2D crystals, which tend to be the main goal of 2D crystallization trials, are treated as described below (except for the peel-blot in Subheading 3.3.2). These types of 2D crystals will have sheet, vesicle, and/or planar-tubular morphology, not uncommonly in a mixture of morphologies. Narrow helical arrays of just a few molecules across the diameter of the tube and typically several hundred nanometers or more in lengths [19], will be ideal for the helical analysis approach. Thin 3D crystals, usually consisting of 2D crystals stacked in register, of approximately 1–2 μm and less than 1 μm thick will be tested for MicroED [10]. Smaller, stacked 2D crystals, as well as crystals found not to be suitable for MicroED, are subjected to the peel-blot approach.

**3.3 Cryo-EM Grid
Preparation
and Cryo-EM**

Single- or double-layered 2D crystals will be subjected to the steps below, except for the peel-blot in Subheading 3.3.1, **step 2**. These 2D crystals are vitrified by the back-injection method in Subheading 3.3.1 [22] (Fig. 2). Smaller, stacked crystals are subjected to the peel-blot (Subheading 3.3.1, **step 2**, but not Subheading 3.3.1), followed by the standard cryo-EM data collection and image processing used for single- and double-layered 2D crystals.

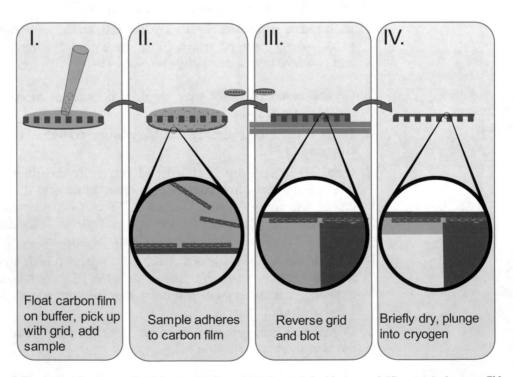

Fig. 2 The back-injection method for preparation of single- and double-layered 2D crystals for cryo-EM grid preparation

3.3.1 Cryo-EM Grid Preparation of Single- Or Double-Layered 2D Crystals

1. Two stacked pieces of Whatman #4 filter paper are placed on a clean bench.

2. A piece of Parafilm is placed next to the filter paper on the bench.

3. Two droplets of 150 µL of trehalose solution are pipetted onto the Parafilm (*see* **Note 20**).

4. A grid is picked up with a pair of anticapillary forceps (*see* **Note 21**).

5. A square piece of carbon-coated mica is cut to a size slightly larger than a grid.

6. The carbon film is floated off on the first drop (*see* **Note 22**).

7. The carbon film is picked up with the anticapillary forceps (*see* **Note 23**).

8. The grid is touched to the surface of the second drop in order to remove excess carbon film at the periphery of the grid.

9. The grid is rotated by 180° (the carbon side is facing down) and the forceps are placed on the bench.

10. The side of a pipette tip is gently drawn across the surface of the grid to remove any remaining broken pieces of carbon film.

11. A volume of 1.3–2 µL of the 2D crystal solution is pipetted onto the grid and distributed in the solution on the grid by pipetting 10×.

12. The solution on the grid is incubated for 1 min.

13. The grid is rotated by 180° and placed on the two layers of filter paper without releasing it from the forceps.

14. The grid is picked up, air-dried for 10–15 s, and hand-plunged into liquid nitrogen, either for immediate or later use.

3.3.2 Peel-Blot

It is not uncommon for 2D crystals to "stack" into many layers, which may impede attempts at structure determination. We have developed a cryo-EM grid preparation (Fig. 3) technique that mechanically separates the layers of these crystals stacks by using the high capillary pressure from blotting cryo-EM grids prepared by back-injection on a filter membrane of submicron pore size. This causes the carbon film to be tightly suctioned onto the surface of the metal grid bars, and any 2D crystals stacks positioned between the carbon film and the grid bars adhere to both surfaces. When these surfaces are subsequently separated by addition of buffer, the stack layers separate, often leaving single layers adhered to the carbon film surface that retain crystallinity and are suitable for EM data collection (Fig. 4). This method is also capable of separating disordered lamellae, such as collapsed lipid vesicles (Fig. 5).

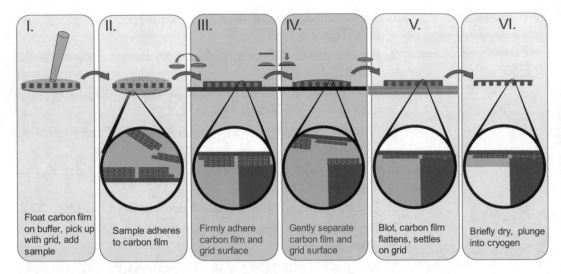

Fig. 3 The back-injection method with peel-blotting for unstacking of multilamellar crystals

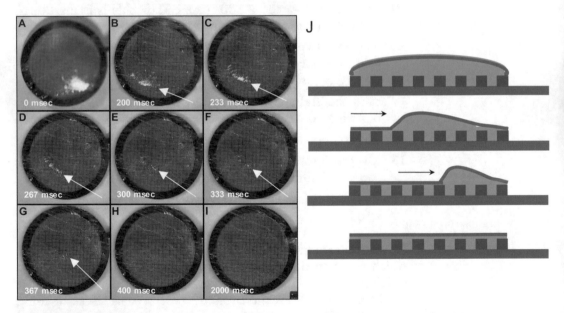

Fig. 4 Peel-blotting and negative staining of DMPC liposomes. DMPC (14:0 PC; 1,2-dimyristoyl-*sn*-glycero-3-phosphocholine) vesicles prepared by dialysis of detergent solubilized lipids, peel-blotted, and negative stained (1% uranyl acetate). (**a**) Low magnification micrograph showing a "grid bar footprint" pattern resulting from peel-blot preparation. (**b**) Higher magnification of boxed region, showing both unpeeled (top) and peeled (bottom) regions. (**c**) Collapsed vesicles after grid preparation without peel-blotting. (**d**) Single-layer vesicle remnants after peel-blotting. (**e**) Large half-peeled vesicle with ordered ("ripple phase") DMPC, with arrows showing peeled (bottom) and unpeeled (top) regions

1. Following positive identification of multilamellae by negative stain, carry out the cryo-EM grid preparation by back-injection according to the previous section, up to **step 12**, in which

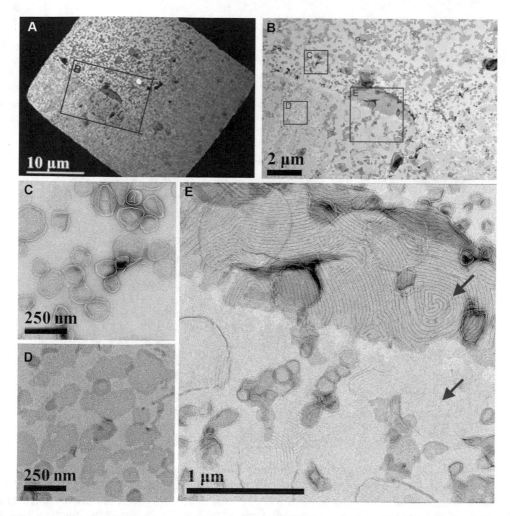

Fig. 5 Peel-blotting of stacked 2D crystals of a human membrane protein. Two-dimensional crystals of human leukotriene C$_4$ synthase (LTC$_4$S) were prepared as described previously [42, 43], and negatively stained (1% uranyl acetate). (**a**) LTC$_4$S 2D crystals can exhibit a strong tendency to stack into multilamellae. Arrows point to regions with small single-layer 2D crystals (bottom) and large multi-layered crystals (top). (**b**) Higher magnification image of negatively stained stacked 2D crystals, with computed power spectra showing a single lattice, which indicates that 2D crystal layers have stacked in register (inset). (**c**) Low magnification image of peeled LTC$_4$S crystals. Arrows point to unpeeled, stacked crystals (right), and peeled region (left), showing micron-sized single-layer regions. (**d**) High-magnification image of a peeled 2D crystal, in which crystalline order is maintained (inset)

sample has been applied to a wetted carbon film on an EM grid (*see* **Note 24**).

2. A 1.7 µL droplet of sample buffer with 4% trehalose is applied onto Parafilm (*see* **Note 25**).

3. The grid is rotated by 180° and firmly placed on a single layer of Millipore DAWP membrane filter paper (0.65 µm pore size)

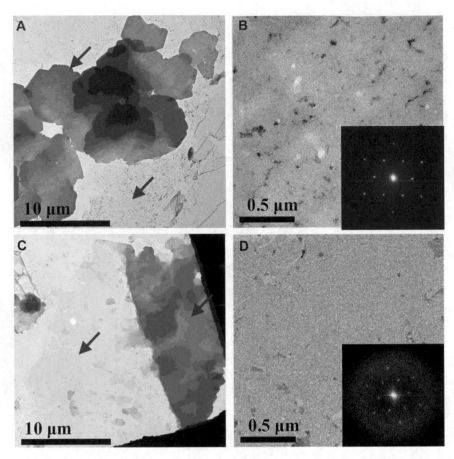

Fig. 6 Blotting an EM grid prepared by back-injection on DAWP membrane as an important step of the peel-blot separation of double-layered and multilamellar membrane. Slow blotting on submicron pore size membrane induces "wicking" and the displacement of carbon film on the grid mesh. Single frames from a video taken at (**a**) initial contact with the membrane, and at (**b–i**) 200, 233, 267, 300, 333, 367, 400, and 2000 ms after contact. The arrows indicate the "wave front" of carbon film being suctioned tightly onto the surface of the grid by capillary pressure. (**j**) Cartoon of carbon film wicking during peel-blot

atop two layers of Whatman #4 filter paper, without releasing it from the forceps (*see* **Note 26**) (Fig. 6).

4. The grid is blotted and held in place ~3 s after carbon film has wicked flat against the EM grid surface (~5–7 s of total blotting time) (*see* **Note 27**).

5. The grid is picked up, and briskly touched to the droplet on Parafilm, until the grid mesh contacts the Parafilm and buffer is forced through the grid mesh to lift the carbon film away from the grid (*see* **Note 28**).

6. (Optional) **steps 2–5** may be repeated, for multiple rounds of peel-blots (*see* **Note 29**).

7. From here, the grid is blotted as in Subheading 3.3.1 above, by being placed on the two layers of Whatman filter paper without releasing it from the forceps.

8. The grid is picked up, air-dried for 10–15 s, and hand-plunged into liquid nitrogen, either for immediate or later use (*see* **Note 30**).

3.3.3   Initial Assessment of Cryo-EM Conditions

Prior to data collection, it is often necessary to refine sample preparation conditions after an initial assessment by cryo-EM. Vitrified samples stored in liquid nitrogen are transferred into a cryogenic transmission electron microscope and evaluated for ice thickness and crystal integrity. If ice is too thick (i.e., the beam does not penetrate the sample), it may be necessary to blot for longer periods of time, or to alter sample buffer components. If the ice is too thin, sample will be dehydrated and freeze-dried, rather than vitrified, resulting in loss of crystalline order, and shorter blot times and/or buffer alterations may be warranted. If ice thickness is ideal and 2D crystals can be located at low magnification and low electron dose (Fig. 7), small rounds of data collection are attempted to screen for crystal quality. Crystalline order may be assessed one of two ways: either by observation of high-resolution reflections via electron diffraction, or by real-space imaging followed by observation of high-resolution lattice spots in computed Fourier power spectra. If crystals are not well ordered, it may be necessary to refine crystallization parameters, such as the lipid-to-protein ratio (LPR), pH, salt concentration, and lipid type.

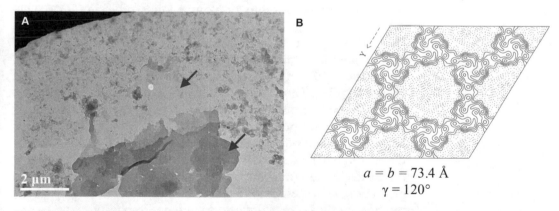

$$a = b = 73.4 \text{ Å}$$
$$\gamma = 120°$$

Fig. 7 Peel-blotting and cryo-EM of stacked 2D crystals of a human membrane protein. Stacked 2D LTC$_4$S crystals were peel-blotted and hand-plunged into liquid nitrogen, and then imaged under low-dose conditions. (**a**) low magnification micrograph of half-peeled crystalline region, arrows showing unpeeled stacked 2D crystals (bottom) and peeled single-layer 2D crystal region (top). (**b**) Projection map of peeled-blotted LTC$_4$S 2D crystal (data truncated to 9 Å resolution). Four unit cells are shown. Imaging conditions: JEOL-1400 with LaB$_6$ filament at 120 kV, Gatan Ultrascan CCD, 0.43 nm/pixel. Processed with 2DX package [44]

When it comes to data collection of 2D crystals, there are two broad strategies. If 2D crystals a well ordered and large (at least micron-scale), it is possible to collect electron diffraction data by cryo-EM, as described in [32]. It is then necessary to obtain phases through indirect means, such as molecular replacement. On the other hand, it is always possible to image in real-space, with the added benefit that both phase and amplitude information is conserved. The later data collection approach has been common for the large majority of 2D crystals. Real-space cryo-EM imaging protocols will be instrument- and facility-specific, but it is vital to image using low-dose procedures, which can be easily utilized using modern high-throughput imaging programs, e.g., Leginon [33], SerialEM [34], or EPU (Thermo Fisher Co.). Under ideal conditions, imaging will be performed using direct electron detection to maximize achievable resolution.

**3.4 Image
Processing**

Initial stages of image processing proceed as with other cryo-EM methods. If a direct electron detector was used during data collection, movies should be aligned to correct for motion blurring using, e.g., MotionCorr [35], MotionCor2 [36], or Unblur [37]. CTF estimation is also necessary, for instance, with CTFFIND4 [38] or gCTF [39].

These initial stages of data processing, as well as later stages, can be easily accomplished using the user-friendly Focus software package [31]. This package, which can be operated in-line with data collection, provides access to the full pipeline of image processing necessary for 2D crystallography, containing the core elements first developed in the MRC 2D crystallography programs, including lattice estimation, tilt estimation, unbending, projection map generation, 2D merging, 3D merging, and final data evaluation [40]. The highest achievable resolution will depend on crystal quality as well as grid preparation, including sample and carbon film flatness [27, 28], as well as careful data collection and image processing [31]. The first important data to provide information on the membrane protein structure is a projection map, which provides information perpendicular to the plane of the phospholipid bilayer. At resolutions that are lower/worse than 10 Å, information on the oligomeric state of largely hydrophobic membrane proteins can be discerned. For these types of proteins at 6–9 Å resolution in projection, α-helices, provided that they are not too highly tilted, can be identified. Projection maps at higher resolutions tend to not allow clear interpretation of α-helices, which may be overcome by reducing the resolution in an additional map of the same data set to the intermediate resolution range of 6–9 Å to reveal projection density typical of α-helices [42]. For membrane proteins with large soluble domains, the projection map will not allow for interpretation of secondary structure. The 3D map will provide critical information perpendicular to the plane of the membrane

(Z direction) such as described in [2, 3, 6, 22, 23, 30], and, under special circumstances of ordered phospholipids [6], conclusions on the importance of the phospholipids to the protein structure and function may be drawn. The 3D maps of samples in different conformational states such as bacteriorhodopsin [24, 26] can reveal substantial information on the detailed reaction mechanism.

4 Notes

1. Lipid stock concentrations of 1 and 10 mg/mL are useful for many lipid-to-protein ratios used during 2D crystallization trials. DMPC has been successfully used for 2D crystallization of a large percentage of membrane proteins. Lipid in chloroform is preferable due to the hygroscopic nature of lipids in powder form.

2. The stream of nitrogen gas is slowly increased to avoid splashing and drying of the lipid on the sides of the round-bottom flask.

3. Lipid aliquots for 2D crystallization trials can frequently be stored for 1–2 years, if not longer.

4. Approximately 4–8 crystallization conditions often provide sufficient and critical information to optimize parameters for the next experiment. It is important to note that simultaneous testing of a larger number of crystallization conditions will require a substantial amount of protein as well as many hours, if not days, of TEM time, which is often not available in one large time block. Thus, it is more efficient to set up fewer crystallization conditions on a frequent basis.

5. An LPR of 0 provides critical information in the first crystallization trial on possible co-purified lipids. Even small amounts of lipids may result in membrane formation; therefore, if membrane formation is seen at an LPR of 0 (no additional lipids added), the protein sample requires further washes during purification.

6. Removal of the air in the tubing is not necessary and can result in sample loss. The air is, in fact, helpful when removing the sample after dialysis as the tubing is less likely to bend upon cutting.

7. The large spacing between the clips has been shown to result in maximal recovery of the dialysate.

8. The $90°$ rotation between the clips will ensure that the bottom clip remains submerged under the surface of the buffer, preventing sample loss and/or drying of the tubing.

9. The dialysis time will depend on the detergent and its concentration. A temperature of 24 °C, which is just above the phase transition temperature is often sufficient for experiments with DMPC.

10. Air bubbles indicate that a substantial amount of detergent remains due to an insufficient length of dialysis. The sample can either be placed in new dialysis tubing (repeat **steps 8–16**), or a new dialysate can be prepared to replicate conditions with a longer dialysis time.

11. Liposomes, proteoliposomes, and 2D crystals will usually adhere to carbon film without the need for glow-discharge.

12. Volumes of 1.5–5 μL of dialysate may be used. The sample may be blotted from the top, with the surface of the filter paper approaching the grid in a parallel manner, especially if the sample volume is either large and/or concentrated.

13. Several laboratories prefer the use of uranyl formate [41] due to the close-to-physiological pH and the smaller grain size of the stain. We prefer uranyl acetate due to the longevity of the grids after preparation.

14. With longer dialysis timeframes, it is most efficient to screen the grid as quickly as possible in order to obtain information for the next dialysis experiment.

15. While 120 kV is standard for screening negatively stained samples in many cryo-EM laboratories, an accelerating voltage anywhere in the range of 80–120 kV will work well for screening.

16. Proteoliposomes can be concentrated at the bottom of an Eppendorf tube by centrifugation at ~1008 × g for 3–5 min.

17. Ordered arrays of membrane proteins with large soluble domains may be identified visually, but FFT assessment will provide important information such as on crystal stacking, amount of stacking, mosaicity, first information on plane group symmetry, and image quality).

18. For assessment of samples before crystallization, it is valuable to evaluate images with FFTs of both smaller and larger areas for the best possible signal-to-noise ratio.

19. Ordered arrays of even a few unit cells will provide critical information for optimization of crystallization conditions. Thus, a large number of membranes need to be carefully evaluated.

20. The drops are placed closer to the center of the Parafilm to avoid them rolling onto the bench when they are manipulated in the subsequent protocol steps.

21. The smooth ("shiny") side of the grid is facing up and the straight leg of the anticapillary forceps is facing down.

22. Only continuous, unbroken carbon film is used for the next step.

23. The smooth side of the grid is still facing up for this step.

24. Success using this grid preparation method depends greatly on using 600-mesh (or finer) EM grids. Standard 400-mesh grids do not provide enough support for the carbon film during capillary pressure adhesion, leading to an unacceptable amount of carbon film breakage. Grids should be pre-cleaned by rinsing in chloroform and dried completely. As this method causes significant stress to the carbon film, it is also imperative that the film be solid "spark-free" carbon film, without any defects that could eventually result in breakage. It can be advantageous to inspect the carbon film carefully as it is initially floated onto buffer, checking the reflection of a light off the carbon/buffer surface to check for cracks, embedded particles, etc.

25. To utilize this method for preparation of negatively stained samples, buffer should contain negative stain, e.g., 1–2% uranyl acetate or uranyl formate.

26. It is important to keep the grid parallel to the membrane surface during placement, and for the EM grid to make full contact with the membrane surface, ensuring a good seal for **step 4** below.

27. Millipore DAWP membrane filters absorb aqueous buffer at a much slower rate than Whatman paper, meaning that somewhat long blotting times are needed. From the time of initial contact with the membrane, it can take 1–5 s for the carbon film to become completely flat on the surface of the grid. In addition, for the capillary pressure to sufficiently bring the EM grid surface and the carbon film into close contact takes a further 2–3 s. The best results were observed by blotting until 3 s after the carbon film had completely flattened.

28. The grid should be quickly moved onto the droplet after lifting from the membrane to ensure that the surface does not dry out. In some cases, the carbon film will adhere tightly to the grid, and it can be helpful to use a second pair of forceps to gently press the grid down onto parafilm, forcing buffer through the grid mesh.

29. This may be helpful when the stacked crystals are many layers thick, with the increased risk of damage that accompanies increased manipulation.

30. For negatively stained samples, air dry only (omit plunging).

Acknowledgments

Part of this work was supported by NIH grant HL090630 (ISK) and an SREB Fellowship (KN).

References

1. Henderson R, Unwin PN (1975) Three-dimensional model of purple membrane obtained by electron microscopy. Nature 257 (5521):28–32

2. Henderson R, Baldwin JM, Ceska TA, Zemlin F, Beckmann E, Downing KH (1990) Model for the structure of bacteriorhodopsin based on high-resolution electron cryomicroscopy. J Mol Biol 213(4):899–929

3. Wang DN, Kühlbrandt W, Fujiyoshi Y (1994) Atomic model of plant light-harvesting complex by electron crystallography. Nature 367 (6464):614–621

4. Nogales E, Wolf SG, Downing KH (1998) Structure of the alpha beta tubulin dimer by electron crystallography. Nature 391:199–203

5. Löwe J, Li H, Downing KH, Nogales E (2001) Refined structure of alpha beta-tubulin at 3.5 A resolution. J Mol Biol 313:1045–1057

6. Gonen T, Cheng Y, Sliz P, Hiroaki Y, Fujiyoshi Y, Harrison SC, Walz T (2005) Lipid-protein interactions in double-layered two-dimensional AQP0 crystals. Nature 438 (7068):633–638

7. Kühlbrandt W (2014) The resolution revolution. Science 343(6178):1443–1444

8. Efremov RG, Gatsogiannis C, Raunser S (2017) Lipid nanodiscs as a tool for high-resolution structure determination of membrane proteins by single-particle cryo-EM. Methods Enzymol 594:1–30

9. Sun C, Gennis RB (2019) Single-particle cryo-EM studies of transmembrane proteins in SMA copolymer nanodiscs. Chem Phys Lipids 221:114–119

10. Nannenga BL, Gonen T (2018) MicroED: a versatile cryoEM method for structure determination. Emerg Top Life Sci 2(1):1–8

11. Martynowycz MW, Zhao W, Hattne J, Jensen GJ, Gonen T (2019) Collection of continuous rotation microed data from ion beam-milled crystals of any size. Structure 27(3):545–548

12. Jap BK, Zulauf M, Scheybani T, Hefti A, Baumeister W, Aebi U, Engel A (1992) 2D crystallization: from art to science. Ultramicroscopy 46:45–84

13. Kühlbrandt W (1992) Two-dimensional crystallization of membrane proteins. Q Rev Biophys 25:1–49

14. Stahlberg H, Fotiadis D, Scheuring S, Rémigy H, Braun T, Mitsuoka K, Fujiyoshi Y, Engel A (2001) Two-dimensional crystals: a powerful approach to assess structure, function and dynamics of membrane proteins. FEBS Lett 504:166–172

15. Mosser G (2001) Two-dimensional crystallogenesis of transmembrane proteins. Micron 32:517–540

16. Kühlbrandt W (2003) In: Schägger H, Hunte C (eds) Membrane protein purification and crystallization: a practical approach, 2nd edn. Academic Press, San Diego, pp 253–284

17. Schmidt-Krey I (2007) Electron crystallography of membrane proteins: two-dimensional crystallization and screening by electron microscopy. Methods 41(4):417–426

18. Signorell GA, Kaufmann TC, Kukulski W, Engel A, Rémigy HW (2007) Controlled 2D crystallization of membrane proteins using methyl-betacyclodextrin. J Struct Biol 157 (2):321–328

19. Vink M, Derr KD, Love J, Stokes DL, Ubarretxena-Belandia I (2007) A high throughput strategy to screen 2D crystallization trials of membrane proteins. J Struct Biol 160(3):295–304

20. Johnson MC, Schmidt-Krey I (2013) Two-dimensional crystallization by dialysis for structural studies of membrane proteins by the cryo-EM method electron crystallography. Methods Cell Biol 113:325–337

21. Uddin YM, Schmidt-Krey I (2015) Inducing two-dimensional crystallization of membrane proteins by dialysis for electron crystallography. Methods Enzymol 557:351–362

22. Kühlbrandt W, Downing KH (1989) Two-dimensional structure of plant light harvesting complex at 3.7 °A resolution by electron crystallography. J Mol Biol 207:823–826

23. Wang DN, Kühlbrandt W (1991) High-resolution electron crystallography of light-harvesting chlorophyll a/b-protein complex in three different media. J Mol Biol 217 (4):691–699

24. Subramaniam S, Faruqi AR, Oesterhelt D, Henderson R (1997) Electron diffraction studies of light-induced conformational changes in the Leu-93 --> Ala bacteriorhodopsin mutant. Proc Natl Acad Sci U S A 94(5):1767–1772

25. Fujiyoshi Y (1998) The structural study of membrane proteins by electron crystallography. Adv Biophys 35:25–80

26. Subramaniam S, Henderson R (1999) Electron crystallography of bacteriorhodopsin with millisecond time resolution. J Struct Biol 128(1):19–25

27. Koning RI, Oostergetel GT, Brisson A (2003) Preparation of flat carbon support films. Ultramicroscopy 94:183–191

28. Gyobu N, Tani K, Hiroaki Y, Kamegawa A, Mitsuoka K, Fujiyoshi Y (2004) Improved specimen preparation for cryo-electron microscopy using a symmetric carbon sandwich technique. J Struct Biol 146:325–333

29. Amos LA, Henderson R, Unwin PN (1982) Three-dimensional structure determination by electron microscopy of two-dimensional crystals. Prog Biophys Mol Biol 39(3):183–231

30. Grigorieff N, Ceska TA, Downing KH, Baldwin JM, Henderson R (1996) Electron-crystallographic refinement of the structure of bacteriorhodopsin. J Mol Biol 259(3):393–421

31. Biyani N, Righetto RD, McLeod R, Caujolle-Bert D, Castano-Diez D, Goldie KN, Stahlberg H (2017) Focus: The interface between data collection and data processing in cryo-EM. J Struct Biol 198(2):124–133

32. Gonen T (2013) The collection of high-resolution electron diffraction data. In: Electron crystallography of soluble and membrane proteins. Humana Press, Totowa, NJ, pp 153–169

33. Suloway C, Pulokas J, Fellmann D, Cheng A, Guerra F, Quispe J, Stagg S, Potter CS, Carragher B (2005) Automated molecular microscopy: the new Leginon system. J Struct Biol 151(1):41–60

34. Mastronarde DN (2005) Automated electron microscope tomography using robust prediction of specimen movements. J Struct Biol 152(1):36–51

35. Li X, Mooney P, Zheng S, Booth CR, Braunfeld MB, Gubbens S, Agard DA, Cheng Y (2013) Electron counting and beam-induced motion correction enable near-atomic-resolution single-particle cryo-EM. Nat Methods 10(6):584–590

36. Zheng SQ, Palovcak E, Armache JP, Verba KA, Cheng Y, Agard DA (2017) MotionCor2: anisotropic correction of beam-induced motion for improved cryo-electron microscopy. Nat Methods 14(4):331–332

37. Grant T, Grigorieff N (2015) Measuring the optimal exposure for single particle cryo-EM using a 2.6 Å reconstruction of rotavirus VP6. elife 4:e06980

38. Rohou A, Grigorieff N (2015) CTFFIND4: Fast and accurate defocus estimation from electron micrographs. J Struct Biol 192(2):216–221

39. Zhang K (2016) Gctf: Real-time CTF determination and correction. J Struct Biol 193(1):1–2

40. Crowther RA, Henderson R, Smith JM (1996) MRC image processing programs. J Struct Biol 116(1):9–16

41. Walz T, Häner M, Wu XR, Henn C, Engel A, Sun TT, Aebi U (1995) Towards the molecular architecture of the asymmetric unit membrane of the mammalian urinary bladder epithelium: a closed "twisted ribbon" structure. J Mol Biol 248(5):887–900

42. Schmidt-Krey I, Kanaoka Y, Mills DJ, Irikura D, Haase W, Lam BK, Austen KF, Kühlbrandt W (2004) Human leukotriene C4 synthase at 4.5 Å resolution in projection. Structure 12(11):2009–2014

43. Zhao G, Johnson MC, Schnell JR, Kanaoka Y, Haase W, Irikura D, Lam BK, Schmidt-Krey I (2010) Two-dimensional crystallization conditions of human leukotriene C4 synthase requiring adjustment of a particularly large combination of specific parameters. J Struct Biol 169(3):450–454

44. Gipson B, Zeng X, Zhang ZY, Stahlberg H (2007) 2dx—user-friendly image processing for 2D crystals. J Struct Biol 157(1):64–72

Chapter 11

Sample Preparation and Data Collection for Electron Crystallographic Studies on Membrane Protein Structures and Lipid–Protein Interaction

Ka-Yi Chan, Chloe Du Truong, Yu-Ping Poh, and Po-Lin Chiu

Abstract

Electron crystallography is a unique tool to study membrane protein structures and lipid–protein interactions in their native-like environments. Two-dimensional (2D) protein crystallization enables the lipids immobilized by the proteins, and the generated high-resolution density map allows us to model the atomic coordinates of the surrounding lipids to study lipid–protein interaction. This protocol describes the sample preparation for electron crystallographic studies, including back-injection method and carbon sandwich method. The protocols of data collection for electron crystallography, including electron imaging and diffraction, of the 2D membrane crystal will be followed.

Key words Electron crystallography, 2D Crystals, Sugar embedding, Back-injection method, Carbon-sandwich method, Electron diffraction

1 Introduction

Membrane proteins are the proteins residing in biological membranes. In most organisms, about 20–30% of the genes encode membrane proteins [1, 2], which play important roles as a gatekeeper, a signal transducer, or a diode in the membranes for living processes of a cell [3]. They serve as the main targets for drug design in promotion or inhibition of a signaling transduction pathway or modulation of the ion channel conductance [4]. Thus, it is essential to understand the structures of these proteins and how they coordinate with their surrounding proteins or lipids to synergistically perform their function. Two-dimensional (2D) protein crystallization combined with electron cryogenic microscopy (cryo-EM), particularly electron crystallography, is one of the powerful techniques that are used to determine the protein structures in the membrane, which is in their native-like environment. A 2D protein array of the membrane protein is crystallized with lipids

Tamir Gonen and Brent L. Nannenga (eds.), *CryoEM: Methods and Protocols*, Methods in Molecular Biology, vol. 2215,
https://doi.org/10.1007/978-1-0716-0966-8_11, © Springer Science+Business Media, LLC, part of Springer Nature 2021

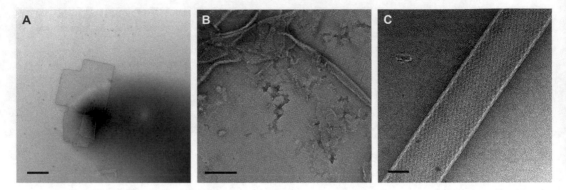

Fig. 1 Membrane crystals in different morphologies in the negative stain. (**a**) Two-dimensional (2D) crystal of aquaporin-0 in a planar sheet. Scale bar indicates 1 μm. (**b**) Vesicular crystal of the MloK1 potassium channel. The crystal was flattened on the carbon film support. Scale bar indicates 100 nm. (**c**) Tubular crystal of prokaryotic glucose transporter, IIC. Scale bar indicates 50 nm

in vivo or in vitro. Some membrane proteins form 2D arrays in their native membranes, such as S-layer protein [5], bacteriorhodopsin (bR) [6, 7], aquaporin-0 (AQP0) [8], porin [9], and photosynthetic apparatus [10]. Electron crystallography can be used to determine the membrane crystal structure and provides a unique way to probe lipid–protein interaction. The technique has been successfully used in the studies of bacteriorhodopsin [6], aquaporins [11–18], S-layer [19], lactose permease [20], prokaryotic potassium channel [21–23], galactose transporter [24], photosystem II [25], glutathione transferase [26], and many other cases. Because cryo-EM can record the structural data of these membrane proteins in the context of a lipid bilayer, which is similar to their native conditions, the electron crystallographic studies of these 2D membrane arrays can thus provide fruitful information of their structures and how they interact with the surrounding lipids.

Membrane crystals may have different forms, such as a 2D planar sheet [12–14, 27], a vesicular crystal [21, 25], or a helical tube [28] (Fig. 1). In some cases, there may be single or multiple forms appear in the same crystallization condition, depending on the nature of the protein or the crystallization buffer conditions. Among all the crystal types, a well-ordered and thin 2D crystal in a planar sheet is ideal for 2D electron crystallography to determine the 3D structure. However, if the sample of the vesicular or helical crystal with a large tube diameter can lie flat on the carbon film support with a size of about a few 100 nm, it is also possible to obtain enough number of unit cells to generate a projection map in good quality [21, 29]. One usually uses negative-stain EM to screen good crystal samples by analyzing the diffraction spots (or Fourier spots) in their image power spectrum [30]. The image power spectrum of a well-ordered crystal will show a crystal lattice with sharp diffraction spots in the second orders or beyond. An automated procedure has been proposed to screen the 2D protein crystals [31]. Once the crystal sample is screened, it will be frozen at liquid-nitrogen temperatures to keep it hydrated and minimize the beam damage for cryo-EM data collection.

Sample preparation for the electron crystallographic study is more critical than that of single-particle cryo-EM. Crystal sample for electron crystallographic study lies on the carbon film support, and an intact and flat carbon film support can help preserving the lattice order of the 2D arrays. If the carbon support is not flat or the crystal does not lie flat, the diffraction spots that are perpendicular to the tilt axis will become blurred when the stage is tilted [32]. Therefore, specimen flatness is usually the key to the success of 2D electron crystallography. Secondly, to prevent the 2D crystal sample from dehydration, sugar embedding is a general method to protect the sample in the high vacuum of the electron microscope (EM) column [33]. The commonly used sugars are trehalose, glucose, tannic acid, and sucrose, which mimic the solvent effects and act as a cryo-protectant [34, 35]. Trehalose has been reported to best preserve the high-resolution information of 2D crystals [34, 36], while the choice of sugar to preserve on a specific type of protein will need to determine empirically. The embedded specimen can be readily plunged into liquid nitrogen without fast cooling. This method was first introduced for electron crystallographic studies on the purple membrane and catalase crystals with glucose [7]. With low-dose imaging, the structure of bR was revealed at high resolution [6, 37].

Because of the small conductivity of biological specimens, when the electron beam illuminates the specimen, the charges are temporarily separated and built up around the specimen, leading to an effect similar to the image drift and resulting in poor image quality. This is often referred to as beam-induced movement or charging [35, 38]. This phenomenon is more noticeable when imaging the tilted specimen under low-dose conditions [39]. To minimize the beam-induced charging, the grid specimens could be prepared with a thicker carbon film, with a 20–30 nm thickness [35]. However, the thick specimen can lead to less elastic scattering and lower the signal contents. Alternatively, spot-scan imaging was designed to image the crystal sample with a small beam size (about 100 nm in diameter), which reduces the beam-induced charging. However, it compromises with a lower spatial coherence of the beam [40, 41]. On the other hand, the development of carbon-sandwich method is an improvement on the sample preparation to minimize beam-induced charging. It was developed to introduce an additional piece of carbon film on the EM grid and help alleviate the charging problem [39]. The idea of carbon-sandwich method is the built-up charges that are symmetrically distributed on both sides of the specimen will supposedly be canceled out when the electron beam illuminates the tilted specimen [39]. Apart from this, the two carbon films sandwich the crystal specimen in the middle, and this provides additional protection and keep the crystals hydrated while imaging in the high vacuum. However, the downside of the carbon-sandwich method is that the specimen thickness is not easy to control. The specimen thickness may be affected by the variation of humidity, sugar concentration, and the thickness of carbon film used.

Once a sugar-embedded 2D crystal is frozen, it will be transferred onto the TEM specimen stage for data collection. The process of the cryo-specimen transfer can be accomplished using a side-entry cryo-specimen holder (Fig. 2) or a modern automated specimen loader system. Because the biological specimen is sensitive to beam radiation damage, a low-dose imaging protocol is applied. Note that the 2D crystal samples are generally more sensitive to the beam damage than single-particle samples [42]. Typically

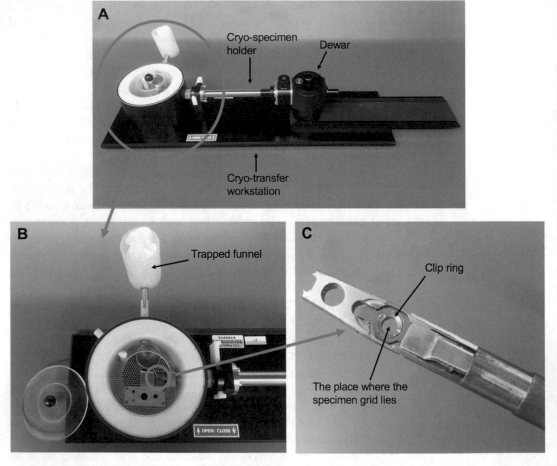

Fig. 2 Side-entry cryo-specimen transfer holder (Oxford CT-3500 cryo-specimen transfer holder) with a transfer station. (**a**) Overall view of the cryo-specimen holder on the transfer station. (**b**) Cryo-specimen transfer station. Liquid nitrogen is added through the trapped funnel of the transfer workstation. (**c**) Loading tip of the cryo-specimen holder. The grid specimen is secured by the clip ring

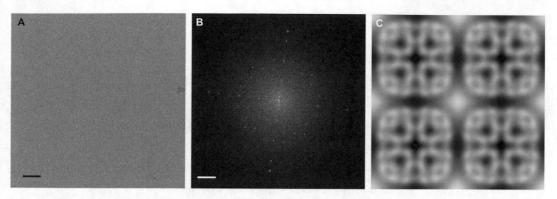

Fig. 3 Electron imaging of a two-dimensional aquaporin-0 crystal. (**a**) Electron image of a two-dimensional (2D) crystal of aquaporin-0 (AQP0). The image was recorded using an FEI Polara TEM with a Gatan Ultrascan CCD camera. Scale bar indicates 50 nm. (**b**) Image power spectrum of the AQP0 2D crystal. Scale bar indicates $1/30$ Å^{-1}. (**c**) 2D crystal projection in 2×2 unit cells at 4 Å resolution. The square length of the shown unit cells is 131 Å

for electron crystallographic data collection, the electron dose is limited to about 10 $e^-/\text{Å}^2$ or less [11, 42] in comparison to 30 $e^-/\text{Å}^2$ for typical single-particle cryo-EM imaging. On the other hand, the choice of using electron diffraction or imaging depends on the crystal quality and the target resolution of the reconstruction. Electron diffraction is usually the option when the 2D crystal is well ordered and large (in the μm size range). With diffraction, the intensities, and therefore the amplitudes, of the reflections are more accurately measured because they are not modulated by the contrast transfer function (CTF) of the TEM. However, with diffraction the initial phases for the map reconstruction will need to be determined separately. On the other hand, imaging on a 2D crystal provides both the amplitudes and phases of the structural factor from the image Fourier transform, allowing us to directly reconstruct a 2D crystal projection [35] (Fig. 3). Meanwhile, the CTF modulates the phase contrast transfer, particularly in the high-spatial frequency region [35]. To minimize this effect, less defocus (about 500–750 nm) is usually used for imaging 2D crystals than what is generally used for single-particle imaging (about 1–3 μm). More details about data processing on the images or diffraction patterns of the 2D crystals were described in the previous studies [43–45].

In this chapter, we focus on the procedures of grid specimen preparation of 2D crystals and electron crystallographic data collection. The goal is to produce a flat and thin grid specimen and generate high-resolution structural data for studying membrane protein structure and lipid–protein interaction. The following procedures provide a guide for performing experiments and can be modified for the user's particular needs and their unique sample.

2 Materials

2.1 Grid Specimen Preparation

1. Crystal sample suspension.
2. Pipettes with tips (1000, 200, 20, and 2 μL).
3. High-grade mica sheets.
4. Graphite rods for carbon evaporation.
5. Carbon evaporator with an oil-free high vacuum.
6. Embedding buffer with sugar only (*see* **Note 1**).
7. Sample mother buffer for dilution.
8. Small reservoirs for holding the buffers (*see* **Note 2**).
9. Molybdenum EM grids (300 meshes) (*see* **Note 3**).
10. Whatman #2 filter paper.
11. A 4–5 mm platinum wire loop.
12. Grid boxes for storing the frozen grids.
13. Anti-capillary self-closing tweezers.
14. Liquid nitrogen.

2.2 Frozen Grid Specimen Transfer

1. A side-entry cryo-specimen holder for the TEM with a side-entry specimen stage (*see* **Note 4**).
2. Liquid nitrogen.
3. Transferring tools (a clip-ring holder and a tweezer).

3 Methods

3.1 Sample Preparation

Before preparing a frozen specimen, screening good crystal samples and optimizing their concentrations using negative-stain EM are strongly recommended. The diffraction spots on the image power spectrum are a good indication of the crystal packing. A detailed protocol for preparing negatively stained samples can be found in the previous study [30]. Once the 2D crystal sample is screened and ready for electron imaging or diffraction, one will prepare the grid specimen with a sugar-embedded 2D crystal on a flat carbon support. The specimen can be readily frozen into liquid nitrogen. The first part of the protocol will describe the procedure for preparing the carbon films. The support carbon film is required to be flat and intact in an appropriate thickness. The second part of the protocol will cover the procedures of back-injection method and carbon-sandwich method, which embeds the crystal sample with the chosen sugar on a flat carbon support.

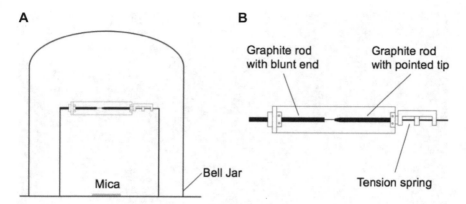

Fig. 4 Schematic of a carbon evaporator. (**a**) Carbon evaporation is performed in a high vacuum within a bell jar. Carbon will be emitted from the graphite rods and deposited on the surface of a piece of mica. (**b**) Enlarged view of the graphite rod assembly. One graphite rod is sharpened as a pencil fixed on the tension spring, and the other with a blunt end

3.1.1 Preparation of the Thin Carbon Films in Spark-Free Conditions

The goal of this procedure is to prepare a high-quality carbon film as flat and intact support for 2D crystals.

1. File one graphite rod to a pencil point and file the other graphite rod with a flat and blunt end (Fig. 4).

2. Carefully mount the two graphite rods into the holders in a carbon evaporator. The graphite rod with the pointed end should be mounted in the spring-loaded holder.

3. Secure the vacuum bell jar on the carbon evaporator and start the vacuum pump.

4. Wait until the vacuum has stabilized at a pressure of less than 1×10^{-4} Pa. The waiting time may vary from instruments to instruments.

5. Slowly turn on the filament and increase the current until the carbon starts to evaporate (*see* **Note 5**).

6. Adjust stepwise to a current that steadily evaporates carbon. Carefully watch for sparks (*see* **Note 6**).

7. Once the sparking has subsided, turn off the filament and vent the bell jar.

8. Place a piece of freshly cleaved mica sheet inside the apparatus. Fold a piece of filter paper and place it to the side of the mica (Fig. 4).

9. Replace the bell jar and restart the vacuum.

10. When the vacuum has stabilized at a pressure of less than 1×10^{-4} Pa, turn on the filament.

11. Very slowly raise the current until the carbon begins to evaporate.

Fig. 5 Examples of the carbon deposition on the mica sheet support. Here show the two Petri dishes with a long and rectangular piece of mica sheet with deposited carbon film. An appropriate thickness of the carbon film is required for a carbon-sandwiched sample. The thickness of the carbon film can be correlated to the contrast of the deposited carbons on the top of the folded filter paper (brown) to where is shaded without deposited carbon (white). Left panel shows a proper thickness of carbon film, and the right shows the carbon film with a larger thickness, which does not apply to use in carbon-sandwiched method

12. Carefully watch for sparks during evaporation. Watch for changes in pressure in parallel. The pressure should not increase abruptly.

13. Watch the contrast by looking at the shadow that is formed on the folded filter paper. Estimate the thickness of the carbon deposition. The carbon film used for support cannot be too thin or too thick, which needs to be determined empirically. Figure 5 shows the examples of the carbon shadows that were formed by the carbon deposition on the filter paper.

14. During the evaporation, the carbon should be steadily evaporated from the graphite rod and the pressure may slightly increase. If the pressure rises at a fast rate, slowly turn off the filament and wait for the vacuum to recover.

15. Once the carbon evaporation has been completed, store the carbon films in a Petri dish in a desiccator until use.

3.1.2 Back-Injection Method

The back-injection method with glucose embedding was successfully applied to the 2D crystal sample of the purple membrane [46].

1. Optimize the crystal concentration using negative-stain EM.

2. If a recycled molybdenum grid is used, clean the grid with plasma cleaning before usage to remove any residue from the grid (*see* **Note 8**).

3. Add the sugar-embedding (usually glucose) buffer into the reservoir and fill it to the top.

4. Cut a section of carbon-coated mica sheet from Subheading 3.1.1 into a size similar or slightly smaller than the EM grid (about $3 \times 3 \ mm^2$). Float the carbon film on the surface of the glucose solution. The floated carbon film should be flat and not split. If the carbon film is split, change the mica and float another piece of the carbon film.

5. Use an anti-capillary self-closing tweezer to pick up a molybdenum grid. Insert the grid deep into the sugar-embedding buffer and then slowly pick up the floated carbon film from the bottom of the carbon film (Fig. 6).

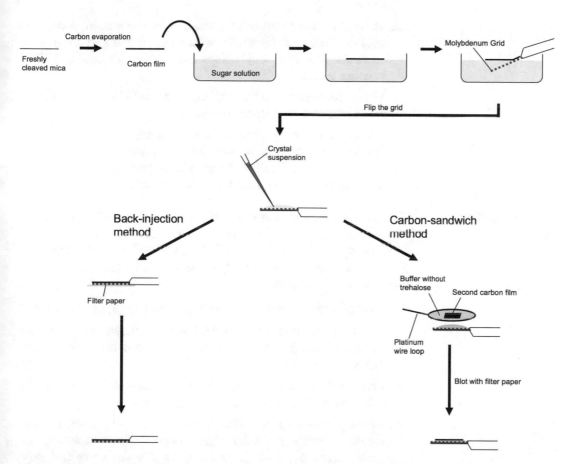

Fig. 6 Schematics of back-injection and carbon-sandwich method. The floated carbon film is picked up by a molybdenum grid from the bottom. The crystal suspension is mixed with the embedding sugar on the opposite of the grid. For back-injection method, a filter paper is used to blot away all the remaining solution. For carbon-sandwich method, a second carbon film is introduced to apply on top of the grid. An additional filter paper is used to blot away all the remaining solution. After the grid specimen is dried, the grid specimen can be directly cooled down in liquid nitrogen

6. Flip the grid upside down and set the tweezer and the grid on the working bench (Fig. 5). A small volume (less than about 1 μL) of the embedding buffer should be maintained on the top of the grid.

7. Apply 2 μL of the crystal suspension on the grid and gently mix the crystal suspension with the remaining buffer on the grid by pipetting (*see* **Note 7**).

8. Wait about 10 s and then remove the solution from the grid using a pipette.

9. Use the filter paper to remove the remaining solution from the grid. With glucose embedding, the grid specimen does not need to freeze into the liquid nitrogen before transferring to the EM stage.

10. If a side-entry cryo-specimen transfer holder is used, secure the grid on the tip of the holder with a clip ring at room temperatures and insert the holder into the TEM. Wait until the TEM vacuum is stable.

11. When the holder is in the TEM column, add liquid nitrogen to the dewar and cool down the holder.

12. Closely monitor the temperature of the cryo-specimen transfer holder. When the tip reaches the liquid nitrogen temperature, wait about 10–15 min until the liquid nitrogen stops boiling. Then start the procedure for data collection.

3.1.3 Carbon-Sandwich Method

The carbon-sandwich method has the advantages of the reduction of the beam-induced charging and the better protection of the crystal sample with two layers of carbon films [39]. The following protocol is for the carbon-sandwich method with trehalose embedding for sample preparation.

1. Optimize the crystal concentration using negative-stain EM.

2. If a recycled molybdenum grid is used, clean the grid with plasma cleaning before use to remove the residuals on the grid (*see* **Note 8**).

3. Prepare two reservoirs, one with trehalose solution and the other with the mother buffer (*see* **Note 9**).

4. Cut a piece of carbon film from the mica sheet with a size similar or slightly smaller than the size of the molybdenum grid. Float a piece of carbon film on the surface of trehalose solution. The floated carbon film should be flat and intact.

5. Float another piece of carbon film on the surface of the sample mother buffer. The size of the carbon film should be smaller than that of the first piece (*see* **Note 10**).

6. Use an anti-capillary self-closing tweezer to pick up an EM grid. Pick up the carbon film that is on the surface of the

trehalose solution by inserting the grid deep in the solution and slowly moving up out to the buffer surface.

7. Flip over the grid and set the tweezer and the grid on the working bench. Mix with 2 μL of the crystal suspension by gentle pipetting (*see* **Note 7**) (Fig. 6). Be careful not to break the carbon film beneath the grid.

8. Let the mixture to settle for 15 s.

9. Use the pipette to remove most of the solution on the grid.

10. Use a platinum wire loop to pick up the second carbon film from the sample buffer reservoir.

11. Gently place the second carbon film on the top of the grid. While the wire loop touches the grid, use a filter paper to touch the loop edge and blot away the excess solution.

12. Once the second carbon film is settled, use another piece of filter paper to remove the remaining solution.

13. Wait for about 1 or 2 min. Directly freeze the grid specimen into the liquid nitrogen and transfer it to the grid box for storage.

14. Store the specimen in the liquid nitrogen dewar for later data collection.

15. If a cryo-specimen transfer holder is used, pre-cool the cryo-transfer holder and wait until the holder is stable at liquid nitrogen temperatures. Transfer the frozen grid to the tip of the holder and secure the grid with a clip ring. Insert the holder to the TEM and wait until the temperature and EM vacuum are stable for data collection.

3.2 Data Collection

For crystallographic data collection, it is important to obtain a parallel electron beam to generate the Bragg reflections on the diffraction pattern or image Fourier transform from the 2D crystalline specimen [35, 47]. The electron beam is required to be well aligned to obtain high-resolution information. Different from the diffraction obtained from a three-dimensional (3D) crystal, the reflections of the 2D crystal pattern are the samples that are intercepted by a plane at a specific Euler angle with the 2D lattice lines [35]. To reconstruct the 2D crystal projections into a 3D reconstruction, we collect the 2D crystal images or diffraction patterns at different tilting angles (Fig. 6). These data will be later extracted and merged into a set of lattice lines for 3D structure determination.

The following procedures are written for a FEI Tecnai Polara F30 TEM. The procedures for beam alignment can be mostly found in the Direct Alignment panel in the vendor software. Details can be modified for different TEMs with a field-emission gun (FEG) source.

3.2.1 Electron Microscope Checkup

1. Several TEM routines are needed to be carefully checked and tuned before the experiments.

   (a) Check the liquid nitrogen levels in the cold-trap. If the liquid nitrogen levels are low, fill up the liquid nitrogen and wait until the stage is stable.

   (b) Check the TEM vacuum status if the pressures are in a normal range.

   (c) Make sure that the stage is pre-cooled to liquid nitrogen temperatures.

   (d) Open the column valve and see if a stable beam can be found in the low magnification.

   (e) Check if the digital camera is cooled down.

   (f) Prepare gain and dark references for the digital recording camera.

2. If cryo-specimen transfer holder will be used for grid transfer, several routines are needed to be prepared:

   (a) Before using the cryo-specimen transfer holder, check if there is any residual grid or contamination on the tip. Remove them before cooling the holder.

   (b) Clean up and check if there is any residual grid or contamination in the transfer station before cooling down.

   (c) Cool down the holder and the transfer station with liquid nitrogen to $-180\ ^{\circ}C$ and wait until it is stable.

   (d) Use fresh liquid nitrogen to cool down the holder and transfer the grid specimen.

3.2.2 Beam Alignment

1. Open the column valve. Find the beam on the screen at low magnification. Adjust the beam intensity and centered the beam by varying beam intensity/brightness (C2 lens current) and beam shift. Keep the objective aperture and selected area diffraction (SAD) aperture out of the beam path at this stage.

2. If the specimen grid is inserted, find an area without any carbon film or specimen.

3. Align the beam conditions for electron diffraction or imaging. Turn on the Low Dose system and switch to the Exposure mode. Keep the mode changing in a sequence from the Search, Focus, and Exposure. First set the magnification to about $5000\times$.

4. Align the gun tilt to maximize the brightness and align the gun shift according to different spot sizes (C1 lens).

5. Choose the target spot size for electron imaging or diffraction (*see* **Note 11**).

6. Insert the proper size of the C2 aperture. For electron diffraction on an FEI Tecnai TEM or a two-condenser lens system, a small C2 aperture (50 μm) is recommended for a better beam coherence.

7. Adjusting the C2 lens currents using the Intensity/Brightness knob. Fine-tune the position of the C2 aperture until the beam shrinks or expands its size to the screen center while adjusting the Intensity/Brightness knob.

8. Align the beam shift/tilt of the condenser lens system.

9. Align the rotation center until the beam shrinks or expands its size to the screen center.

10. Set a higher magnification to about 50,000×. Go to **step 8** and repeat the alignment procedure a few cycles until the alignment is stable.

11. Insert the objective aperture if one plans to perform electron imaging on the 2D crystal samples. Switch to diffraction mode and align the position of the objective aperture.

12. If the specimen is present, find an area with a flat carbon film and adjust the eucentric height. Normalize the objective lenses using the Eucentric Focus knob. Perform the coma-free alignment.

3.2.3 Electron Diffraction

Electron diffraction is generally used when a well-ordered and large 2D crystal (in μm size range) is available. The intensity data can be accurately recorded without CTF modulation.

1. Align the beam condition to obtain a parallel illumination. In a two-condenser lens system, a small condenser aperture (50 μm) and the smallest spot size (spot size **step 11**) are recommended to use.

2. In the Low Dose system, go to the Exposure mode. Find the right focus and centered the beam. Make sure that the objective aperture is out of the path.

3. Set the magnification to about 50,000× and spread out the beam to obtain a target dose rate of about 0.3 $e^-/Å^2/s$.

4. Switch to the diffraction mode. Set an appropriate camera length for diffraction experiments (*see* **Note 12**).

5. Align the beam to the screen center using a diffraction lens shift. Adjust the beam shape to a circle with the diffraction lens astigmatism.

6. Focus the beam using the Diffraction/Focus knob.

7. Go to Search mode and set the magnification to the smallest in the SA mode. Set the Search mode in the diffraction mode. Use the same camera length as that in the Exposure mode. Overfocus to spread the beam until the crystal can be recognized.

8. Cycle the Search and Exposure modes in sequence for a few times until the beam is stable and not moving anymore.

9. Go to the Search mode and remove the beam stop if it is in the beam path. Search for the crystal target and wobble the stage to find the eucentric *z*-height.

10. Find an adjacent area without a crystal and then go to the Exposure mode. Normalize the lenses using the Eucentric Focus button. Focus the beam and move it to the screen center.

11. Cycle between the Search and Exposure a few times until the beam is stable.

12. In the Search mode, locate the crystal target. Tilt the stage to the target tilting angle.

13. Insert the SAD aperture to select the crystal area. Insert the beam stop. Make sure that the origin of the diffracted beam is under the beam stop.

14. Turn on the beam blanker before switching to the Exposure mode.

15. Switch to the Exposure mode. Raise the screen and record the diffraction pattern on the digital camera for 30 s.

16. Tilt the stage back to 0° and repeat the steps for recording the diffraction pattern of another crystal. If moving to another grid square, repeat from **step 10**.

3.2.4   Electron Imaging

1. Insert the C2 aperture. Turn on the Low Dose system and align the three modes, which are Search, Focus, and Exposure mode. The spot sizes used for three modes are recommended to be the same.

2. In the Exposure mode, set the spot size and beam intensity for imaging. If one uses the counting or super-resolution mode in direct electron detector (DED) camera, the intensity will need to be adjusted to a counting rate that minimizes the coincidence loss.

3. Insert the objective aperture with a larger size, for example, 70 μm in diameter. Align the position of the objective aperture in the diffraction mode.

4. Set the magnification to have a resulting pixel size for obtaining an appropriate target resolution.

5. Align the images and stage position of the target area by adjusting the image shifts in Search mode and stage shifts in Exposure mode.

6. In Focus mode, set the Focus distance and Focus angle along the tilt axis to be away from the target imaging area in about 1.5 μm. Set a higher magnification than that of the Exposure

mode. The texture of the carbon film should be visualized in the image.

7. Cycle the Search, Focus, and Exposure modes in sequence for a few rounds until the beam is stable.

8. In Search mode, locate an area with intact carbon and an appropriate density of crystals. Find a target crystal for imaging.

9. Use the wobbler tool to set the eucentric z-height of the stage. Tilt the stage to the desired tilting angle.

10. Switch to Focus mode. To minimize the CTF modulation on the image amplitudes and phases, less defocus will be applied. Switch to Focus mode and use the focus knob to set a nominal defocus to a value between −400 and −1200 nm.

11. While in focus mode, correct any astigmatism of the objective lens and centered the beam using the beam-shift trackball. Wait a few seconds for stabilization if the stage is moving.

12. Switch to Exposure mode, insert the digital recording media and record the image. Set an exposure time that gives an accumulating dose of about 10 $e^-/Å^2$.

13. Switch back to Search mode, identify another crystal for imaging and repeat steps.

3.3 Initial Data Quality Assessment

The quality of the recorded intensities can be evaluated from the diffraction pattern or the power spectrum of the crystal image. The diffraction peaks should be sharp and symmetric without splitting or skewed. Also, the distribution of the diffraction peaks should be isotropic, particularly the diffraction patterns or the images of tilted crystal specimens. The highest resolution of the diffraction peaks can be roughly estimated by the known unit cell parameters and their Miller indices. For the study of lipid–protein interaction, one will need high-resolution information (higher than 2.5 Å resolution) to accurately model the atomic coordinates of the lipids.

4 Notes

1. Glucose is non-volatile and often used to embed crystal by the back-injection method [34]. However, it was reported that glucose embedding may not maintain the protein in its native condition if the preparation is not carefully taken [48]. Trehalose was reported to preserve high-resolution information of the 2D crystal samples [34, 49]. The concentration of trehalose used for embedding is usually 1–7% (w/v) [11, 14, 27, 29], which can be varied in different cases. Other sugars that are commonly used are sucrose and tannic acid [35].

2. The small reservoir that provides a large water surface can be used to hold the embedding solution or mother buffer. The larger surface area will help one to pick up the carbon film.

3. Molybdenum grids are usually used in electron crystallography to reduce the cryo-crinkling effect [50], which is a phenomenon that the support carbon film wrinkles when the grid is cooled down to the liquid nitrogen temperatures. The wrinkling is due to different rates of thermal expansion between the grid metal and the carbon film. The thermal expansion coefficient of the molybdenum is two to three times less than that of copper, and molybdenum is thus a better material than copper for reducing the wrinkling of carbon film [50].

4. In this protocol, we will use an Oxford CT3500 side-entry or a Gatan 626 cryo-specimen holder as an example. The actual procedure of the specimen transfer depends on the stage design of the TEM used.

5. The quality of the carbon film is critical to the success of grid specimen preparation. When evaporating the carbon, make sure to slowly and continuously deposit the carbon on top of the mica sheet.

6. Sparking usually occurs when the current increases too fast and leads to a strong discharging from the carbon source. If the sparks are observed, the carbon film is possibly broken and will be needed an inspection under a light microscope.

7. Before applying the 2D crystal sample onto the grid, one may need to pipet a few times in the crystal suspension because the membrane crystals are easy to cluster while sitting in the refrigerator. Gently pipetting ensures an appropriate amount of the crystal sample when preparing the EM grid specimen.

8. If one reuses the molybdenum grids, the grids are needed to be cleaned before use. The used grids can be cleaned in chloroform, ethanol, and deionized water with sonication. Before working on specimen preparation, one or two rounds of plasma cleaning on the grid is recommended.

9. Sugar may not be evenly distributed on the grid and may form a small size of the crystals that are distinguishable especially in the diffraction pattern. It shows a diffraction pattern of a hexagonal lattice overlaid on the diffraction pattern of the 2D protein crystal. Mixing the crystal sample thoroughly with the embedding solution may help to alleviate this problem. On the other hand, the intensities from the sugar crystalline are strong and can be visible in the diffraction pattern (Fig. 7). These reflections can be masked out before the background subtraction and have a minimal effect on the integration of the diffraction intensities of the 2D crystal.

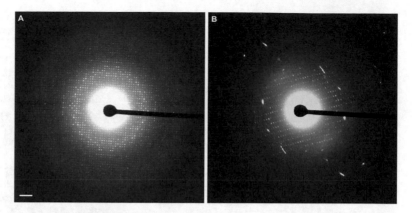

Fig. 7 Diffraction patterns of the two-dimensional crystal of aquaporin-0 at different stage tilting angles. Diffraction patterns of the AQP0 2D crystal at (**a**) 0°- and (**b**) 65°-tilting angles. The strong amorphous diffraction on the pattern (**b**) was scattered from the embedded sugar crystals. Scale bar indicates $1/6\ \text{Å}^{-1}$. The unit cell parameters are $a = 65.5\ \text{Å}$, $b = 65.5\ \text{Å}$, and $\gamma = 90°$

10. In the carbon sandwich method, the size of the second carbon film should not be too large. If the carbon film is too large, it will fold to the other side of the grid and increase the thickness of the grid specimen.

11. Spot size will need to be set to a step that can generate an appropriate dose rate during the period of exposure time. One should also consider about the size of the condenser aperture in parallel when setting the spot size.

12. The camera length used for electron diffraction needs to be determined by crystal quality and target resolution. An ideal camera length provides a diffraction pattern with the highest resolution reflection near the image edge.

Acknowledgments

We thank the Eyring Materials Center (EMC) at Arizona State University (ASU) for providing the resources. Electron crystallographic data of the AQP0 and MloK1 2D crystals was the work in Dr. Thomas Walz and Dr. Henning Stahlberg laboratories supported by the NIH grants, respectively.

References

1. Sonnhammer EL et al (1998) A hidden Markov model for predicting transmembrane helices in protein sequences. Proc Int Conf Intell Syst Mol Biol 6:175–182

2. Wallin E, von Heijne G (1998) Genome-wide analysis of integral membrane proteins from eubacterial, archaean, and eukaryotic organisms. Protein Sci 7(4):1029–1038

3. Sachs JN, Engelman DM (2006) Introduction to the membrane protein reviews: the interplay of structure, dynamics, and environment in

membrane protein function. Annu Rev Biochem 75:707–712

4. Yin H, Flynn AD (2016) Drugging membrane protein interactions. Annu Rev Biomed Eng 18:51–76

5. Schuster B et al (2008) S-layer stabilized lipid membranes (review). Biointerphases 3(2):FA3

6. Henderson R et al (1990) Model for the structure of bacteriorhodopsin based on high-resolution electron cryo-microscopy. J Mol Biol 213(4):899–929

7. Unwin PNT, Henderson R (1975) Molecular structure determination by electron microscopy of unstained crystalline specimens. J Mol Biol 94(3):425–440

8. Zampighi G et al (1982) On the structural organization of isolated bovine lens fiber junctions. J Cell Biol 93(1):175–189

9. Kessel M et al (1988) Naturally crystalline porin in the outer membrane of Bordetella pertussis. J Mol Biol 203(1):275–278

10. Jungas C et al (1999) Supramolecular organization of the photosynthetic apparatus of Rhodobacter sphaeroides. EMBO J 18 (3):534–542

11. Chiu P-L et al (2015) Evaluation of super-resolution performance of the K2 electron-counting camera using 2D crystals of aquaporin-0. J Struct Biol 192(2):163–173

12. Gonen T et al (2004) Aquaporin-0 membrane junctions reveal the structure of a closed water pore. Nature 429(6988):193–197

13. Gonen T et al (2005) Lipid-protein interactions in double-layered two-dimensional AQP0 crystals. Nature 438(7068):633–638

14. Hite RK et al (2015) Effect of lipid head groups on double-layered two-dimensional crystals formed by aquaporin-0. PLoS One 10 (1):e0117371

15. Murata K et al (2000) Structural determinants of water permeation through aquaporin-1. Nature 407(6804):599–605

16. Schenk AD et al (2005) The 4.5 a structure of human AQP2. J Mol Biol 350(2):278–289

17. Tani K et al (2009) Mechanism of aquaporin-4's fast and highly selective water conduction and proton exclusion. J Mol Biol 389 (4):694–706

18. Viadiu H et al (2007) Projection map of aquaporin-9 at 7 a resolution. J Mol Biol 367 (1):80–88

19. Norville JE et al (2007) 7A projection map of the S-layer protein sbpA obtained with trehalose-embedded monolayer crystals. J Struct Biol 160(3):313–323

20. Zhuang J et al (1999) Two-dimensional crystallization of Escherichia coli lactose permease. J Struct Biol 125(1):63–75

21. Chiu PL et al (2007) The structure of the prokaryotic cyclic nucleotide-modulated potassium channel MloK1 at 16 a resolution. Structure 15(9):1053–1064

22. Kowal J et al (2014) Ligand-induced structural changes in the cyclic nucleotide-modulated potassium channel MloK1. Nat Commun 5:3106

23. Li HL et al (1998) Two-dimensional crystallization and projection structure of KcsA potassium channel. J Mol Biol 282(2):211–216

24. Zheng H et al (2010) The prototypical H+/galactose symporter GalP assembles into functional trimers. J Mol Biol 396(3):593–601

25. Nakazato K et al (1996) Two-dimensional crystallization and cryo-electron microscopy of photosystem II. J Mol Biol 257(2):225–232

26. Holm PJ et al (2006) Structural basis for detoxification and oxidative stress protection in membranes. J Mol Biol 360(5):934–945

27. Hite RK et al (2010) Principles of membrane protein interactions with annular lipids deduced from aquaporin-0 2D crystals. EMBO J 29(10):1652–1658

28. Tsiotis G et al (1996) Tubular crystals of a photosystem II core complex. J Mol Biol 259 (2):241–248

29. Kalbermatter D et al (2015) 2D and 3D crystallization of the wild-type IIC domain of the glucose PTS transporter from Escherichia coli. J Struct Biol 191(3):376–380

30. Ohi M et al (2004) Negative staining and image classification - powerful tools in modern Electron microscopy. Biol Proced Online 6:23–34

31. Kim C et al (2010) An automated pipeline to screen membrane protein 2D crystallization. J Struct Funct Genom 11(2):155–166

32. Vonck J (2000) Parameters affecting specimen flatness of two-dimensional crystals for electron crystallography. Ultramicroscopy 85 (3):123–129

33. Hite RK et al (2010) Collecting electron crystallographic data of two-dimensional protein crystals. Methods Enzymol 481:251–282

34. Chiu PL et al (2011) The use of trehalose in the preparation of specimens for molecular electron microscopy. Micron 42(8):762–772

35. Glaeser RM (2007) Electron crystallography of biological macromolecules. Oxford University Press, Oxford

36. Hirai T et al (1999) Trehalose embedding technique for high-resolution electron

crystallography: application to structural study on bacteriorhodopsin. J Electron Microsc 48 (5):653–658

37. Henderson R, Unwin PN (1975) Three-dimensional model of purple membrane obtained by electron microscopy. Nature 257 (5521):28–32

38. Russo CJ, Henderson R (2018) Charge accumulation in electron cryomicroscopy. Ultramicroscopy 187:43–49

39. Gyobu N et al (2004) Improved specimen preparation for cryo-electron microscopy using a symmetric carbon sandwich technique. J Struct Biol 146(3):325–333

40. Bullough P, Henderson R (1987) Use of spot-scan procedure for recording low-dose micrographs of beam-sensitive specimens. Ultramicroscopy 21(3):223–230

41. Downing KH, Glaeser RM (1986) Improvement in high resolution image quality of radiation-sensitive specimens achieved with reduced spot size of the electron beam. Ultramicroscopy 20(3):269–278

42. Grant T, Grigorieff N (2015) Measuring the optimal exposure for single particle cryo-EM using a 2.6 Å reconstruction of rotavirus VP6. elife 4:e06980

43. Schenk AD et al (2013) A pipeline for comprehensive and automated processing of electron diffraction data in IPLT. J Struct Biol 182 (2):173–185

44. Gipson B et al (2007) 2dx_merge: data management and merging for 2D crystal images. J Struct Biol 160(3):375–384

45. Gipson B et al (2007) 2dx--user-friendly image processing for 2D crystals. J Struct Biol 157 (1):64–72

46. Wall J et al (1985) Films that wet without glow discharge, 35th EMSA meeting. San Francisco Press, Louisville

47. Kittel C, McEuen P (1996) Introduction to solid state physics. Wiley, New York

48. Perkins GA et al (1993) Glucose alone does not completely hydrate bacteriorhodopsin in glucose-embedded purple membrane. J Microsc 169(Pt 1):61–65

49. Kimura Y et al (1997) Surface of bacteriorhodopsin revealed by high-resolution electron crystallography. Nature 389(6647):206–211

50. Booy FP, Pawley JB (1993) Cryo-crinkling: what happens to carbon films on copper grids at low temperature. Ultramicroscopy 48 (3):273–280

Chapter 12

Single Particle Analysis for High-Resolution 2D Electron Crystallography

Ricardo Righetto and Henning Stahlberg

Abstract

Electron crystallography has been used for decades to determine three-dimensional structures of membrane proteins embedded in a lipid bilayer. However, high-resolution information could only be retrieved from samples where the 2D crystals were well ordered and perfectly flat. This is rarely the case in practice. We implemented in the FOCUS package a module to export transmission electron microscopy images of 2D crystals for 3D reconstruction by single particle algorithms. This approach allows for correcting local distortions of the 2D crystals, yielding much higher resolution reconstructions than otherwise expected from the observable diffraction spots. In addition, the single particle framework enables classification of heterogeneous structures coexisting within the 2D crystals. We provide here a detailed guide on single particle analysis of 2D crystal data based on the FOCUS and FREALIGN packages.

Key words Electron crystallography, 2D Crystals, Single particle analysis, Membrane proteins, MloK1, Ion channels, Cryo-electron microscopy, Image processing, 3D Reconstruction, 3D Classification

1 Introduction

Historically, the structures of membrane proteins have been much more difficult to determine than those of their soluble counterparts. This is mainly due to the hydrophobic nature of the transmembrane domains, which render these proteins challenging both for crystallization and solubilization. However, a special type of crystallography, in which the proteins form ordered two-dimensional (2D) periodical arrays in a lipid membrane, is suitable for gathering information on their three-dimensional (3D) structures. It was from a naturally occurring 2D crystal, the purple membrane of *Halobacterium salobium*, that the first 3D structure of a membrane protein was obtained [1]. This breakthrough made the transmission electron microscope (TEM) a promising instrument for studying membrane proteins, as long as 2D crystals could be obtained. In fact, after it was demonstrated

Tamir Gonen and Brent L. Nannenga (eds.), *CryoEM: Methods and Protocols*, Methods in Molecular Biology, vol. 2215, https://doi.org/10.1007/978-1-0716-0966-8_12, © Springer Science+Business Media, LLC, part of Springer Nature 2021

that embedding the sample in vitreous ice eliminated drying and fixation artifacts and reduced the impact of radiation damage [2], electron crystallography became the method of choice for solving membrane protein structures, reaching near-atomic resolution in a number of cases, such as bacteriorhodopsin [3], the plant light-harvesting complex [4], αβ-tubulin [5], and aquaporin-0- [6]. Unfortunately, it was also realized that satisfying the sample quality requirements for achieving high-resolution diffraction, namely the perfect ordering and flatness of the 2D crystals, in most cases was very difficult or even impossible.

While it is possible in some cases to employ X-ray crystallography [7, 8] or nuclear magnetic resonance (NMR) [9] to solve structures of membrane proteins, these techniques also have their own shortcomings, namely the need for suitably large 3D crystals, or being constrained to small molecular weights, respectively. Furthermore, in both techniques, the membrane protein environment is often not comparable to the native cellular membrane. In single particle cryo-electron microscopy (cryo-EM), however, most of these technical challenges are overcome by performing 3D reconstructions from proteins randomly oriented in solution [10, 11]. Membrane proteins, specifically, can be solubilized if surrounded by detergent micelles [12] or a lipid nanodisc [13].

Nevertheless, there are specific cases where working with 2D crystals is necessary or even advantageous compared to these other techniques. One example is when such periodical arrangements occur natively in the cell membrane [1, 3], or when their natural formation is associated with a biological specific function [14]. Another example is when the 2D array is designed artificially, enabling studies of proteins and other molecules that would be difficult otherwise [15, 16]. Other advantages are that 2D crystallography offers the highest possible efficiency in terms of number of particles in the field of view and allows reconstructions from arbitrarily small proteins, which can be challenging for conventional single particle analysis (SPA). These are particularly interesting aspects if combined with the single particle approach outlined in this chapter.

2D electron crystallography can be performed both in the diffraction or the imaging mode of the TEM [17]. In this chapter, we will cover only the imaging mode, in which several projection images of 2D crystals at different orientations are recorded. As in single particle analysis, the principle of 3D reconstruction is based on the central section theorem [10, 18]. The periodical nature of the two-dimensional array in real space appears as diffraction spots (Bragg peaks) in reciprocal space. These spots can then be indexed, measured, and corrected for the contrast transfer function (CTF) of the microscope. After merging the detected spots in 3D reciprocal space, a real space map of the protein can be obtained by Fourier inversion [17, 19]. Algorithms for performing these operations have been implemented in the MRC package [20], and subsequently in the 2dx [21] and FOCUS [22] packages, among others [23]. A technical limitation of 2D electron crystallography is that samples in the microscope can typically only be tilted up to about 60° for geometric reasons, therefore leaving a cone of missing information in 3D Fourier space. Because of this, the reconstructed maps may appear elongated in the z-direction in real space. Furthermore, only if the 2D crystals are perfectly flat and well-ordered, high-resolution diffraction spots will be observed. Because this is rarely the case in practice, image processing algorithms were developed to correct for the crystal distortions in silico [24, 25].

One such algorithm, the so-called *image unbending* [26], has been particularly successful [3]. This method works by moving small patches of the 2D crystal image by iteratively comparing cross-correlation peaks indicating the locations of unit cells in real space with their predicted positions. However, this approach is intrinsically limited to distortions in the image plane and cannot account for azimuth angle variations that are present when 2D crystals with larger in-plane distortions are imaged at higher tilts. In order to be able to correct for out-of-plane distortions, such as "bumps" in the 2D crystal, the patches need to be compared (e.g., cross-correlated) against a reference in 3D space, which is essentially what single particle refinement and reconstruction algorithms do [10]. First attempts in this direction showed promise [27, 28], but were still limited to low-resolution reconstructions. This was in part because the datasets analyzed in these works had not yet been collected on direct electron detectors (DED) [29], and partially because the algorithms implemented did not account for the very low signal-to-noise (SNR) ratios of cryo-EM images in a probabilistic manner [30].

We have since then implemented a module in the FOCUS package that allows the user to export a 2D crystallography project for processing with standard single particle analysis software. This not only brings to electron crystallography the maturity that these packages have achieved in terms of robustness and performance [31–33], but, more importantly, it opens the possibility of classifying heterogeneous structures [34, 35] coexisting within the 2D crystals. In the *Methods* section, we will describe in detail the workflow for processing 2D crystal data within the single particle framework, with an emphasis on practical tips and tricks. The approach presented here is based on the one we used to process poorly diffracting 2D crystals of MloK1, a 160 kDa prokaryotic potassium channel [36, 37]. Using single particle refinements, we were able to obtain a map of MloK1 at 4 Å resolution and to observe distinct conformations of its cyclic-nucleotide binding domain (CNBD), helping to elucidate the mechanism of gating for this channel [38]. The method is generally applicable to other 2D crystal samples as well.

2 Materials

2.1 Software and Hardware

In this work, we refer to the software package FOCUS (http://www.focus-em.org) [22] and a version of the FREALIGN package [39] extended with additional features useful for the processing of 2D crystal data, hereby called frealign-2dx to avoid confusion with the original implementation (http://github.com/C-CINA/frealign-2dx) [38]. Both, FOCUS and frealign-2dx are freely available as open-source software and run on Linux-based operating systems.

A typical computing workstation to carry out the data processing steps described ahead will have the following hardware components (*see* **Note 1**):

1. 2× 12-core CPU or better.

2. 256 GB RAM.

3. >50 TB HD storage.

4. 1 TB SSD storage (used as a fast "scratch" disk).

5. 2× NVIDIA GTX 1080 GPU card or better (*see* **Note 2**).

With this setup, it is possible to carry out real-time data pre-processing (e.g., drift correction and CTF estimation, *see* Subheading 3.1), classical 2D crystal processing (*see* Subheading 3.2), and single particle refinements (Subheading 3.3.1 onwards). However, for the processing of large datasets, in particular with large box sizes, a high-performance computing (HPC) cluster may be required.

3 Methods

We here describe the computational steps required to process 2D crystal data with single particle software. For details and protocols on the growth of 2D crystals and their sample preparation for TEM imaging, please refer to [40–43]. After exporting the data from FOCUS [22], we will use frealign-2dx, an extended version of the FREALIGN package [39] (*see* Supplementary Notes 1 and 2 of [38]) although any other single particle analysis package can also be used, in principle (*see* **Note 3**).

3.1 Data Acquisition and Initial Processing of Movies

To ensure a good spectral SNR (SSNR) in the high-resolution range, we recommend the acquisition of movies at a microscope magnification corresponding to a pixel size <1 Å at the sample level, to benefit from the improved detective quantum efficiency (DQE) of DEDs at lower detector resolutions [29]. Total exposures in the range of 40–50 e$^-$/Å^2 per movie are known to work well. Multiple 2D crystals should be imaged, applying tilt angles ranging from 0° to ±60° in steps of 5°–10°, usually (a full exposure is acquired at a given tilt angle). The finer the angular sampling the better. A typical dataset will contain ~100 to ~1000 movies, which should be spread across the tilt range. Note that in contrast to electron tomography, where unique samples are imaged in a tilt series, in the here described approach each 2D crystal is imaged only once and at one specific tilt angle. However, different 2D crystals should then be imaged at different fixed tilt angles.

Movies of 2D crystals should be corrected for beam-induced drift [44] and for beam-induced resolution loss by applying dose-dependent temperature factors to each frame before computing frequency-weighted averages from all frames [45], generating a drift-corrected average image. Also, defocus and astigmatism have to be estimated for each image [46, 47]. Using FOCUS, these steps can be carried out in real time during the microscopy session and the dataset can be conveniently pruned based on parameters such as total drift, measured defocus and estimated CTF resolution, among others [22].

3.2 Conventional 2D Crystallographic Reconstruction

Before starting with the single particle approach, it is necessary to process the data in conventional 2D electron crystallography fashion. For each 2D crystal image in the dataset, the crucial parameters to be obtained at this stage are:

- The defocus and astigmatism at the center of the image.
- The tilt geometry of the specimen plane, defined by the angles TLTANG and TLTAXIS.
- The orientation of the 2D crystal on the specimen plane, defined by the angle TAXA.
- The phase origin (translational shifts).
- The unit cell positions provided by the unbending algorithm [26].

Ideally, the best possible 3D map should be obtained in this way by iterative 3D merging and refinement before proceeding, although an initial, low-resolution map may suffice (*see* **Note 4**). For more information and protocols for the processing of 2D crystal data, please refer to [17, 19, 21, 43, 48].

3.3 Exporting a 2D crystal Project for Single Particle Analysis

Having performed conventional 2D crystallography data processing, we can proceed to the single particle analysis of 2D crystals. The single particle scripts are available under the **Particles** tab on the top bar of the **FOCUS GUI**. The first script, **Pick & export particles**, is the core of our approach, allowing the user to extract the particles and export these data in a way that is understood by single particle programs. The script will operate on the images that have been selected via the FOCUS library GUI (*see* **Note 5**). The

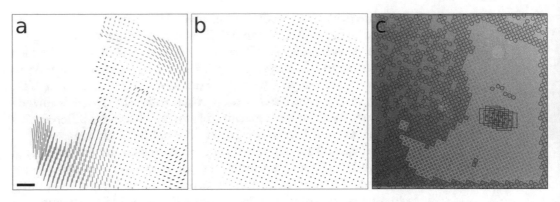

Fig. 1 Particle picking from unbending profile. (**a**) The unbending algorithm compares the observed and predicted unit cell positions. Displacement vectors are 10× exaggerated for visualization, with colors indicating the displacement direction. (**b**) CC peaks indicating unit cell positions after unbending. Darker dots indicate stronger peaks. (**c**) Particle picking based on the information from (**b**). Green circles are picked particles (i.e., above user-selected threshold) and magenta circles are ignored particles. The blue squares indicate the size (416 Å in this case) of the overlapping patches to be extracted, with the crosses indicating the box centers. Scale bar: 500 Å

script presents a number of parameters to the user, the most important of which will be explained below.

3.3.1 Particle Picking

The classical unbending algorithm [26] generates a cross-correlation (CC) profile (Fig. 1). The peaks in this CC profile indicate the position of the unit cells, available after running the Unbend II script during the 2D crystal processing in Subheading 3.2 (Fig. 1a, b). This information, stored in the file <**image_name**>**_profile.dat** for each micrograph, will be used to pick the particles from the 2D crystals (Fig. 1c).

Optionally, a **Phase Origin Shift** can be applied to the CC peaks. This is a translational shift, here presented in degrees in relation to the unit cell dimensions. A phase origin shift of 180° in one direction means shifting by half the length of the unit cell. Such a shift may be required when the center of the protein does not coincide with the center of the crystal unit cell, as shown in Fig. 2. The shift is calculated accordingly for non-tilted crystals, based on the current estimate of the tilt geometry.

The **Box size** parameter defines the side length of the (squared) box into which the particles will be windowed, in pixels. The particles, in this context, are patches of the 2D crystals, and this parameter defines how large such patches should be.

Differently from conventional SPA, it is not possible here to have a single protein unit isolated in the box. Also, in the tilted views the projection of the neighboring proteins overlaps with that of the central protein, which has to be taken into account during the particle alignments (*see* Fig. 4). For this reason, the box size should be large enough to contain the protein of interest plus at

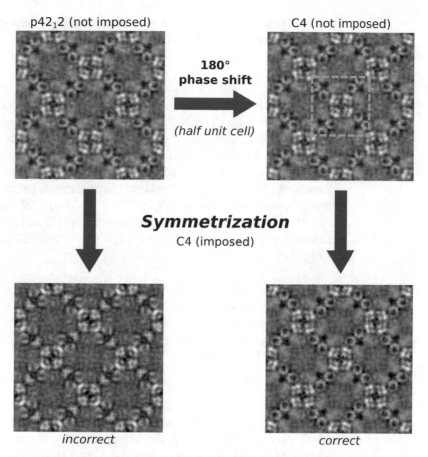

p42₁2 (not imposed)

C4 (not imposed)

180° phase shift

(half unit cell)

Symmetrization
C4 (imposed)

incorrect

correct

Fig. 2 Shifting the phase origin. In single particle analysis, it is desired to have the protein of interest at the center of the reconstruction box. This is especially important when the object is symmetrical in 3D. Crystallographic reconstructions may have *point-group* symmetry, while single particle reconstructions may have *space-group* symmetry. In the case of MloK1, the phase origin of non-tilted views has to be shifted by half a unit cell (180°), to bring one tetramer (green dashed square) to the center of the particle box and then be able to apply C4 symmetry. The procedure is illustrated for the projection map of a non-tilted crystal. The tilted views are also shifted accordingly by the cosine of the tilt angle

least one neighbor protein on each side, implying an overlap between the boxes of adjacent particles, as shown by the blue boxes in Fig. 1c. This also favors the smoothness of the alignment parameters across the 2D crystal because of the natural correlations existing among adjacent particles (proteins located close to each other in the 2D crystal tend to appear in similar orientations). Furthermore, larger boxes contain more signal, rendering the alignments more reliable. On the other hand, a box too large detracts from the goal of correcting local crystal distortions as accurately as possible. The optimal box size will depend mainly on the actual size and molecular weight of the protein. Smaller proteins will certainly require larger patches comprising more neighboring units, and vice-versa.

The **Threshold to include particles** parameter allows to define how strict the picking should be:

$$CC_{thr} \geq \mu_{CC} + z\sigma_{CC}, \tag{1}$$

where CC_{thr} is the threshold, above which candidate particles defined by the CC peaks in a given 2D crystal will be picked, μ_{CC} is the average of all CC peaks in that crystal, and σ_{CC} their standard deviation. z is a multiplier of the standard deviation to make the threshold more or less selective. Higher (positive) values of z make the picking more stringent and in theory selects "better" particles. Conversely, lower (negative) values will pick more, potentially also "worse", particles. The default for **Threshold to include** particles is $z = 0$, meaning only CC peaks above average will be boxed, which in most cases is a good compromise (*see* green and magenta circles in Fig. 1c).

3.3.2 CTF Correction

Each particle will be assigned a defocus value. For perfectly non-tilted images, the defocus will be the same for all particles. However, for tilted images there will be a defocus gradient. The defocus at the center of each particle box is computed according to the CTFTILT equation [49]:

$$\Delta f = \Delta f_0 + [(x-x_0)\sin(\phi)-(y-y_0)\cos(\phi)]\tan(\gamma), \tag{2}$$

where Δf is the defocus at the center of the particle box, Δf_0 is the defocus at the center of the image, (x, y) and (x_0, y_0) are the coordinates of the box and the image centers, respectively, ϕ is the angle between the tilt axis and the X-axis (i.e., TLTAXIS) and γ is the tilt angle (TLTANG). Astigmatism is assumed to be constant across the whole image. The particles are boxed without CTF correction, which will be performed later by frealign-2dx. However, in addition to boxing uncorrected particles, it is possible and recommended to generate also a CTF-corrected particle stack. There are three options for this CTF correction in FOCUS: *phase flipping*, *CTF multiplication*, or *Wiener filtering*. The latter requires a user-defined *ad hoc* constant proportional to the inverse of the dataset's SNR (the default value is set to 0.4). Any of the three options should yield similar results for the purposes of calculating crystal averages (Subheading 3.4.1).

3.3.3 Exporting the Metadata

When generating the particle metadata files, the orientation for each particle will be inherited from the 2D crystal it was picked from. As frealign-2dx uses the SPIDER angle convention [50], the crystallographic tilt geometry is converted to Euler angles as follows [48]:

$$\psi = 270^{\circ} - \text{TLTAXIS}$$

$$\theta = \text{TANGL} \tag{3}$$

$$\phi = 90^{\circ} - \text{TAXA}$$

where ψ is an in-plane rotation (2D), θ is the tilt angle, and ϕ is a 3D rotation. These values, along with defocus information and other metadata, are stored in a text file called **particles.par**. The particle stack is stored in the file **particles.mrcs**. If CTF-corrected stacks are generated, additional similar files will also be created. All these files are located in the directory **stacks/** inside the location specified by the **General Single-Particle directory** parameter. If an absolute path is not specified, this location will be relative to the **merge/** directory within the FOCUS project directory. Optionally, figures indicating the included and ignored picking coordinates overlaid on the micrographs can also be saved (inside the **picking/** directory), and the metadata can also be saved in **.star** format for RELION (*see* **Note 6**). The underlying Python script that performs particle picking and metadata creation can be efficiently run in parallel, with a user-defined number of threads.

It is important to notice that while creating the particle stacks and respective metadata, the script will by default ensure that particles extracted from the same 2D crystal stay assigned to the same "half-set." This is required to prevent inflated resolution estimates because of the large overlap between adjacent particles (Fig. 1c) [51]. In frealign-2dx, this is accomplished by interleaving the 2D crystals in the particle stack and in the **.par** file. In RELION, this is controlled by the **rlnRandomSubset** and **rlnHelicalTubeID** labels in the **.star** file (*see* **Note 7**).

3.4 Pre-Refinement

After picking and exporting the particles, it should be possible to start a single particle refinement straight away (Subheading 3.5). It is fast and useful, however, to run a "pre-refinement" (*see* below), using crystal averages prior to the full refinement with individual particles.

3.4.1 Crystal Averages

The next script in the pipeline, **Generate crystal averages**, will calculate crystal averages, using the CTF-corrected particle stacks (Subheading 3.3.2). These averages are analogous to 2D class averages in SPA. They provide a straightforward way of visually assessing the picking parameters chosen (e.g., phase origin shift, box size, etc.). This procedure is also known as the correlation averaging method [25]. The Fourier ring correlation (FRC) for each crystal is also computed; its plots can be consulted under the **FRC/** directory. As a validation measure, the crystal averages should be highly similar to their corresponding projection maps, such as those shown in Fig. 2 (except for the phase origin, if a shift was applied as described in Subheading 3.3.1). The difference is that the crystal averages are computed by averaging the boxed unit cells directly, while the projection maps are computed from the information in the diffraction spots alone (which is also a form of averaging). Examples of crystal averages are shown in Supplementary Fig. 3 of [38].

3.4.2 Running the Pre-Refinement

The script **Prepare pre-refinement** offers a graphical interface to generate an **mparameters** file for frealign-2dx [39]. It will also put the files required by frealign-2dx into the **pre-refine** directory, by default. The default values should be reasonable, but users are encouraged to fine-tune the parameters to their needs. We draw the attention here to the **Auto-refinement** parameters available in our modified version of frealign-2dx:

- **FSC threshold for resolution limit? (thresh_fsc_ref)**: the resolution limit used in refinement (for preventing overfitting) will be taken as the resolution where the Fourier shell correlation (FSC) curve crosses this threshold (default: 0.800).

- **FSC threshold for map improvement evaluation? (thresh_fsc_eval)**: any map improvement between refinement cycles will be assessed by looking at the resolution where the FSC curve crosses this threshold (default: 0.143).

- **Minimum resolution limit? (res_min)**: this parameter prevents the auto-refiner from using a resolution too low, which may cause the refinement to diverge (default: 40 Å).

- **Stay away from current map resolution? (ref_stay_way)**: this parameter prevents the resolution limit from getting too close to the current map resolution, which would potentially introduce bias in the refinement (default: 2 Å).

- **Try different combinations of parameter mask? (change_p-mask)**: if enabled, this parameter makes the auto-refiner try different combinations among the active refinement parameters (ϕ, θ, ψ, x, y), before declaring convergence, which helps avoiding local minima (default: *yes*).

- **Keep the tilt angle fixed? (no_theta)**: if active, the tilt angle (θ) will be kept fixed. Sometimes this is helpful for 2D crystal data (default: *no*).

For more information, please *see* Supplementary Note 2 of [38]. After preparing the pre-refinement, it can be launched via the **Run pre-refinement** script. Alternatively, it can be launched from the command line by running the **frealign_run_refine_auto** program inside the pre-refinement directory. The crystal averages are only a few hundred for a typical dataset and they have a high SNR. The pre-refinement should therefore complete in only a few minutes and noticeably improve the tilt geometry estimation (now converted to Euler angles), as shown in Fig. 3. Plots of the angular

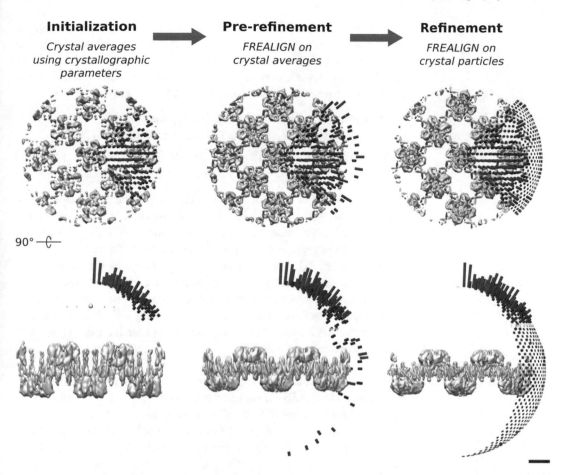

Fig. 3 Single particle refinement from 2D crystals. Left: The initial reconstruction uses parameters from 2D crystallographic data processing. Center: The map is then improved by doing a "pre-refinement" with frealign-2dx using crystal averages. Right: Using particles (local patches) extracted from the 2D crystals starting with parameters obtained in the pre-refinement, the map is refined to high resolution with frealign-2dx. Both the height and the color of the cylinders (here shown for the asymmetric unit only) indicate the proportion of particles or crystal averages in the corresponding orientation, relative to each dataset (red: high particle density, blue: low particle density; a.u.). Maps shown are not sharpened. Scale bar: 50 Å

changes can also be obtained by running the **Analyze pre-refinement results** script.

3.5 Consensus Refinement

After the pre-refinement is complete, the full refinement using the extracted particles can take place. This refinement finally corrects for the local distortions of the 2D crystals. As in the pre-refinement, the FOCUS script **Prepare refinement** will generate the **mparameters** file and generate the proper file structure for the frealign-2dx refinement. At this stage, the extracted particles will inherit the pre-refined alignment parameters from their "parent" 2D crystals (Subheading 3.4). This is done internally by the Python script **SPR_FrealignParameterInheritance**. The extracted particles have a much lower SNR than the crystal averages, and their alignment parameters are expected to not deviate too much from the crystal average. Therefore, besides the auto-refinement parameters (Subheading 3.4.2), it is now recommended to activate the alignment restraints available in frealign-2dx (*see* **Note 8**):

– **Restraint for Euler angles, in degrees (sigma_angles)**: standard deviation for a Gaussian restraint on the Euler angles.

– **Restraint for x,y shifts, in pixels (sigma_shifts)**: standard deviation for a Gaussian restraint on the x, y shifts.

More information about these restraints can be found in the Supplementary Note 1 of [38]. Upon convergence, a higher quality map should have been obtained, and a more diverse set of views should be observed, as shown in Fig. 3. This evidences that particles from the same 2D crystal are not exactly always in the same orientation. This map, commonly referred to as a "consensus map" because no 3D classification has been performed (yet), can now be post-processed as described in Subheading 3.7. The user is also encouraged to experiment with more advanced features, such as defocus refinement [32, 39].

3.6 3D Classification   Perhaps the most interesting feature of the single particle method is its ability to classify heterogeneous data into conformationally homogeneous classes [35, 52]. Structural variability is also a source of disorder in 2D crystals, thus limiting the achievable resolution by conventional crystallographic methods. We have used 3D classification to detect distinct conformations of the MloK1 CNBD in relation to the transmembrane domain (TMD) [38].

Differently from conventional SPA, however, in 2D crystal particles the protein of interest is always surrounded by neighboring proteins. This means that, in tilted views, the projection of the neighbors overlaps with the projection of the classification target. This problem is illustrated in Fig. 4.

Therefore, 3D classification will only work properly if the signal from the neighbor proteins is subtracted from the particle images. For more accurate results in signal subtraction of 2D crystal data, the user should calculate a completely unmasked reconstruction in frealign-2dx (*see* Fig. 4). This can be done by setting a reconstruction radius larger than the particle box in the **mparameters** file and then running the command **frealign_calc_reconstructions** from the refinement directory (*see* **Note 9**). Based on this reconstruction, the user should then define a soft mask focused on the protein of interest (e.g., the central MloK1 tetramer), and invert this mask to select everything else that should be subtracted from the experimental projections (i.e., the particles). The **focus.postprocess** tool can be used to create such masks (*see* Subheading 3.7). The signal

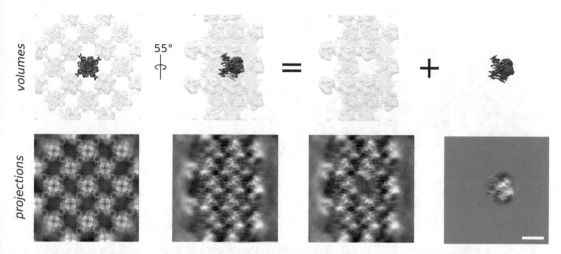

Fig. 4 Signal subtraction for 3D classification. For 2D crystal data, the projection of the neighboring proteins (transparent yellow densities) overlap with the projection of the classification target (red density) in the tilted views. In order to perform 3D classification, it is necessary to subtract the signal of the neighboring proteins from the particles. Shown here is an example from the MloK1 dataset. Scale bar: 50 Å

subtraction itself can be accomplished from the command line using RELION [53]. An example command line for this would be:

```
relion_project --i particles_9_r1-subtract.mrc --subtract_exp --angpix
2.68 --ctf --ang particles_9_r1.star --o subtracted_particles
```

The **mparameters** file can then be edited to point to the new subtracted **particles.mrcs** stack. If required, the new particle stack and the subtracted 3D reference can be down-sampled to a coarser pixel size for computational speedup. Also, the user should set the **nclasses** parameter to the number of classes desired. In order to run the 3D classification, the user should start immediately after the refinement cycle in which the consensus refinement converged (e.g., 10, if the auto-refinement converged in cycle 9). Because the signal-subtracted particles may contain too little signal for reliable alignments, it is recommended to disable the optimization of any alignment parameters at this stage and run the classification for a pre-defined number of cycles (e.g., 40). In this case, the auto-refiner will not be used and the 3D classification can be launched with the command **frealign_run_refine**. The occupancies of the particles in each class will be randomized to initialize the classification. Convergence of the 3D classification can be assessed by the user with the **frealign_calc_stats**.

3.7 Assessing Resolution and Post-Processing

Resolution is typically assessed by computing the FSC curves [54, 55] between independently (beyond a certain resolution) refined half-maps. As explained in Subheading 3.3.3, FOCUS ensures that particles from the same 2D crystal are assigned to the

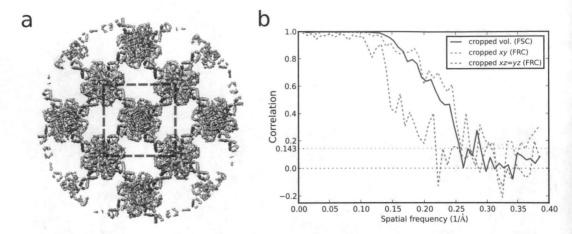

Fig. 5 Resolution estimation. (**a**) The full, sharpened consensus map of MloK1, with the central tetramer, which was cropped for resolution estimation, highlighted. (**b**) FSC curve for the central tetramer, after adjusting for the protein volume, and FRC curves for the *xy* and *xz* = *yz* planes

same half-sets to prevent an inflation of the FSC curve. Normally, only a sub-volume is of interest, for example, the central tetramer in the case of MloK1, as shown in Fig. 5.

We developed a program called **focus.postprocess** that conveniently offers a number of volume cropping, masking, resolution estimation, and sharpening operations. A typical command line is:

```
focus.postprocess map1.mrc map2.mrc --angpix 1.3 --crop_center
0,0,-12 --crop_size 104,104,104 --mtf data_mtf_k2_300kv.star
--mw 160.0 --mask_radius 42 --out consensus
```

The **focus.postprocess** tool has features that allow the user to assess the resolution anisotropy of the map due to the missing cone (options --**cone_aperture**, --**xy_only**, --**xz_only**, and --**yz_only**). All options can be found by typing **focus.postprocess** --**help** on the command line.

If maps from multiple 3D classes have been obtained and refined, the same procedures should be applied on them. Finally, the post-processed maps can be used for modeling the atomic structure of the protein [56].

3.8   Conclusions

Single particle refinement algorithms can obtain 3D reconstructions from disordered 2D crystals at resolutions well beyond the range of observable diffraction spots. A dedicated module in the FOCUS package and the frealign-2dx program, a modified version of FREALIGN, enable this method. The approach is not limited to these specific implementations, though. A likely reason for so many 2D crystals being disordered is conformational heterogeneity. The single particle approach to 2D crystals allows signal subtraction and 3D classification, which can be used to disentangle different conformations as demonstrated for the MloK1 potassium channel dataset. In summary, this method presents a viable alternative to conventional 2D crystallography, when the 2D crystal form is related to protein function, or when 2D crystals were intentionally designed.

4   Notes

1. For other possible hardware examples, *see* the official online FOCUS documentation: http://focus.c-cina.unibas.ch/wiki/doku.php?id=1_0:hardware_linux or http://focus-em.org

2. Depending on which packages you use, GPU cards may not be required. Please consult the specific documentation of the software packages present in your data processing pipeline.

3. We have also successfully used cisTEM [32], RELION [52] and cryoSPARC [31] with 2D crystal data, although these packages have not been as extensively tested and customized as frealign-2dx in this context.

4. Actually, just merging in 2D may be enough in some cases, as long as phase origins are sufficiently accurate. In this case, however, an initial 3D map needs to be obtained by other means. For example, we have successfully used stochastic gradient descent (SGD) [31] to this end.

5. It is assumed the dataset has been pruned accordingly before and/or during the conventional 2D crystal reconstruction workflow.

6. The program **par2star.py** can also be called from the command line as a standalone tool.

7. If using cryoSPARC, please note that, at present, it ignores these fields and always randomizes the order of the particles when assigning its half-sets. In this case, it is up to the user to ensure that the resolution estimation is unbiased by, for example, re-splitting the data based on the 2D crystals after the refinement and calculating half-maps using FREALIGN or RELION.

8. If using RELION, similar behavior would be accomplished by using local searches, i.e., setting the initial angular sampling equal to the sampling from which to perform local searches.

9. For the signal subtraction and 3D classification tasks, the user should work from the command line, as these options are not available from the FOCUS GUI.

Acknowledgments

Calculations were performed at sciCORE (http://scicore.unibas.ch/) scientific computing center at University of Basel. RDR acknowledges funding from the Fellowships for Excellence program sponsored by the Werner-Siemens Foundation and the University of Basel. This work was supported by the Swiss National Science Foundation through grant 205320_166164 and the NCCR TransCure.

References

1. Henderson R, Unwin PNT (1975) Three-dimensional model of purple membrane obtained by electron microscopy. Nature 257:28–32

2. Adrian M, Dubochet J, Lepault J, McDowall AW (1984) Cryo-electron microscopy of viruses. Nature 308:32–36

3. Henderson R, Baldwin JMM, Ceska TAA et al (1990) Model for the structure of bacteriorhodopsin based on high-resolution electron cryomicroscopy. J Mol Biol 213:899–929

4. Kühlbrandt W, Wang DN, Fujiyoshi Y (1994) Atomic model of plant light-harvesting complex by electron crystallography. Nature 367:614–621

5. Nogales E, Wolf SG, Downing KH (1998) Structure of the αβ tubulin dimer by electron crystallography. Nature 391:199–203

6. Gonen T, Cheng Y, Sliz P et al (2005) Lipid-protein interactions in double-layered two-dimensional AQP0 crystals. Nature 438:633–638

7. Deisenhofer J, Michel H (1989) The photosynthetic reaction center from the purple bacterium Rhodopseudomonas viridis. Science 245:1463–1473

8. Landau EM, Rosenbusch JP (1996) Lipidic cubic phases: a novel concept for the crystallization of membrane proteins. Proc Natl Acad Sci U S A 93:14532–14535

9. Liang B, Tamm LK (2016) NMR as a tool to investigate the structure, dynamics and function of membrane proteins. Nat Struct Mol Biol 23:468

10. Frank J (2006) Three-dimensional Electron microscopy of macromolecular assemblies, 2nd edn. Oxford University Press, Oxford

11. Kühlbrandt W (2014) The resolution revolution. Science 343:1443–1444

12. Liao M, Cao E, Julius D, Cheng Y (2013) Structure of the TRPV1 ion channel determined by electron cryo-microscopy. Nature 504:107–112

13. Gao Y, Cao E, Julius D, Cheng Y (2016) TRPV1 structures in nanodiscs reveal mechanisms of ligand and lipid action. Nature 534:347–351

14. Goni GM, Epifano C, Boskovic J et al (2014) Phosphatidylinositol 4,5-bisphosphate triggers activation of focal adhesion kinase by inducing clustering and conformational changes. Proc Natl Acad Sci 111:E3177–E3186

15. Gonen S, DiMaio F, Gonen T, Baker D (2015) Design of ordered two-dimensional arrays mediated by noncovalent protein-protein interfaces. Science 348:1365–1368

16. Subramanian RH, Smith SJ, Alberstein RG et al (2018) Self-assembly of a designed nucleoprotein architecture through multimodal interactions. ACS Cent Sci 4:1578–1586

17. Schenk AD, Castaño-Díez D, Gipson B et al (2010) 3D reconstruction from 2D crystal image and diffraction data. In: Enzymology GJJBT-M in Cryo-EM, part B: 3-D reconstruction. Academic Press, New York, pp 101–129

18. De Rosier DJ, Klug A (1968) Reconstruction of three dimensional structures from Electron micrographs. Nature 217:130–134

19. Stahlberg H, Biyani N, Engel A (2015) 3D reconstruction of two-dimensional crystals. Arch Biochem Biophys 581:68–77

20. Crowther RA, Henderson R, Smith JM (1996) MRC image processing programs. J Struct Biol 116:9–16

21. Gipson B, Zeng X, Zhang ZY, Stahlberg H (2007) 2dx—user-friendly image processing for 2D crystals. J Struct Biol 157:64–72

22. Biyani N, Righetto RD, McLeod R et al (2017) Focus: the interface between data collection and data processing in cryo-EM. J Struct Biol 198:124–133

23. van Heel M, Harauz G, Orlova EV et al (1996) A new generation of the IMAGIC image processing system. J Struct Biol 116:17–24

24. van Heel M, Hollenberg J (1980) On the stretching of distorted images of two-dimensional crystals. In: Baumeister W, Vogell W (eds) Electron microscopy at molecular dimensions: state of the art and strategies for the future. Springer, Berlin, Heidelberg, pp 256–260

25. Saxton WO, Baumeister W (1982) The correlation averaging of a regularly arranged bacterial cell envelope protein. J Microsc 127:127–138

26. Henderson R, Baldwin JM, Downing KH et al (1986) Structure of purple membrane from halobacterium halobium: recording, measurement and evaluation of electron micrographs at 3.5 Å resolution. Ultramicroscopy 19:147–178

27. Scherer S, Arheit M, Kowal J et al (2014) Single particle 3D reconstruction for 2D crystal images of membrane proteins J Struct Biol 185:267–277

28. Kuang Q, Purhonen P, Pattipaka T et al (2015) A refined single-particle reconstruction procedure to process two-dimensional crystal images from transmission Electron microscopy. Microsc Microanal 21:876–885

29. McMullan G, Faruqi AR, Henderson R (2016) Direct Electron detectors. In: the resolution revolution: recent advances in cryoEM, 1st edn. Elsevier Inc., Amsterdam, pp 1–17

30. Sigworth FJ, Doerschuk PC, Carazo J M, Scheres SHW (2010) An introduction to maximum-likelihood methods in Cryo-EM. In: enzymology GJJBT-M in Cryo-EM, part B: 3-D reconstruction. Academic Press, New York, pp 263–294

31. Punjani A, Rubinstein JL, Fleet DJ, Brubaker MA (2017) cryoSPARC: algorithms for rapid unsupervised cryo-EM structure determination. Nat Methods 14:290–296

32. Grant T, Rohou A, Grigorieff N (2018) cis-TEM, user-friendly software for single-particle image processing. elife 7:e35383

33. Zivanov J, Nakane T, Forsberg BO et al (2018) New tools for automated high-resolution cryo-EM structure determination in RELION-3. elife 7:e42166

34. Scheres SHW, Gao H, Valle M et al (2007) Disentangling conformational states of macromolecules in 3D-EM through likelihood optimization. Nat Methods 4:27–29

35. Lyumkis D, Brilot AF, Theobald DL, Grigorieff N (2013) Likelihood-based classification of cryo-EM images using FREALIGN. J Struct Biol 183:377–388

36. Kowal J, Chami M, Baumgartner P et al (2014) Ligand-induced structural changes in the cyclic nucleotide-modulated potassium channel MloK1. Nat Commun 5:3106

37. Kowal J, Biyani N, Chami M et al (2018) High-resolution Cryoelectron microscopy structure of the cyclic nucleotide-modulated Potassium Channel MloK1 in a lipid bilayer. Structure 26:20–27.e3

38. Righetto RD, Biyani N, Kowal J et al (2019) Retrieving high-resolution information from disordered 2D crystals by single-particle cryo-EM. Nat Commun 10:1722

39. Grigorieff N (2016) Frealign: an exploratory tool for single-particle Cryo-EM. In: enzymology BT-M in (ed). Academic Press, New York, pp 191–226

40. Abeyrathne PD, Chami M, Pantelic RS et al (2010) Preparation of 2D crystals of membrane proteins for high-resolution Electron crystallography data collection. In: enzymology GJJBT-M in (ed) Cryo-EM part a sample preparation and data collection. Academic Press, New York, pp 25–43

41. Abeyrathne PD, Arheit M, Kebbel F et al (2012) Analysis of 2-D crystals of membrane proteins by Electron microscopy. In: Comprehensive biophysics. Elsevier, Amsterdam, pp 277–310

42. Goldie KN, Abeyrathne P, Kebbel F et al (2014) Cryo-electron microscopy of membrane proteins. Humana Press, Totowa, NJ

43. Schmidt-Krey I, Cheng Y (2013) Electron crystallography of soluble and membrane proteins: methods and protocols. Humana Press, Totowa, NJ

44. Brilot AF, Chen JZ, Cheng A et al (2012) Beam-induced motion of vitrified specimen on holey carbon film. J Struct Biol 177:630–637

45. Grant T, Grigorieff N (2015) Measuring the optimal exposure for single particle cryo-EM using a 2.6 Å reconstruction of rotavirus VP6. elife 4:e06980

46. Rohou A, Grigorieff N (2015) CTFFIND4: fast and accurate defocus estimation from electron micrographs. J Struct Biol 192:216–221

47. Zhang K (2016) Gctf: real-time CTF determination and correction. J Struct Biol 193:1–12

48. Biyani N, Scherer S, Righetto RD et al (2018) Image processing techniques for high-resolution structure determination from badly ordered 2D crystals. J Struct Biol 203:120–134

49. Mindell JA, Grigorieff N (2003) Accurate determination of local defocus and specimen tilt in electron microscopy. J Struct Biol 142:334–347

50. Frank J, Radermacher M, Penczek P et al (1996) SPIDER and WEB: processing and visualization of images in 3D Electron microscopy and related fields. J Struct Biol 116:190–199

51. He S, Scheres SHW (2017) Helical reconstruction in RELION. J Struct Biol 198:163–176

52. Scheres SHW (2012) A Bayesian view on Cryo-EM structure determination. J Mol Biol 415:406–418

53. Bai X, Rajendra E, Yang G et al (2015) Sampling the conformational space of the catalytic subunit of human γ-secretase. elife 4:e11182

54. Harauz G, van Heel M (1986) Exact filters for general geometry three dimensional reconstruction. Optik (Stuttg) 78:146–156

55. Rosenthal PB, Henderson R (2003) Optimal determination of particle orientation, absolute hand, and contrast loss in single-particle Electron Cryomicroscopy. J Mol Biol 333:721–745

56. Brown A, Long F, Nicholls RA et al (2015) Tools for macromolecular model building and refinement into electron cryo-microscopy reconstructions. Acta Crystallogr Sect D Biol Crystallogr 71:136–153

Part IV

Microcrystal Electron Diffraction (MicroED)

Chapter 13

MicroED Sample Preparation and Data Collection For Protein Crystals

Guanhong Bu and Brent L. Nannenga

Abstract

Microcrystal Electron Diffraction (MicroED) enables structure determination of very small crystals that are much too small to be of use for other conventional diffraction techniques. MicroED has been used to determine the structures of many proteins and small organic molecules, and the technique can be performed on most standard cryo-TEM instruments equipped with high-speed detectors capable of collecting electron diffraction data. Here, we present protocols for MicroED sample preparation and data collection for protein microcrystals.

Key words Microcrystal electron diffraction, MicroED, Protein crystallography, Structural biology, Sample preparation

1 Introduction

Microcrystal electron diffraction (MicroED) is a method for collecting electron diffraction data using a cryo-transmission electron microscope (cryo-TEM) [1, 2]. MicroED makes it possible to collect high-resolution electron diffraction data sets from extremely small three-dimensional crystals that would otherwise be unusable for conventional X-ray diffraction methods [3, 4]. Since its initial development, MicroED has been successfully used on a variety of protein [5–10], peptide [11–16], and small molecule targets [17–21]. Sample preparation methods can vary depending on the type of crystal being studied, and here we will present methods for preparing protein crystals samples for MicroED analysis.

When working with protein samples, it is critical that the crystals remain hydrated; therefore, protein microcrystals are vitrified in liquid ethane in order to ensure they are preserved in a frozen hydrated state within the cryo-TEM. Sample preparation equipment and procedures for MicroED are similar to those used in single particle cryo-electron microscopy (cryo-EM) [22, 23]. Briefly, microcrystals are applied onto a carbon-coated

Tamir Gonen and Brent L. Nannenga (eds.), *CryoEM: Methods and Protocols*, Methods in Molecular Biology, vol. 2215,
https://doi.org/10.1007/978-1-0716-0966-8_13, © Springer Science+Business Media, LLC, part of Springer Nature 2021

electron microscopy (EM) grid (e.g., Quantifoil) and excess solution is blotted away using filter paper. Following blotting, the crystal-loaded EM grid is immediately vitrified in liquid ethane and either loaded into the cryo-TEM or stored under liquid nitrogen for future use. While the protocols presented here will focus on more standard sample preparation methods, there are also other procedures that can be used including crystal fragmentation [6] and cryo-focus ion beam (cryo-FIB) milling [24, 25]. Readers are encouraged to refer to other sources for details on how to use these other methods for sample preparation.

Once the sample is inserted into the microscope, the quality of the sample is screened, and data is collected from the best diffracting crystals on the grid. While there are recent methods for automated data collection [26–28] and new user interfaces for MicroED data collection, this chapter will focus on the manual method for MicroED data collection that does not require additional software besides the standard cryo-TEM control interface. Data collection begins by screening the grid for the presence of suitable microcrystals on the grid. Suitable crystals are those which have a defined shape and are not too close to the EM grid bars or other crystals, which would limit the range of data that could be collected (*see* Fig. 1). Following the identification of a suitable crystal, an initial diffraction pattern is collected, and its quality is assessed. If the diffraction is of high quality (*see* Fig. 2), a continuous rotation data set [7] should be collected from this crystal. This involves tilting the stage of the cryo-TEM to high-tilt and then starting a steady and continuous rotation of the stage. Diffraction data are recorded using a high-speed detector as the crystal continuously tilts in the beam. These diffraction data sets are then processed using standard data processing workflows [29, 30].

2   Materials

1. Holey carbon or continuous carbon-coated EM grids (copper; 200–400 mesh).

2. Protein microcrystals.

3. Glow discharge system.

4. Clean glass microscope slides.

5. Plunge freezing apparatus (*see* **Note 1**).

6. Locking fine-tip tweezers for plunge freezing apparatus.

7. Blotting paper for plunge freezing apparatus.

8. Liquid Nitrogen.

9. Liquid Ethane.

10. Pipettes capable of dispensing in the range of 1–20 µL.

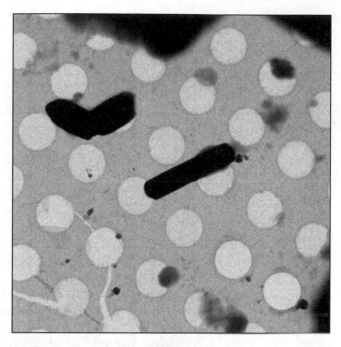

Fig. 1 Images of protein microcrystals as seen under a typical medium-magnification search. Crystals show well-defined edges and are well-dispersed on the grid. Holes in the carbon film are approximately 2 μm

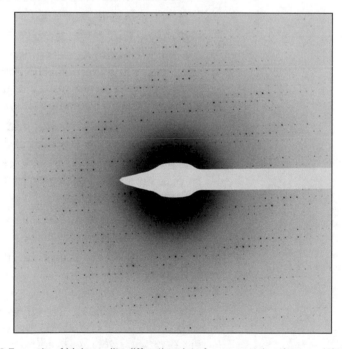

Fig. 2 Example of high-quality diffraction data from a protein microcrystal. Spots are sharp and well separated and extend to high resolution

11. Cryo grid box.

12. Cryo-TEM (*see* **Note 2**).

13. High-speed detector for the collection of electron diffraction data (*see* **Note 3**).

3 Methods

Here, we will present the generic workflow for MicroED sample preparation and data collection when working with protein crystals. The procedures start after the identification of protein microcrystals [22, 31–33] and ends with the collection of continuous rotation MicroED data sets that can be processed and used for structure determination [22]. It is recommended that first time users of MicroED attempt these protocols with known standards in order to become familiar with the process. Suggested protein standards are lysozyme and proteinase K and there are several publications where these samples have been analyzed by MicroED [6, 7, 9, 34].

3.1 Sample Preparation

1. Use clean tweezers to place carbon-coated EM grids onto a clean glass microscope slide.

2. Use a glow discharge unit and follow the manufacturer's protocol to clean the surface of the carbon.

3. Use tweezers to flip the grids over on the glass slide and repeat **step 2** to treat the opposite surface of the grid.

4. Assemble the components of the plunge freezing coolant container according to the specific manufacturer's instructions. Fill container with liquid nitrogen and allow the temperature to stabilize. Add additional liquid nitrogen as needed.

5. Fill the reservoir where the grid will be plunged with liquid ethane according to manufacturer's instructions.

6. Use the locking fine-tip tweezers for plunge freezing apparatus to pick up an EM grid prepared following **step 3** and place tweezers on the lab bench.

7. Use a pipette to aspirate 2–3 μL of solution containing microcrystals (*see* **Note 4**). Dispense solution onto the EM grid.

8. Allow crystal solution to rest on the surface of the grid for 30 s to 1 min.

9. In some cases, microcrystal solution is applied to both sides. If crystals will only be deposited on one side of the grid, skip this step. To deposit crystals on the other side of the grid as well, flip over the tweezers holding the EM grid that has just had microcrystal solution applied to it, and repeat **steps 7** and **8** to deposit on to the other side of the grid.

10. Attach fine-tip tweezers holding the EM grid and protein microcrystal sample to the plunge freezing device.

11. Blot and vitrify the sample using the plunge freezing device (*see* **Note 5**). Place the vitrified sample into the cryo grid box that is under liquid nitrogen. Ensure that once the grid has been vitrified it always remains at liquid nitrogen temperatures.

12. Repeat **steps 6–11** to make multiple samples. Additional samples should be made to screen around blotting time, volume of crystals dispensed on grid, and microcrystal treatment strategies (e.g., cryo-protectants, dilutions).

13. Once all samples have been made, store the cryo grid boxes containing the prepared EM grids under liquid nitrogen until loading into a cryo-TEM for analysis.

3.2 Screening Microcrystal Diffraction Quality

1. Follow protocols for the cryo-TEM to align the microscope in both imaging and diffraction mode (*see* **Note 6**).

2. Load grids into a cryo-TEM equipped with a high-speed detector for diffraction data collection (*see* **Notes 2** and **3**).

3. When the samples have been loaded in the cryo-TEM, examine the grid under low magnification ($\sim$100–200$\times$). This step is to examine the overall thickness of the grid and verify presence of protein microcrystals. It is common to observe an uneven distribution of crystals and overall thickness. If there are no visible regions of the grid because grid is too thick everywhere, or if there are no crystals visible on the grid, either load a new grid or repeat the sample preparation steps to prepare new samples with different blotting or sample handling conditions.

4. Note areas of the grid which are promising (i.e., thin enough for the beam to penetrate and holding crystals). Move the stage to one of these grid squares and increase the magnification to a medium level (approximately 500–1000$\times$). The field of view should be approximately one grid square (*see* Fig. 1).

5. Locate promising crystals on the grid square. If no crystals are visible when zooming in after **step 4**, return to **step 4**, and move to a new grid square.

6. When a promising crystal is identified (well-defined edges and well-separated from other crystals or the grid bars) on the grid square, move the crystal to the center of the detector. The eucentric height should be accurately set so that the crystal does not move as the stage is rotated.

7. Insert the selected area aperture.

8. Insert the beam stop, unless the detector equipped on the cryo-TEM does not require the use of a beam stop.

9. Blank the beam and switch the microscope into diffraction mode that has been set up to collect low-dose diffraction patterns.

10. Unblank the beam and collect a still exposure diffraction pattern from the single crystal.

11. Assess the quality of the diffraction pattern. If the reflections are sharp, well separated, and extend to high resolution (*see* Fig. 2), proceed to the next section to collect a continuous rotation data set from this crystal. If the crystal shows poor diffraction, or no diffraction, switch the microscope back to imaging mode, remove the selected area aperture and beam stop, and return to **step 5**.

3.3 MicroED Data Collection

1. Once a crystal has shown high-quality diffraction from a single exposure. Return the microscope to imaging to visualize the crystal on the grid square again (500–1000× magnification). Remove the selected area aperture and beam stop.

2. Tilt the stage in both the positive and negative directions to ensure the crystal remains centered while rotating and to determine the maximum possible tilt range that can be collected without the crystal being obscured by other crystals or the bars of the grid. If the crystals move while rotating, correct the Z-height to return the crystal to the eucentric position.

3. Once the tilt range to be collected has been confirmed in **step 2**, tilt the grid to the maximum tilt angle of this range. Put in the selected area aperture and beam stop. Blank the beam.

4. Based on the manufacturer and model, set up the stage rotation such that when data collection begins, the stage can be rotated slowly from the starting tilt angle to the end tilt angle (*see* **Note 7**).

5. According to manufactures direction, set the high-speed detector up for data to be collected in a movie mode.

6. Return the microscope to diffraction mode used for collecting low-dose diffraction patterns.

7. Unblank the beam and begin the rotation of the stage. Once the stage begins to rotate, begin recording with the high-speed detector.

8. Once the data set has been collected, return the stage to 0° tilt, remove the beam stop and the selected area aperture.

9. Repeat the procedures presented in Subheadings 3.2 and 3.3 to collect data sets from multiple crystals on the grid. Generally, multiple crystals are needed for structure determination. Therefore, it is recommended to collect as many high-quality data sets as possible from grid that shows high-quality crystals. Following the successful use of these methods on suitable

samples, the user will have collected several high-quality MicroED data sets that can be processed by standard programs and procedures [29, 30, 35].

4 Notes

1. There is not one specific plunge freezing device required for MicroED. Many types of automated or manual plunge freezing devices can be used for vitrification of samples for MicroED. Users are encouraged to use whichever device they prefer.

2. In principle, any cryo-TEM capable of low-dose data collection can be used for MicroED. Much of the previous work on MicroED has used Thermo Fisher cryo-TEMs (e.g., TF-20, Titan Krios); however, other manufactures such as JEOL can be used as well. One of the required features of the cryo-TEM is the ability to load vitrified samples into the grid either with an autoloader system or a compatible cryo-holder. The second required feature is the ability to collect data under very low-dose conditions and to be able to switch between the three aligned low-dose modes: low-magnification search, medium-magnification search, and diffraction data collection mode. Low-magnification search should be approximately $100–200\times$ and will be used to visualize large areas of the grid at once. Medium-magnification mode will be around $500–1000\times$ (enough to see one grid square) and is used to identify crystals on the grid. Alternatively, medium-magnification search can be performed in over-focused diffraction mode. Both search modes should be set up to have the absolute minimum amount of dose required to visualize crystals on the grid. The third mode, diffraction data collection mode, is performed while the microscope is in diffraction mode and should be set such that the exposure is on the order of 0.01 $e^-/\text{Å}^2/s$. It is critical that the diffraction data collection mode is very well aligned with the medium-magnification search mode, so that when a crystal is centered in the search mode, it is also centered in the beam for diffraction data collection mode.

3. A requirement for continuous rotation MicroED data collection is that the detector used must have a high dynamic range, be sensitive, and be able to collect high-speed data. Several camera systems have been used for MicroED including high-speed scintillator-based CMOS detectors [7, 8, 16, 17], direct electron detectors [35], and hybrid-pixel detectors [10, 36, 37].

4. In some cases, crystals must be diluted using the precipitant solution. In the cases where the crystals grow in highly viscous

solutions (e.g., high-molecular weight PEGs), dilution in the presence of less viscous (e.g., low-molecular weight PEGs) is required. This is generally a trial-and-error process where the user optimizes the ability to blot the solution with the stability of the microcrystals in these solutions. For sensitive crystals, pipetting should be kept as gentle as possible. If crystals are stable yet too large, there are protocols for fragmenting by pipetting, and other means, available [6].

5. Blotting time is another variable that must be screened to obtain optimal samples. Ideally, the crystals will have a thin layer of vitrified solution surrounding them, and the rest of the grid will be easily penetrated by the electron beam. If blotting time is too long, too much solution may be removed, and crystal quality can be affected. If blotting time is too short, the grid may be too thick and the electron beam will not penetrate. It is recommended to try a wide range of blotting time (e.g., 2–20 s) and multiple blotting procedures, where the sample is blotted more than once before plunge freezing.

6. It is critical that the microscope be well aligned for the collection of high-quality MicroED data. Also, the medium-magnification search mode must be aligned with the diffraction data collection mode so that when a crystal is located in medium-magnification search, it can be exposed by the diffraction mode and data can be collected.

7. The rotation rate of the microscope is a parameter that can be adjusted depending on the crystal samples being studied. By adjusting the rotation rate along with the integration time on the high-speed detector, the user can control how many degrees of rotation occurs for each diffraction frame in the data set. For example, if the rotation rate on the stage is $0.1°/s$ and the frame rate of the detector is set to 4 s, each frame in the data set will consist of a $0.4°$ sampling of reciprocal space.

Acknowledgments

The Nannenga lab is supported by the National Institutes of Health grant R01GM124152 and R21GM135784, the Air Force Office of Scientific Research grant FA9550-18-1-0012, and the National Science Foundation award 1942084. We acknowledge the use of facilities within the LeRoy Eyring Center for Solid State Science at Arizona State University, Tempe, AZ, specifically the use of the Titan Krios and the funding of this instrument by NSF MRI 1531991.

References

1. Nannenga BL (2020) MicroED methodology and development. Struct Dyn 7(1):014304. https://doi.org/10.1063/1.5128226

2. Nannenga BL, Gonen T (2019) The cryo-EM method microcrystal electron diffraction (MicroED). Nat Methods 16(5):369–379. https://doi.org/10.1038/s41592-019-0395-x

3. Wolff AM, Young ID, Sierra RG, Brewster AS, Martynowycz MW, Nango E, Sugahara M, Nakane T, Ito K, Aquila A, Bhowmick A, Biel JT, Carbajo S, Cohen AE, Cortez S, Gonzalez A, Hino T, Im D, Koralek JD, Kubo M, Lazarou TS, Nomura T, Owada S, Samelson AJ, Tanaka T, Tanaka R, Thompson EM, van den Bedem H, Woldeyes RA, Yumoto F, Zhao W, Tono K, Boutet S, Iwata S, Gonen T, Sauter NK, Fraser JS, Thompson MC (2020) Comparing serial X-ray crystallography and microcrystal electron diffraction (MicroED) as methods for routine structure determination from small macromolecular crystals. IUCrJ 7(Pt 2):306–323. https://doi.org/10.1107/S205225252000072X

4. Zatsepin NA, Li C, Colasurd P, Nannenga BL (2019) The complementarity of serial femtosecond crystallography and MicroED for structure determination from microcrystals. Curr Opin Struct Biol 58:286–293. https://doi.org/10.1016/j.sbi.2019.06.004

5. Purdy MD, Shi D, Chrustowicz J, Hattne J, Gonen T, Yeager M (2018) MicroED structures of HIV-1 Gag CTD-SP1 reveal binding interactions with the maturation inhibitor bevirimat. Proc Natl Acad Sci 115(52):13258–13263. https://doi.org/10.1073/pnas.1806806115

6. de la Cruz MJ, Hattne J, Shi D, Seidler P, Rodriguez J, Reyes FE, Sawaya MR, Cascio D, Weiss SC, Kim SK, Hinck CS, Hinck AP, Calero G, Eisenberg D, Gonen T (2017) Atomic-resolution structures from fragmented protein crystals with the cryoEM method MicroED. Nat Methods 14(4):399–402. https://doi.org/10.1038/nmeth.4178

7. Nannenga BL, Shi D, Leslie AG, Gonen T (2014) High-resolution structure determination by continuous-rotation data collection in MicroED. Nat Methods 11(9):927–930. https://doi.org/10.1038/nmeth.3043

8. Nannenga BL, Shi D, Hattne J, Reyes FE, Gonen T (2014) Structure of catalase determined by MicroED. elife 3:e03600. https://doi.org/10.7554/eLife.03600

9. Shi D, Nannenga BL, Iadanza MG, Gonen T (2013) Three-dimensional electron crystallography of protein microcrystals. elife 2:e01345. https://doi.org/10.7554/eLife.01345

10. Xu H, Lebrette H, Clabbers MTB, Zhao J, Griese JJ, Zou X, Högbom M (2019) Solving a new R2lox protein structure by microcrystal electron diffraction. Sci Adv 5(8):eaax4621. https://doi.org/10.1126/sciadv.aax4621

11. Warmack RA, Boyer DR, Zee C-T, Richards LS, Sawaya MR, Cascio D, Gonen T, Eisenberg DS, Clarke SG (2019) Structure of amyloid-β (20-34) with Alzheimer's-associated isomerization at Asp23 reveals a distinct protofilament interface. Nat Commun 10(1):3357. https://doi.org/10.1038/s41467-019-11183-z

12. Krotee P, Griner SL, Sawaya MR, Cascio D, Rodriguez JA, Shi D, Philipp S, Murray K, Saelices L, Lee J, Seidler P, Glabe CG, Jiang L, Gonen T, Eisenberg DS (2018) Common fibrillar spines of amyloid-beta and human islet amyloid polypeptide revealed by micro-electron diffraction and structure based inhibitors. J Biol Chem 293(8):2888–2902. https://doi.org/10.1074/jbc.M117.806109

13. Gallagher-Jones M, Glynn C, Boyer DR, Martynowycz MW, Hernandez E, Miao J, Zee CT, Novikova IV, Goldschmidt L, McFarlane HT, Helguera GF, Evans JE, Sawaya MR, Cascio D, Eisenberg DS, Gonen T, Rodriguez JA (2018) Sub-angstrom cryo-EM structure of a prion protofibril reveals a polar clasp. Nat Struct Mol Biol 25(2):131–134. https://doi.org/10.1038/s41594-017-0018-0

14. Krotee P, Rodriguez JA, Sawaya MR, Cascio D, Reyes FE, Shi D, Hattne J, Nannenga BL, Oskarsson ME, Philipp S, Griner S, Jiang L, Glabe CG, Westermark GT, Gonen T, Eisenberg DS (2017) Atomic structures of fibrillar segments of hIAPP suggest tightly mated beta-sheets are important for cytotoxicity. elife 6:e19273. https://doi.org/10.7554/eLife.19273

15. Sawaya MR, Rodriguez J, Cascio D, Collazo MJ, Shi D, Reyes FE, Hattne J, Gonen T, Eisenberg DS (2016) Ab initio structure determination from prion nanocrystals at atomic resolution by MicroED. Proc Natl Acad Sci U S A 113(40):11232–11236. https://doi.org/10.1073/pnas.1606287113

16. Rodriguez JA, Ivanova MI, Sawaya MR, Cascio D, Reyes FE, Shi D, Sangwan S, Guenther EL, Johnson LM, Zhang M, Jiang L, Arbing MA, Nannenga BL, Hattne J, Whitelegge J, Brewster AS, Messerschmidt M, Boutet S, Sauter NK, Gonen T, Eisenberg DS

(2015) Structure of the toxic core of alpha-synuclein from invisible crystals. Nature 525 (7570):486–490. https://doi.org/10.1038/nature15368

17. Levine AM, Bu G, Biswas S, Tsai EHR, Braunschweig AB, Nannenga BL (2020) Crystal structure and orientation of organic semiconductor thin films by microcrystal electron diffraction and grazing-incidence wide-angle X-ray scattering. Chem Commun (Camb) 56 (30):4204–4207. https://doi.org/10.1039/d0cc00119h

18. van Genderen E, Clabbers MT, Das PP, Stewart A, Nederlof I, Barentsen KC, Portillo Q, Pannu NS, Nicolopoulos S, Gruene T, Abrahams JP (2016) Ab initio structure determination of nanocrystals of organic pharmaceutical compounds by electron diffraction at room temperature using a Timepix quantum area direct electron detector. Acta Crystallogr A Found Adv 72(Pt 2):236–242. https://doi.org/10.1107/S2053273315022500

19. Gruene T, Wennmacher JTC, Zaubitzer C, Holstein JJ, Heidler J, Fecteau-Lefebvre A, De Carlo S, Muller E, Goldie KN, Regeni I, Li T, Santiso-Quinones G, Steinfeld G, Handschin S, van Genderen E, van Bokhoven JA, Clever GH, Pantelic R (2018) Rapid structure determination of microcrystalline molecular compounds using electron diffraction. Angew Chem Int Ed Eng 57 (50):16313–16317. https://doi.org/10.1002/anie.201811318

20. Ting CP, Funk MA, Halaby SL, Zhang Z, Gonen T, van der Donk WA (2019) Use of a scaffold peptide in the biosynthesis of amino acid–derived natural products. Science 365 (6450):280. https://doi.org/10.1126/science.aau6232

21. Jones CG, Martynowycz MW, Hattne J, Fulton TJ, Stoltz BM, Rodriguez JA, Nelson HM, Gonen T (2018) The cryoEM method microED as a powerful tool for small molecule structure determination. ACS Central Sci 4 (11):1587–1592. https://doi.org/10.1021/acscentsci.8b00760

22. Shi D, Nannenga BL, de la Cruz MJ, Liu J, Sawtelle S, Calero G, Reyes FE, Hattne J, Gonen T (2016) The collection of MicroED data for macromolecular crystallography. Nat Protoc 11(5):895–904. https://doi.org/10.1038/nprot.2016.046

23. Carragher B, Cheng Y, Frost A, Glaeser RM, Lander GC, Nogales E, Wang HW (2019) Current outcomes when optimizing 'standard' sample preparation for single-particle cryo-EM. J Microsc 276(1):39–45. https://doi.org/10.1111/jmi.12834

24. Martynowycz MW, Zhao W, Hattne J, Jensen GJ, Gonen T (2019) Collection of continuous rotation MicroED data from ion beam-milled crystals of any size. Structure 27(3):545–548. e542. https://doi.org/10.1016/j.str.2018.12.003

25. Duyvesteyn HME, Kotecha A, Ginn HM, Hecksel CW, Beale EV, de Haas F, Evans G, Zhang P, Chiu W, Stuart DI (2018) Machining protein microcrystals for structure determination by electron diffraction. Proc Natl Acad Sci U S A 115(38):9569–9573. https://doi.org/10.1073/pnas.1809978115

26. de la Cruz MJ, Martynowycz MW, Hattne J, Gonen T (2019) MicroED data collection with SerialEM. Ultramicroscopy 201:77–80. https://doi.org/10.1016/j.ultramic.2019.03.009

27. Wang B, Zou X, Smeets S (2019) Automated serial rotation electron diffraction combined with cluster analysis: an efficient multi-crystal workflow for structure determination. IUCrJ 6 (Pt 5):854–867. https://doi.org/10.1107/S2052252519007681

28. Cichocka MO, Angstrom J, Wang B, Zou X, Smeets S (2018) High-throughput continuous rotation electron diffraction data acquisition via software automation. J Appl Crystallogr 51(Pt 6):1652–1661. https://doi.org/10.1107/S1600576718015145

29. Hattne J, Reyes FE, Nannenga BL, Shi D, de la Cruz MJ, Leslie AG, Gonen T (2015) MicroED data collection and processing. Acta Crystallogr A Found Adv 71(Pt 4):353–360. https://doi.org/10.1107/S2053273315010669

30. Clabbers MTB, Gruene T, Parkhurst JM, Abrahams JP, Waterman DG (2018) Electron diffraction data processing with DIALS. Acta Crystallogr D Struct Biol 74(Pt 6):506–518. https://doi.org/10.1107/S2059798318007726

31. Stevenson HP, Lin G, Barnes CO, Sutkeviciute I, Krzysiak T, Weiss SC, Reynolds S, Wu Y, Nagarajan V, Makhov AM, Lawrence R, Lamm E, Clark L, Gardella TJ, Hogue BG, Ogata CM, Ahn J, Gronenborn AM, Conway JF, Vilardaga JP, Cohen AE, Calero G (2016) Transmission electron microscopy for the evaluation and optimization of crystal growth. Acta Crystallogr D Struct Biol 72(Pt 5):603–615. https://doi.org/10.1107/S2059798316001546

32. Barnes CO, Kovaleva EG, Fu X, Stevenson HP, Brewster AS, DePonte DP, Baxter EL, Cohen AE, Calero G (2016) Assessment of

microcrystal quality by transmission electron microscopy for efficient serial femtosecond crystallography. Arch Biochem Biophys 602:61–68. https://doi.org/10.1016/j.abb.2016.02.011

33. Stevenson HP, Makhov AM, Calero M, Edwards AL, Zeldin OB, Mathews II, Lin G, Barnes CO, Santamaria H, Ross TM, Soltis SM, Khosla C, Nagarajan V, Conway JF, Cohen AE, Calero G (2014) Use of transmission electron microscopy to identify nanocrystals of challenging protein targets. Proc Natl Acad Sci U S A 111(23):8470–8475. https://doi.org/10.1073/pnas.1400240111

34. Hattne J, Shi D, de la Cruz MJ, Reyes FE, Gonen T (2016) Modeling truncated pixel values of faint reflections in MicroED images. J Appl Crystallogr 49(Pt 3):1029–1034. https://doi.org/10.1107/S1600576716007196

35. Hattne J, Martynowycz MW, Penczek PA, Gonen T (2019) MicroED with the Falcon III direct electron detector. IUCrJ 6 (Pt 5):921–926. https://doi.org/10.1107/S2052252519010583

36. Tinti G, Frojdh E, van Genderen E, Gruene T, Schmitt B, de Winter DAM, Weckhuysen BM, Abrahams JP (2018) Electron crystallography with the EIGER detector. IUCrJ 5 (Pt 2):190–199. https://doi.org/10.1107/S2052252518000945

37. Clabbers MTB, van Genderen E, Wan W, Wiegers EL, Gruene T, Abrahams JP (2017) Protein structure determination by electron diffraction using a single three-dimensional nanocrystal. Acta Crystallogr D Struct Biol 73 (Pt 9):738–748. https://doi.org/10.1107/S2059798317010348

Chapter 14

Detection of Microcrystals for CryoEM

Simon Weiss, Sandra Vergara, Guowu Lin, and Guillermo Calero

Abstract

Here, we present a strategy to identify microcrystals from initial protein crystallization screen experiments and to optimize diffraction quality of those crystals using negative stain transmission electron microscopy (TEM) as a guiding technique. The use of negative stain TEM allows visualization along the process and thus enables optimization of crystal diffraction by monitoring the lattice quality of crystallization conditions. Nanocrystals bearing perfect lattices are seeded and can be used for MicroED as well as growing larger crystals for X-ray and free electron laser (FEL) data collection.

Key words Protein crystallization, Optimization, Brightfield microscopy, UV microscopy, Negative staining TEM, FFT-calculation, Granular aggregate, Microcrystal, Crystal lattice quality

---

1 Introduction

Crystallization of protein targets remains the most significant challenge in the process of structure determination by macromolecular crystallography. When setting up crystallization trays experimenters can obtain three different negative outcomes: (a) conditions with amorphous precipitate, (b) few or no conditions with apparent crystals or (c) crystals that do not diffract or diffract to low resolution. To tackle these problems most recent efforts have focused mainly on improving sample "crystallizability" by implementing techniques such as alanine scanning mutagenesis [1–3] or the use of chimeric proteins to promote/improve crystal packing [4–8], which has led to structures of important targets. However, less effort has been applied towards the discovery, evaluation, and optimization of crystals and nanocrystals. Here, we present the use of transmission electron microscopy (TEM) as an easy, reproducible technique to guide the optimization of protein crystallization [9–12]. We will describe reproducible protocols to use TEM for visualizing lattices from microcrystals, as well as fragmented larger crystals and to study details of the crystallization process. Our experiments have shown that for all protein crystals tested,

Tamir Gonen and Brent L. Nannenga (eds.), *CryoEM: Methods and Protocols*, Methods in Molecular Biology, vol. 2215,
https://doi.org/10.1007/978-1-0716-0966-8_14, © Springer Science+Business Media, LLC, part of Springer Nature 2021

TEM images reveal details of the crystal lattices that prove to be accurate qualitative indicators of the potential diffraction of the crystal. In general, the detection by negative stain TEM methods of well-ordered lattices with third order Bragg spots was a good predictor of X-ray diffraction, while samples with disorganized lattices yielded no diffraction [12]. Additionally, we show that TEM can play an important role to identify crystal pathologies that contribute to poor X-ray diffraction data [11]. Among them are: (a) crystal lattice defects; (b) anisotropic diffraction; and (c) crystal "polluting" by heavy protein aggregates and microcrystal nuclei. Detection of lattice defects in some crystals could point to the presence of samples containing protein contaminants, aggregates or partially proteolyzed protein as well as discrepancies in the stoichiometry of the sample. Negative stain TEM analysis was able to identify crystals that possess anisotropic diffraction. In these cases, performing steps to improve crystal contacts, such as altering or adding reagents to the crystallization conditions or modification of the protein itself, may be advisable. It is important to note that this information cannot be observed with other techniques before performing X-ray diffraction experiments. The use of TEM has also enabled qualitative estimation of crystal solvent content and allowed the study of lattice dehydration on crystal diffraction [11]. This application was particularly noteworthy since (a) crystal dehydration protocols have proven very useful in the improvement of X-ray diffraction [13–15], and (b) negative staining with uranyl acetate provides very high contrast between solvent channels and biological macromolecules. Overall, information obtained by TEM experiments could provide critical advice to the experimenter about which crystallization conditions to be pursued and would also allow monitoring of crystal optimization protocols.

2   Materials

2.1   Microscopy

1. Brightfield and light microscope.

2. Ultraviolet (UV) microscope.

3. Electron microscope (e.g., FEI TECNAI T12 operating at 120 kV with a single-tilt specimen holder and 2 k × 2 k Gatan UltraScan 1000 CCD camera).

2.2   Negative Staining Components

1. 2% (w/v) uranyl acetate in 15 mL plastic Falcon tube, covered in aluminum foil.

2. 0.22 μm sterile filter.

3. Carbon-coated CF400-CU grids.

4. Glow discharge unit.

5. Filter paper.

6. Optional: Vortex Mixer.

7. Optional: glass beads 0.5 mm or 1 mm.

3   Methods

1. Set up hanging drop vapor diffusion crystallization screens of your target protein (*see* **Note 1**).

2. Identify conditions with granular aggregate and micro or nanocrystals by screening the crystallization trays under a Brightfield (BF) microscope to select conditions for the negative stain TEM analysis (*see* **Note 2** and Fig. 1).

3. Confirm the presence of protein in the conditions identified with BF microscopy by taking UV microscope images (*see* **Note 3** and Fig. 1).

4. Prepare a 2% (w/v) uranyl acetate solution in dH_2O, dissolving the UA by rocking it in a 15 mL Falcon tube covered with aluminum foil at RT for at least half an hour before filtering with a 0.22 μm filter immediately before use (*see* **Note 4**).

5. Negatively glow discharge the TEM grids (e.g., CF400-CU, Electron Microscopy Sciences) for 1 min at 25 mV and 0.2 mbar (EmiTech KX 100) with the carbon side up, not more than 20 min before loading your sample (*see* **Note 5**).

6. Using a razor blade, remove the coverslip of a well, which bears UV-positive granular aggregates from the crystallization tray. Place it upside down on the transparent cover of the crystallization tray in the path of the microscope. Then carefully add about 2 μL of mother liquor (or stabilizing solution; *see* **Notes 6** and **7**) to the crystal drop and gently mix by pipetting up and down several times (*see* **Note 6**).

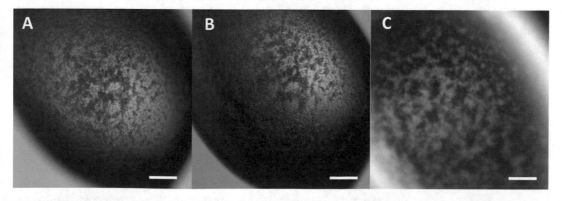

Fig. 1 The "shiny" effect of the granular aggregates under the BF microscope may not be visible in regular settings (**a**) but when the polarization and contrast are carefully adjusted the aggregates are clearly identified (**b**). UV image of the "shiny" sample (**c**). The scale bar in (**a**) and (**b**) represents 50 μm and in (**c**) 30 μm

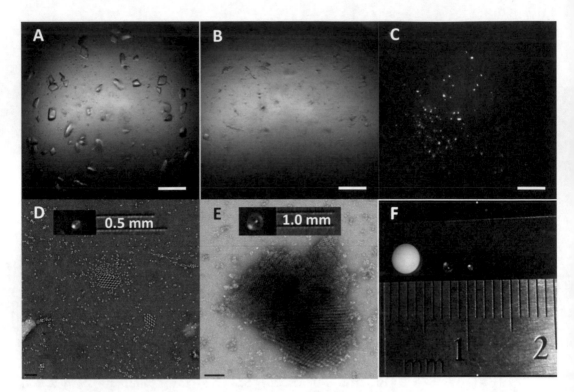

Fig. 2 Fragmentation of microcrystals for negative stain TEM analysis. (**a**) BF image of non-fragmented crystals, (**b**) BF image of crystals after vortexing with 0.5 mm beads, (**c**) UV image of crystals after vortexing with 0.5 mm beads, (**d**) negative stain TEM image of crystals after vortexing with 0.5 mm beads and (**e**) negative stain TEM image of crystals after vortexing with 1.0 mm beads. (**f**) Comparison of 5 mm Teflon ball (white, left) and glass beads of 1 mm (center) or 0.5 mm (right) used for microcrystals fragmentation. Panels **a–c**, **d** reproduced from Lin et al. (2019) [16] with permission from Elsevier. Panels **d**, **e**, reproduced from Stevenson et al. (2016) [12] with permission of the International Union of Crystallography

7. Optional: Fragmentation of larger crystals, i.e., bigger than about 2 μm for the smallest side. First add approximately 10 μL of glass beads (Research Products International) to a 1.5 mL Eppendorf tube and wash them with 500 μL of 20% (v/v) ethanol, followed by washing with 500 μL dH$_2$O. Then the beads are equilibrated with about 30 μL of stabilizing buffer (e.g., reservoir buffer). Pipette the collected crystals into the Eppendorf tube with the glass beads. The crystals are fragmented by vortexing (about 2 s to 2 min) until UV microscopy images reveal a homogeneous slurry of high density crystal fragments with edge lengths of around 1–5 μm (*see* **Note 7** and Fig. 2).

8. Pipette approximately 2 μL of the granular mix or the fragmented crystals onto the carbon film side of the electron microscopy grid and incubate for 1 min. Then remove the excess liquid by blotting the grid with P2 filter paper (Fisherbrand) from the side (*see* **Note 8** and Fig. 3).

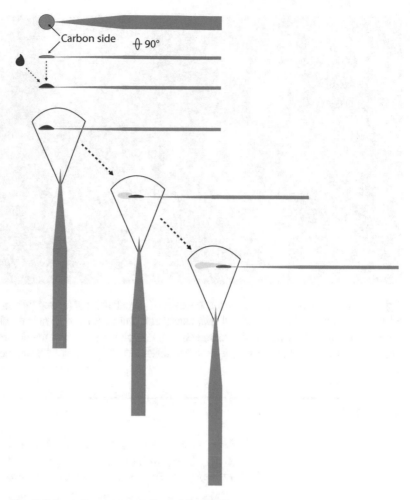

Fig. 3 Schematic representation of the blotting process

9. Immediately after blotting place the grid (carbon side down) on top of a drop of 200 μL of a 2% (w/v) uranyl acetate solution for 40 s. Transfer the grid to a second drop of 200 μL of a 2% (w/v) uranyl acetate solution and stain for another 40 sec without blotting in between the two staining steps. After staining, the grid is blotted from the back with filter paper in order to remove the staining solution and dried on air for at least 20 min prior to further use (*see* **Note 9**).

10. Inspect the nature of the granular aggregates and lattice quality of the (fragmented) microcrystals by collecting TEM images (*see* **Note 10** and Fig. 4).

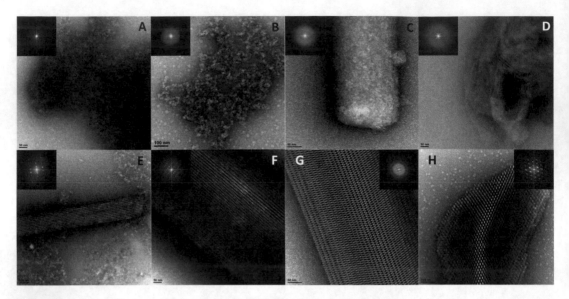

Fig. 4 Negative stain images and corresponding FFT calculations of (**a**) and (**b**) amorphous granular aggregate, (**c**) crystal without lattice, (**d**) crystal with disordered lattice, (**e**) and (**f**) anisotropic crystals, (**g**) well-ordered crystals, and (H) fragmented well-ordered crystal. Complete datasets of the (fragmented) biologically relevant multi-protein-complex crystals (**g**, **h**) could be collected at Synchrotron beamlines with resolutions of about 3.5 Å

4   Notes

1. The final volume of the crystallization drop should be around 1 μL, therefore about 1.0 μL protein stock solution and 1.0 μL reservoir buffer should be used to set up each initial crystallization drop due to evaporation process during the crystallization experiment. The exact values and ratio between protein and reservoir solution will depend on the specific "crystallizability" of each protein of interest.

2. The initial screening of the crystallization tray should identify conditions displaying a significant amount of "shiny" aggregate (*see* Fig. 1 for examples). This screening is best performed at magnifications of 30–100×. In order to obtain the best possible image quality and differentiate between shiny and non-shiny aggregate the experimenter might need to change the polarization and contrast settings of the microscope. It is advised to especially look for any type of shape (e.g., needle like or other symmetric structures), which are likely indicators for the presence of nanocrystals.

3. Crystal drops containing granular aggregates under BF microscopy are selected for further evaluation using ultraviolet (UV) microscopy (Jansi, Molecular Dimensions). UV positive granular aggregates can be easily discerned with the use of a

10× objective (*see* Fig. 1). As with the BF microscopy, it is essential to adjust the focus planes to obtain the most contrasting UV-images possible. For very fine granular aggregates use of a 40× objective might be required. False UV-positive granular aggregates can be observed in crystallization conditions that include calcium salts.

4. The 2% (w/v) uranyl acetate solution can be used for up to 1 month but should be filtered with a 0.22 μm filter on the day of grid preparation to obtain best crystal staining results.

5. We routinely use carbon-coated grids with a 400-mesh. These grids result in less broken areas compared to larger mesh grids (e.g., 300-mesh, 200-mesh) when blotting by hand from the back with the P2 filter paper.

6. When preparing the sample to load onto the grid be sure to visually check that the microcrystals are soaked into the pipette. In some cases, the sample might stick to the cover slid. In those circumstances, use the pipette tip or a crystallization tool to carefully perform slow stirring motions to detach the microcrystals or aggregate from the cover slid. If the reservoir buffer has a very high viscosity (e.g., a PEG-10000 concentration above 10% (w/v) or other similar precipitants), it is recommended to use a stabilizing solution containing methyl pentanediol (MPD) instead of the highly viscous reservoir solution (*see also* **Note 7**).

7. The size of glass beads (0.5 mm or 1.0 mm diameter) should be selected depending on the stability of the protein crystal; smaller beads result in harsher crystal fragmentation (*see* Fig. 2) [11]. For fragile crystals (e.g., needles and thin plates), we recommend using 1.0 mm beads; for sturdier and "chunkier" crystals, 0.5 mm beads are generally needed. Similarly, the vortexing time has to be chosen depending on the crystal sturdiness. In case of temperature-sensitive protein crystals, the fragmentation process has been carried out at 4 °C while cooling the sample on ice in between vortex intervals to prevent protein denaturation. For scarce material containing fewer granular aggregates, it is recommended to use stainless steel beads that can be removed from the tube using a strong magnet. Concentration of the slurry can be achieved by removing the magnetic beads, followed by low speed centrifugation (524 rcf) to pellet down crystal fragments followed by removal of excess stabilizing solution. Crystallization conditions that have highly viscous solutions (e.g., 25% polyethylene glycol 8000 (peg8K)) are difficult to handle or visualize on EM grids. For such conditions, it is convenient to find a stabilizing solution with low viscosity where aggregates do not dissolve. A solution containing 25% peg8K could be exchanged with a

stabilizing solution containing 30–40% methyl pentanediol (MPD) during several rounds of concentrating, removing excess liquid and slowly adding the new stabilizing solution.

8. Hold the grid at a perpendicular angle relative to the filter paper and slowly move it along the filter paper with only its edge touching the filter paper. Make sure to visually check the filter paper during to blotting process for signs of the blotted liquid (*see* Fig. 3). When instead blotting from the front, the crystals might be inadvertently removed from the grid as well, while blotting from the back might not be possible with a non-holey grid due to the high buffer viscosity. Generally, the more viscous the reservoir solution, the slower the blotting process will be and it will take about 2–10 s to fully blot away the excess liquid.

9. Within a couple of seconds after blotting, place the blotted grid sample side down on top of the first drop of uranyl acetate solution in order to keep the sample from completely drying out before the staining. It is important to fully blot away excess liquid but not to let the crystals crack due to dehydration when trying to optimize staining results, therefore a thorough blotting and fast transfer are essential.

10. For the TEM screening, we have predominantly used an FEI TECNAI T12 electron microscope operating at 120 kV with a single-tilt specimen holder. Images are acquired with a 2 k × 2 k Gatan UltraScan 1000 CCD camera, typically at magnifications between 11,000 and 52,000×. Fast Fourier Transform (FFT) calculations of the lattices are used to determine the crystal quality. The experimenter should be prepared to spend up to 2 h looking for crystal lattices on each grid. When looking for crystals, it is important to focus on objects with sharp symmetric edges, which show a dark fringe and a brighter color in its central parts. Some crystals are radiation sensitive and the lattice might "melt" during the focus adjustment. In these cases, it might be necessary to use the "Low Dose" mode of the microscope for imaging purposes.

Acknowledgments

This work was supported by NIH grant R01GM112686 (G.C.), P50GM082251 (G.C. and S.W.), and BioXFEL-STC1231306 (S. W). S.V. acknowledges support from grant R01GM097082. The contents of this publication are solely the responsibility of the authors and do not necessarily represent the official views of NIGMB or NIH.

References

1. Wiskerchen M, Muesing MA (1995) Identification and characterization of a temperature-sensitive mutant of human immunodeficiency virus type 1 by alanine scanning mutagenesis of the integrase gene. J Virol 69(1):597–601

2. Williams PF, Mynarcik DC, Yu GQ, Whittaker J (1995) Mapping of an NH2-terminal ligand binding site of the insulin receptor by alanine scanning mutagenesis. J Biol Chem 270 (7):3012–3016

3. Blaber M, Baase WA, Gassner N, Matthews BW (1995) Alanine scanning mutagenesis of the alpha-helix 115 123 of phage T4 lysozyme: effects on structure, stability and the binding of solvent. J Mol Biol 246(2):317–330

4. Manglik A, Kruse AC, Kobilka TS, Thian FS, Mathiesen JM, Sunahara RK, Pardo L, Weis WI, Kobilka BK, Granier S (2012) Crystal structure of the micro-opioid receptor bound to a morphinan antagonist. Nature 485 (7398):321–326

5. Rasmussen SG, Choi HJ, Rosenbaum DM, Kobilka TS, Thian FS, Edwards PC, Burghammer M, Ratnala VR, Sanishvili R, Fischetti RF, Schertler GF, Weis WI, Kobilka BK (2007) Crystal structure of the human beta2 adrenergic G-protein-coupled receptor. Nature 450(7168):383–387

6. Cherezov V, Rosenbaum DM, Hanson MA, Rasmussen SG, Thian FS, Kobilka TS, Choi HJ, Kuhn P, Weis WI, Kobilka BK, Stevens RC (2007) High-resolution crystal structure of an engineered human beta2-adrenergic G protein-coupled receptor. Science 318 (5854):1258–1265

7. Rasmussen SG, DeVree BT, Zou Y, Kruse AC, Chung KY, Kobilka TS, Thian FS, Chae PS, Pardon E, Calinski D, Mathiesen JM, Shah ST, Lyons JA, Caffrey M, Gellman SH, Steyaert J, Skiniotis G, Weis WI, Sunahara RK, Kobilka BK (2011) Crystal structure of the beta2 adrenergic receptor-Gs protein complex. Nature 477(7366):549–555

8. Kobilka B, Schertler GF (2008) New G-protein-coupled receptor crystal structures: insights and limitations. Trends Pharmacol Sci 29(2):79–83

9. Stevenson HP, DePonte DP, Makhov AM, Conway JF, Zeldin OB, Boutet S, Calero G, Cohen AE (2014) transmission electron microscopy as a tool for nanocrystal characterization pre- and post-injector. Philos Trans R Soc Lond Ser B Biol Sci 369(1647):20130322

10. Stevenson HP, Makhov AM, Calero M, Edwards AL, Zeldin OB, Mathews II, Lin G, Barnes CO, Santamaria H, Ross TM, Soltis SM, Khosla C, Nagarajan V, Conway JF, Cohen AE, Calero G (2014) Use of transmission electron microscopy to identify nanocrystals of challenging protein targets. Proc Natl Acad Sci U S A 111(23):8470–8475

11. Barnes CO, Kovaleva EG, Fu X, Stevenson HP, Brewster AS, DePonte DP, Baxter EL, Cohen AE, Calero G (2016) Assessment of microcrystal quality by transmission electron microscopy for efficient serial femtosecond crystallography. Arch Biochem Biophys 602:61–68

12. Stevenson HP, Lin G, Barnes CO, Sutkeviciute I, Krzysiak T, Weiss SC, Reynolds S, Wu Y, Nagarajan V, Makhov AM, Lawrence R, Lamm E, Clark L, Gardella TJ, Hogue BG, Ogata CM, Ahn J, Gronenborn AM, Conway JF, Vilardaga JP, Cohen AE, Calero G (2016) Transmission electron microscopy for the evaluation and optimization of crystal growth. Acta Crystallogr D Struct Biol 72(Pt 5):603–615

13. Heras B, Martin JL (2005) Post-crystallization treatments for improving diffraction quality of protein crystals. Acta Crystallogr D Biol Crystallogr 61(Pt 9):1173–1180

14. Newman J (2006) A review of techniques for maximizing diffraction from a protein crystal in stilla. Acta Crystallogr D Biol Crystallogr 62 (Pt 1):27–31

15. Russo Krauss I, Sica F, Mattia CA, Merlino A (2012) Increasing the X-ray diffraction power of protein crystals by dehydration: the case of bovine serum albumin and a survey of literature data. Int J Mol Sci 13(3):3782–3800

16. Lin G, Weiss S, Vergara S, Calero G (2019) Transcription with a laser: radiation-damage-free diffraction of RNA polymerase II crystals. Methods 159–160:23

Chapter 15

Low-Dose Data Collection and Radiation Damage in MicroED

Johan Hattne

Abstract

Microcrystal electron diffraction (MicroED) is a technique for structure determination that relies on the strong interaction of electrons with a minuscule, crystalline sample. While some of the electrons used to probe the crystal interact without altering the crystal, others deposit energy which changes the sample through a series of damage events. It follows that the sample cannot be observed without damaging it, and the frames obtained at the beginning of data collection reflect a crystal that differs from the one that yields the last frames of the dataset. Data acquisition at cryogenic temperatures has been found to reduce the rate of damage progression and is routinely used to increase the dose tolerance of the crystal, allowing more useful data to be obtained before the sample is destroyed. Low-dose data collection can further prolong the lifetime of the crystal, such that less damage is inflicted over the course of data acquisition. Ideally, lower doses increase the measurable volume of a single-crystal lattice by reducing the damage caused by probing electrons. However, the information that can be recovered from a diffraction image is directly related to the number of electrons used to probe the sample. The signal from a weakly exposed crystal runs the risk of being lost in the noise contributed by solvent, crystal disorder, and the electron detection process. This work focuses on obtaining the best possible data from a MicroED measurement, which requires considering several aspects such as sample, dose, and camera type.

Key words MicroED, Data collection, Dose, Radiation damage, Cryo-EM

1 Introduction

Microcrystal electron diffraction (MicroED) is a diffraction method in cryo-electron microscopy (cryo-EM) [1]. It exploits the strong interaction of electrons with matter to recover the signal diffracted from tiny, three-dimensional crystals. One of the key insights that made MicroED possible is low-dose data collection [2] and its implication on radiation damage. In earlier work on electron diffraction from two-dimensional crystals, it was often found that crystals were destroyed by radiation damage before an interpretable dataset could be obtained. Crystals would frequently only last for a single diffraction image, which even with today's data-processing algorithms results in a formidable data-analysis challenge when the crystal's unit cell is unknown with hundreds or even thousands of

Tamir Gonen and Brent L. Nannenga (eds.), *CryoEM: Methods and Protocols*, Methods in Molecular Biology, vol. 2215, https://doi.org/10.1007/978-1-0716-0966-8_15, © Springer Science+Business Media, LLC, part of Springer Nature 2021

diffraction images collected from equally many crystals, all in different orientations [3]. However, reducing the exposure lowers the signal, which depends on the number of scattered electrons, and it does not necessarily change the noise. For low-dose data acquisition to be successful, noise must be controlled and the camera that is used to record the diffraction pattern must be sensitive enough to extract information from weak signals. To maximize useful data obtained within the dose tolerance of the sample, it is important to carefully plan data collection to optimally distribute the probing electrons across the sample.

In cryo-EM, the sample is exposed to electrons that are accelerated in the column of an electron microscope. The electrons that contribute useful information in the diffraction spots on the detector interact with the sample elastically, without loss of total kinetic energy. A small fraction of the elastic scattering events is destructive and can directly dislocate atoms from their chemical bonds, but the cross-sections of these knock-on events are so low at voltages common for cryo-EM that they have historically been ignored [4]. For every elastic interaction, an estimated three electrons interact inelastically [5], where the deposited kinetic energy is ultimately transformed to heat. Most damage occurs due to electrons that lose between 5 and 100 eV during the interaction [6], by mechanisms that are assumed to be similar to those that cause damage under X-ray radiation. The deposited energy predominately excites or ionizes valence electrons, which breaks chemical bonds and produces electrons and free radicals. The electrons and their associated Auger electrons liberated during these primary damage events are mobile even at 77 K and can continue to break bonds, even at cryogenic temperatures [7]. Free radicals, on the other hand, generate cascades of secondary chemical reactions [8], and spread through the crystal by thermal diffusion; the damage caused by these events can be controlled by lowering the data-collection temperature. The absorbed energy, which is related to radiation damage, depends on the chemical composition of the crystal and its surrounding mother liquor as well as the energy of the incident electrons. For instance, the two disulfide bonds in proteinase K have been observed to exhibit breakage at different rates, even though all data collection parameters are identical [9, 10].

The individual molecules in the crystal lattice are affected as soon as the first electrons enter the sample, but since every atom in the crystal contributes to every reflection on the diffraction pattern, these effects may not be immediately apparent during data collection. In contrast, the gradual deterioration of the lattice is readily apparent in the diffraction patterns: as the accumulated exposure increases, fine lattice features are disrupted, and spots become fainter, beginning with the weak spots at high resolution. As exposure continues, such global radiation damage causes spots at lower resolution to dim until eventually all spots disappear into the noise of the background. During data processing, global damage is often modeled as increasing mosaicity, even though the crystal is unlikely to break into mosaic blocks under exposure to electrons.

The challenge of data collection is to recover as much information as possible from the sample within the lifetime of the crystal and the limitations of the instrument. Ideally, the final dataset measures all reciprocal space, such that each reflection is observed multiple times, and the resolution is only limited by the diffractive power of the crystal. One of the most critical decisions concerns the exposure, i.e., the number of electrons that impinge on the sample. If well-diffracting and isomorphous crystals are abundant, this decision can in part be deferred to the data-processing stage by collecting several datasets where the exposure is adjusted such that the observable resolution is limited by the diffractive power of each crystal. This is not a panacea, as real crystals exhibit some degree of non-isomorphism and merging data from many, slightly different crystals, introduces complications of its own. Assuming the immediate goal of data acquisition is to obtain the most complete dataset, a generally applicable strategy is to keep the exposure as low as possible, as this will maximize the lifetime of the crystal in the beam. The beam cannot be arbitrarily attenuated; if the data are too weak to be processed, they will not be of any use to the experiment. Finding the optimal compromise to the challenges that arise during low-dose data collection are central to current methods development in MicroED.

2   Sample Preparation

A well-ordered crystal diffracts electrons better than a poorly ordered one, and a large crystal with many unit cells yields stronger diffraction than a small crystal with fewer molecules. However, crystals that are both large and well-ordered are notoriously difficult to come by. The success of MicroED largely stems from its ability to obtain strong diffraction from small crystals, which addresses a major bottleneck in X-ray crystallography: that of obtaining large crystals suitable for high-resolution diffraction measurement. Furthermore, the resolution-limiting macrocrystal disorder may be due to mosaicity, and it has been suggested that MicroED can circumvent this problem by collecting data from individual mosaic blocks [11]. Smaller crystals may also be less susceptible to radiation damage, as they allow secondary electrons to escape before they have deposited all their energy to the sample [12].

The strong interaction between electrons and matter not only enables MicroED to measure diffraction data from small crystals, but also limits the thickness of the sample. This introduces complications for sample preparation. Once the path a probing electron takes through the sample becomes too long, multiple scattering events or absorption will either contaminate the diffracted intensities or prevent their observation altogether [13]. MicroED data are now routinely collected under continuous rotation of the sample, which mitigates dynamical artifacts due to multiple scattering [14, 15]. As crystal milling is becoming more commonplace, it is increasingly possible to carefully prepare the sample to its desired thickness [16, 17]. Even so, owing to their unique chemical composition and unit cell size, the exposure will have to be tuned to each sample depending on its susceptibility to radiation damage.

3   Data Collection

Data acquisition is the final experimental stage in crystal structure determination; all subsequent stages involve calculations that are ultimately limited by the quality of the data. Computational steps may be modified and repeated as necessary, whereas data collection is constrained by sample availability and access to the instrument. Since no data processing program can rescue an irredeemably bad dataset, it is necessary to plan data collection such that it yields the most suitable data for the purpose at hand.

To recover the information in a diffraction pattern, the spots must first be indexed, which entails finding their coordinates in the corresponding three-dimensional reciprocal lattice. Once the identity of each spot is known, their intensities are calculated, either by summing the pixel values in a region around their predicted central location, or by fitting a suitable spot profile, often derived from strong, nearby spots. This reduces a stack of hundreds, or even thousands, of diffraction images to a single indexed table of intensities and associated error estimates. Low-dose data collection inherently implies that data processing software must deal with weak reflections, barely visible over the background.

MicroED often requires data collected from multiple crystals. This is in part because crystal rotation is limited to a single, fixed axis, and its accessible rotation range may be further constrained by obstructing grid bars or other crystals. When processing multi-crystal data, each individual dataset must be strong enough to be indexed on its own; otherwise, it will not be possible to register the datasets with respect to each other. If the unit cell is known before data collection, it has recently become possible to index each frame individually [18]. For a sample with unknown unit cell, indexing usually requires several good diffraction patterns, all at different orientations [19, 20]. It follows that the exposure must generally be adjusted such that each crystal gives sufficiently many reflections distributed over a wide enough rotation range for a dataset to be indexed. For macromolecules, a dataset spanning 10–20° with ~50 clearly visible spots on each frame can readily be indexed with standard data reduction programs.

3.1 Cryo-EM

Ultimately, the amount of data that can be measured from a single crystal is limited by radiation damage. Primary damage from absorption of electrons is inevitable, and its rate has been found to be independent of temperature. Secondary damage, on the other hand, relies on diffusion of reactive species throughout the crystal, which is slowed down at cryogenic temperatures. Low-temperature data collection may also serve to mechanically restrain molecular fragments, preventing the disruption of global order [30]. Cryo-cooling has repeatedly been found to prolong the lifetime of the crystal in the beam and allows data acquisition to proceed under extensive exposure [21–24]. In X-ray crystallography, radiation damage has been observered to be minimized at temperatures closer to 50 K [25], >20 °C below the temperature in a typical, liquid nitrogen-cooled instrument. Because the underlying chemical and physical damage mechanisms are assumed to be identical in cryo-EM, it may be possible to realize a three-fold increase in dose tolerance in cryo-EM as well. Nevertheless, even a helium-cooled sample will eventually be destroyed by radiation damage.

3.2 Resolution Effects

Damage to the fine features of the structure in the crystal can be inferred from the decrease of the average intensity of the high-resolution reflections, even before the data has been interpreted in terms of an atomic model [26]. For instance, for proteinase K, it was found that the exponential fading of the average intensity of reflections in the 2.00–2.04 Å interval due to global damage was 2.5-fold faster than for reflections in the 5.15–21.0 Å interval [9]. In practice, tuning the exposure will trade resolution against completeness: increasing the exposure may raise the faint high-resolution reflections over the noise of the background but will also decrease the number of frames that can be recorded.

Specific damage, in the form of decarboxylation of acidic residues, breakage of disulfide bonds, and various conformational changes of amino acid side chains, occurs as soon as the first electrons impinge on the sample. The effects of site-specific damage are difficult to observe in both single-particle cryo-EM and MicroED, as both methods rely on extensive averaging to boost the signal over the noise. Only when a site is affected in a substantial fraction of the averaged molecules and the data are good enough to resolve the mean effect of the changes, can the damage be seen in the final density, but then, these changes are nefarious. Limited damage may appear as disproportionately high temperature factors indicating disorder or increased mobility of the affected sites. In severe cases specific damage may lead to misinterpretation of structural features and biologically important functional results.

3.3 Exposure and Dose

In electron microscopy, dose and exposure are often used interchangeably, whereas strictly, they refer to different quantities [4]. Exposure measures the number of electrons delivered to the sample, usually counted in electrons per unit area, while dose refers to the actual energy absorbed by the sample per unit mass. In electron microscopy, the exposure can be estimated using a Faraday cup, the microscope's phosphor screen, or an evenly exposed image recorded on a well-calibrated camera. Dose is much more difficult to quantify due to the complex interactions of chemical composition, crystal size, and beam shape. In X-ray crystallography, the dose can be calculated using, e.g., *RADDOSE* [27, 28], but there are no equivalent programs for electron diffraction yet. If the elemental composition and solvent content of the sample is known, the dose in MicroED can instead be approximated from the exposure at a given acceleration voltage.

Ignoring the effects of the crystalline structure of the sample and the vitrified solvent around it, the collision stopping power of a typical protein sample [12] for a 200 keV electron is 2.76 MeV cm^2 g^{-1} [29]. In a typical protein sample with density $\rho = 1.17$ g cm^{-3}, a 200 keV electron loses 323 eV μm^{-1} to the sample. The relative energy loss of the electron during its traversal of the sample is small, even for the thickest protein crystals, and the change in collision stopping power is negligible. Because both total deposited energy and mass increase linearly with the thickness of the sample, the dose does not dependent on crystal thickness, and a single 200 keV electron hitting a 1 Å^2 surface of a crystal will deliver a dose of 4.41×10^6 J kg^{-1} (4.41 MGy). For 300 keV electrons, an exposure of 1 e$^-$ Å^{-2} corresponds to 3.72 MGy; this reflects the smaller number of scattering events at higher acceleration voltages, such that a higher exposure is required to maintain the same signal when the energy of the incident electrons is increased [5].

Several estimates of the dose tolerances of biomolecules are available in the literature [30, 31]. These numbers often constitute limits which are rarely attained in practice. The actual tolerance will vary significantly from one sample to another [32, 33], depend on precisely how damage is quantified [34], and additionally change with the resolution interval under consideration. Finding the ideal exposure can be posed as an optimization problem with respect to information content: if the dose is too small, the signal may not be strong enough to be integrated accurately; if it is too large, the sample may succumb to radiation damage before the desired amount of information has been recovered. The optimum dose will lie somewhere between these extrema. For MicroED, the proper exposure will often have to be established empirically.

3.4 Multicrystal Data Collection

Even within a dataset recorded from a fully illuminated, single crystal, the intensities on frames recorded at different timepoints during data collection are not on a common scale. Effects due to anisotropic absorption, radiation damage, beam intensity variations, and temporal non-uniformity in the response of the detector all cause intensities to exhibit non-random variations over the course of the measurement [35]. Ideally, all these effects are accounted for during scaling, before intensities integrated from the first observations are merged with those measured during later stages. However, scaling a multicrystal dataset assumes that all individual data originate from identical crystals, which is never the case in practice. The effect of lattice discrepancies on the scaled and merged intensities depends on the degree of non-isomorphism.

Multicrystal merging presents an opportunity to obtain a minimally damaged dataset, as frames recorded from a highly exposed crystal can be discarded in favor of the first frames recorded from a different sample. If enough datasets are available, merging a high-multiplicity dataset from short exposures can thus reduce errors and lead to overall better results [36]. Even though MicroED often requires data from a 10°–20° wedge for accurate indexing and integration, not all frames need to be merged into the final dataset. A good strategy for multicrystal data collection is often to alternate between collecting data starting at high and low tilt. In cases of preferred crystal orientation, this will increase the probability that reciprocal space is evenly exposed, and that damage is not concentrated to specific wedges of reciprocal space. In the limit, only the first frame from each dataset would be merged, which requires data from very many crystals. Increasing automation now has the potential to make such a strategy feasible [37].

4   Camera Considerations

The quality of the recorded diffraction images is paramount in MicroED, as they constitute the only experimental measurement on which the final atomic model is based. Weak exposures generally yield noisy images that cannot be accurately integrated, whereas high exposures may fall outside the linear range of the camera. An ideal camera for diffraction measurement would account for every electron that interacts with the sample; it must have high dynamic range in order to accurately record both the intense, low-resolution reflections as well as the faint, high-resolution spots on the same image.

The ability to accurately detect very low electron counts is critical when the exposure is attenuated in the interest of reducing radiation damage. A particularly attractive camera option is provided by electron-counting, which owing to the absence of read-out noise promises to accurately record weak, high-resolution reflections. However, care must be taken not to cause pile-up effects in strong, low-resolution reflections, whereby electrons arriving on the same area in rapid succession will be undercounted because they exceed the count rate of the detector [38]. A possible solution is to collect data in two passes [39]: an initial low-resolution pass using low exposure is aimed at recording only those reflections that would cause overflows at higher exposure. A subsequent high-resolution pass will then record the high-resolution reflections using a higher exposure, without concern for simultaneously preserving the integrity of the low-resolution information.

An additional requirement on a MicroED camera is speed. The absolute position of the stage in an electron microscope can usually not be set accurately, which makes it impossible to arbitrarily orient the crystal before each frame is recorded. Furthermore, accelerating the stage to the desired rotation rate and precisely timing frame capture to the appropriate angular interval requires a level of synchronization currently not available to electron microscopes. These limitations are sidestepped by collecting data in "rolling shutter" mode [40], such that the detector is constantly active while the crystal is continuously rotated in the beam [14]. This is only possible on fast cameras with short dead-times during readout, lest reflections that pass through their diffractive conditions while the camera is blind will be lost and introduce unwanted gaps in the measurement of reciprocal space. Indeed, this shutterless mode of data acquisition has become prevalent in synchrotron X-ray crystallography as well [41].

5   Conclusion and Outlook

Choosing a suitable exposure and rotation range for a given crystal requires information not only about the lattice orientation of the sample currently in the beam, but integrated results from all previously collected datasets. Only then can the experimenter make informed decisions on how to distribute the probing electrons across the rotation range accessible to the instrument, such that data acquisition eventually leads to the best possible model. This is the purpose of data collection strategy [42]. As integrated data collection, processing, and automated strategy optimization tools are not yet commonplace in MicroED, these decisions must currently be made by the experimenter.

Here, techniques to limit the total exposure, without unduly compromising completeness or resolution have been outlined. The direct relationship between exposure, dose, and radiation damage suggest that exposure reduction is a sensible approach for structure determination by MicroED. This is further motivated by the observation that the most interesting parts of a structure are often the ones most susceptible to damage. A possible reason could be that these parts are often in strained geometries that destabilize the amino acid structure [4].

Deciding how to optimally distribute electrons across the sample during data collection requires a good understanding of the sample's susceptibility to damage and the camera's response to electrons. Ideally, this information can be obtained under minimal exposure of the sample to the beam. Even then, all strategies reflect trade-offs because information in MicroED can only be gained by exposing a crystal to electrons which invariably will deposit some of their energy to the sample and damage it.

In practice, a certain amount of radiation damage will have to be accepted, the merged data will exhibit some degree of incompleteness, and the highest-resolution observations will not necessarily contribute to a better atomic interpretation of the data [43]. The data may contain over, as well as underexposed reflections, as it is generally impossible to optimize the exposure such that all reciprocal space is measured at its most suitable exposure. Instead of the perfect dataset, the experimenter is forced to seek the optimal compromise.

Acknowledgments

Gerd Rosenbaum (Advanced Photon Source, Argonne, IL) is acknowledged for helpful discussions on dose in MicroED. Michael Martynowycz (UCLA, Los Angeles, CA) critically read the manuscript and provided insightful comments.

References

1. Shi D, Nannenga BL, Iadanza MG, Gonen T (2013) Three-dimensional electron crystallography of protein microcrystals. Elife [Internet] 2:e01345. http://www.ncbi.nlm.nih.gov/pubmed/24252878

2. Nannenga BL, Gonen T (2019) The cryo-EM method microcrystal electron diffraction (MicroED). Nat Methods [Internet] 16 (5):369–379. http://www.nature.com/articles/s41592-019-0395-x

3. Martynowycz MW, Gonen T (2018) From electron crystallography of 2D crystals to MicroED of 3D crystals. Curr Opin Colloid Interface Sci [Internet] 34:9–16. https://linkinghub.elsevier.com/retrieve/pii/S1359029417301589

4. Baker LA, Rubinstein JL (2010) Radiation damage in electron cryomicroscopy. In: Methods in enzymology [Internet]. Elsevier, Masson SAS, pp 371–388. https://linkinghub.elsevier.com/retrieve/pii/S0076687910810158

5. Henderson R (1995) The potential and limitations of neutrons, electrons and X-rays for atomic resolution microscopy of unstained biological molecules. Q Rev Biophys [Internet] 28(2):171–193. https://www.cambridge.org/core/product/identifier/S003358350000305X/type/journal_article

6. Langmore JP, Smith MF (1992) Quantitative energy-filtered electron microscopy of biological molecules in ice. Ultramicroscopy [Internet] 46(1–4):349–373. https://linkinghub.elsevier.com/retrieve/pii/030439919290024E

7. Jones GDD, Lea JS, Symons MCRR, Taiwo FA (1987) Structure and mobility of electron gain and loss centres in proteins. Nature [Internet] 330(6150):772–773. https://doi.org/10.1038/330772a0

8. Garman EF (2010) Radiation damage in macromolecular crystallography: what is it and why should we care? Acta Crystallogr Sect D Biol Crystallogr [Internet] 66(4):339–351. http://scripts.iucr.org/cgi-bin/paper?S0907444910008656

318   Johan Hattne

9. Hattne J, Shi D, Glynn C, Zee C-T, Gallagher-Jones M, Martynowycz MW et al (2018) Analysis of global and site-specific radiation damage in cryo-EM. Structure [Internet] 26 (5):759–766.e4. https://linkinghub.elsevier.com/retrieve/pii/S0969212618301254

10. Hattne J, Martynowycz MW, Penczek PA, Gonen T (2019) MicroED with the falcon III direct electron detector. IUCrJ [Internet] 6 (5):921–926. http://scripts.iucr.org/cgi-bin/paper?S2052252519010583

11. Nederlof I, Li Y-WW, Van Heel M, Abrahams JP (2013) Imaging protein three-dimensional nanocrystals with cryo-EM. Acta Crystallogr Sect D Biol Crystallogr 69(Pt 5):852–859

12. Nave C, Hill MA (2005) Will reduced radiation damage occur with very small crystals? J Synchrotron Radiat [Internet] 12(Pt 3):299–303. http://www.ncbi.nlm.nih.gov/pubmed/15840914

13. Subramanian G, Basu S, Liu H, Zuo J-M, Spence JCH (2015) Solving protein nanocrystals by cryo-EM diffraction: multiple scattering artifacts. Ultramicroscopy [Internet] 148:87–93. http://linkinghub.elsevier.com/retrieve/pii/S0304399114001715

14. Nannenga BL, Shi D, Leslie AGWW, Gonen T (2014) High-resolution structure determination by continuous-rotation data collection in MicroEDED. Nat Methods [Internet] 11 (9):927–930. http://www.ncbi.nlm.nih.gov/pubmed/25086503

15. Wang B, Zou X, Smeets S (2019) Automated serial rotation electron diffraction combined with cluster analysis: an efficient multi-crystal workflow for structure determination. IUCrJ 6 (5):1–14

16. Duyvesteyn HME, Kotecha A, Ginn HM, Hecksel CW, Beale EV, de Haas F et al (2018) Machining protein microcrystals for structure determination by electron diffraction. Proc Natl Acad Sci [Internet] 115(38):9569–9573. https://doi.org/10.1073/pnas.1809978115

17. Martynowycz MW, Zhao W, Hattne J, Jensen GJ, Gonen T (2019) Qualitative analyses of polishing and precoating FIB milled crystals for MicroED. Structure [Internet] 27 (10):1594–1600.e2. https://doi.org/10.1016/j.str.2019.07.004

18. Gevorkov Y, Barty A, Brehm W, White TA, Tolstikova A, Wiedorn MO et al (2020) PinkIndexer—a universal indexer for pink-beam X-ray and electron diffraction snapshots. Acta Crystallogr Sect A Found Adv [Internet] 76 (2):121–131. http://scripts.iucr.org/cgi-bin/paper?S2053273319015559

19. Hattne J, Reyes FE, Nannenga BL, Shi D, de la Cruz MJ, Leslie AGWW et al (2015) MicroED data collection and processing. Acta Crystallogr Sect A Found Adv [Internet] 71 (4):353–360. http://scripts.iucr.org/cgi-bin/paper?S2053273315010669

20. Clabbers MTBB, Gruene T, Parkhurst JM, Abrahams JP, Waterman DG (2018) Electron diffraction data processing with DIALS. Acta Crystallogr Sect D Struct Biol [Internet] 74 (6):506–518. http://scripts.iucr.org/cgi-bin/paper?S2059798318007726

21. Henderson R, Unwin PNT (1975) Three-dimensional model of purple membrane obtained by electron microscopy. Nature [Internet] 257(5521):28–32. http://www.nature.com/articles/257028a0

22. Hayward SB, Glaeser RM (1979) Radiation damage of purple membrane at low temperature. Ultramicroscopy [Internet] 4 (2):201–210. https://linkinghub.elsevier.com/retrieve/pii/S0304399179902110

23. Uyeda N, Kobayashi T, Ishizuka K, Fujiyoshi Y (1980) Crystal structure of Ag TCNQ. Nature [Internet] 285(5760):95–97. http://www.nature.com/articles/285095b0

24. Jeng T-W, Chiu W (1984) Quantitative assessment of radiation damage in a thin protein crystal. J Microsc [Internet] 136(1):35–44. https://doi.org/10.1111/j.1365-2818.1984.tb02544.x

25. Meents A, Gutmann S, Wagner A, Schulze-Briese C (2010) Origin and temperature dependence of radiation damage in biological samples at cryogenic temperatures. Proc Natl Acad Sci U S A 107(3):1094–1099

26. Glaeser RM (1971) Limitations to significant information in biological electron microscopy as a result of radiation damage. J Ultrastruct Res [Internet] 36(3–4):466–482. https://linkinghub.elsevier.com/retrieve/pii/S0022532071801181

27. Paithankar KS, Owen RL, Garman EF (2009) Absorbed dose calculations for macromolecular crystals: improvements to RADDOSE. J Synchrotron Radiat [Internet] 16 (2):152–162. http://scripts.iucr.org/cgi-bin/paper?S0909049508040430

28. Zeldin OB, Gerstel M, Garman EF (2013) RADDOSE-3D: time- and space-resolved modelling of dose in macromolecular crystallography. J Appl Crystallogr [Internet] 46 (4):1225–1230. http://gateway.webofknowledge.com/gateway/Gateway.cgi?GWVersion=2&SrcAuth=mekentosj&SrcApp=Papers&DestLinkType=FullRecord&DestApp=WOS&KeyUT=000322032300053

29. Berger MJ, Coursey JS, Zucker MA, Chang J (2005) Stopping-power & range tables for electrons, protons, and helium ions [Internet]. http://physics.nist.gov/Star

30. Henderson R (1990) Cryo-protection of protein crystals against radiation damage in electron and X-ray diffraction. Proc R Soc London Ser B Biol Sci [Internet] 241(1300):6–8. https://doi.org/10.1098/rspb.1990.0057

31. Owen RL, Rudiño-Piñera E, Garman EF (2006) Experimental determination of the radiation dose limit for cryocooled protein crystals. Proc Natl Acad Sci U S A 103 (13):4912–4917

32. Zeldin OB, Brockhauser S, Bremridge J, Holton JM, Garman EF (2013) Predicting the X-ray lifetime of protein crystals. Proc Natl Acad Sci [Internet] 110(51):20551–20556. https://doi.org/10.1073/pnas.1315879110

33. Liebschner D, Rosenbaum G, Dauter M, Dauter Z (2015) Radiation decay of thaumatin crystals at three X-ray energies. Acta Crystallogr Sect D Biol Crystallogr 71:772–778

34. Kmetko J, Husseini NS, Naides M, Kalinin Y, Thorne RE (2006) Quantifying X-ray radiation damage in protein cryotals at cryogenic temperatures. Acta Crystallogr D Biol Crystallogr [Internet] 62(Pt 9):1030–1038. http://www.ncbi.nlm.nih.gov/pubmed/16929104

35. Faruqi A, Henderson R (2007) Electronic detectors for electron microscopy. Curr Opin Struct Biol [Internet] 17(5):549–555. https://linkinghub.elsevier.com/retrieve/pii/S0959440X07001212

36. Liu Z-J, Chen L, Wu D, Ding W, Zhang H, Zhou W et al (2011) A multi-dataset data-collection strategy produces better diffraction data. Acta Crystallogr Sect A Found Crystallogr [Internet] 67(6):544–549. http://scripts.iucr.org/cgi-bin/paper?S0108767311037469

37. de la Cruz MJ, Martynowycz MW, Hattne J, Gonen T (2019) MicroED data collection with SerialEM. Ultramicroscopy [Internet] 201:77–80. https://linkinghub.elsevier.com/retrieve/pii/S0304399119300026

38. Trueb P, Dejoie C, Kobas M, Pattison P, Peake DJ, Radicci V et al (2015) Bunch mode specific rate corrections for PILATUS3 detectors. J Synchrotron Radiat [Internet] 22 (3):701–707. http://scripts.iucr.org/cgi-bin/paper?S1600577515003288

39. Dauter Z (2017) Collection of X-ray diffraction data from macromolecular crystals. In: New comprehensive biochemistry [Internet]. Springer, New York, pp 165–184. http://link.springer.com/10.1007/978-1-4939-7000-1_7

40. Stumpf M, Bobolas K, Daberkow I, Fanderl U, Heike T, Huber T et al (2010) Design and characterization of 16 megapixel fiber optic coupled CMOS detector for transmission electron microscopy. Microsc Microanal 16:856–857

41. Powell HR (2017) X-ray data processing. Biosci Rep [Internet] 37(5):BSR20170227. https://doi.org/10.1042/BSR20170227

42. Popov AN, Bourenkov GP (2003) Choice of data-collection parameters based on statistic modelling. Acta Crystallogr - Sect D Biol Crystallogr [Internet] 59(7):1145–1153. http://scripts.iucr.org/cgi-bin/paper?S0907444903008163

43. Karplus PA, Diederichs K (2012) Linking crystallographic model and data quality. Forensic Sci Int 336(6084):1030–1033. https://doi.org/10.1126/science.1218231

Chapter 16

Automation of Continuous-Rotation Data Collection for MicroED

M. Jason de la Cruz

Abstract

Automated coordination of microscope and camera functions for MicroED data collection simplifies the procedure for robust dataset acquisition and enables unattended sequential collection of many crystal targets. This chapter discusses the prerequisites for an algorithm of data collection automation for continuous-rotation MicroED and presents a practical protocol for achieving this goal using the popular TEM control software program SerialEM.

Key words MicroED, CryoEM, Electron microscopy, Microcrystal, Electron diffraction, 3D, Electron crystallography

1 Introduction

Data collection for continuous-rotation microcrystal electron diffraction (MicroED) involves (a) microscope and software setup; (b) nanocrystal screening on transmission electron microscope (TEM) grids (Fig. 1a, b); (c) testing crystals for diffraction (Fig. 1c, d); (d) preparing the crystals for data collection; and (e) coordination of the microscope's beam blanker/column valves, stage rotation, and camera recording for data acquisition. Automating these actions by computer requires these steps to occur in a strictly linear fashion. An experienced microscopist/crystallographer may prefer to identify, prepare for, and collect each crystal manually because crystals on the same grid could have different sizes and available tilt range. For a regular user, automation of data acquisition advantageously reduces the chances for operator error during data collection and frees up the user's own time, particularly when multiple datasets are collected.

Several publicly available software packages exist for the automated collection of images for transmission electron microscopy [1], but many have limited provisions for full microscope control in diffraction mode. As of July 2019, the following software packages

Tamir Gonen and Brent L. Nannenga (eds.), *CryoEM: Methods and Protocols*, Methods in Molecular Biology, vol. 2215, https://doi.org/10.1007/978-1-0716-0966-8_16, © Springer Science+Business Media, LLC, part of Springer Nature 2021

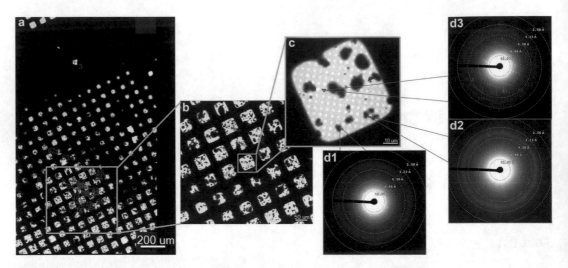

Fig. 1 The progression of proteinase K microcrystal screening through diffraction using automated SerialEM processes on a Thermo Fisher Talos Arctica with Thermo Fisher CetaD camera. (**a**) Low-magnification (155×) grid atlas showing an overview of the TEM grid. (**b**) Closer view of the same grid atlas in (**a**) showing a selected square. (**c**) Defocused-diffraction view of the grid square in (**b**) showing targeted crystals for diffraction. (**d1–3**) Diffraction patterns of the targeted crystals in (**c**)

have the ability to record continuous-rotation MicroED data: SerialEM [2] and ParallEM [3]. Advances are being made in proprietary software from the microscope and camera manufacturers, independent research groups, as well as in established software, e.g., Leginon [4]. In this chapter, continuous-rotation MicroED data collection automation as it relates to SerialEM is described, based on a recently published method [5].

2 Materials

2.1 Microscope

For automated microscopy, the TEM must utilize electronically controlled, servo-driven motors to move and rotate the stage, as well as the ability to vary lens magnetism and deflectors through software control. It must also have the ability to receive instructions from, and send feedback to, control software such as SerialEM. The procedure discussed here using SerialEM is currently compatible with TEMs produced by Thermo Fisher Scientific, with the computer-controlled sample goniometer known as the Compu-Stage; this includes models of the Tecnai, Titan, Spirit, Talos, and Glacios ranges. Late-model TEMs from the Japanese manufacturer JEOL have recently acquired support for the advanced SerialEM commands required for the particular script mentioned in this work.

2.2 Electron Detector

The electron detector, which serves as the camera for image recording purposes, should have the ability to detect and record images with high dynamic range for accurate representation of the crystallographic diffraction patterns. It should also have a high acquisition frame rate in order to capture as much data through the rotation data collection as possible to minimize angular gaps in the rotation.

For this method, cameras must be compatible with SerialEM, such as charge-coupled diode (CCD) and complementary metal-oxide semiconductor (CMOS) models from Gatan, TVIPS, Thermo Fisher, AMT, Direct Electron, and the JEOL Ruby camera. Cameras that have been tested to collect data for MicroED using SerialEM include the TVIPS TemCam-F416, TVIPS TemCam-XF416, Thermo Fisher Ceta (16M, 2, and D models), Thermo Fisher Falcon III, and the Direct Electron DE-20. The Falcon III and DE-20 are direct electron detectors; for these cases, a beamstop was used to mask the incident beam at all times while in diffraction mode to protect the sensor.

2.3 Image Acquisition Software

In general, image acquisition software for the collection of continuous-rotation MicroED data should be compatible with both TEM and camera in order to script concerted actions for automation. SerialEM natively supports diffraction mode and is compatible with a wide range of microscopes and cameras. It also includes a Low Dose Mode, which enables the user to customize the camera settings of a set microscope lens configuration for a given task. In this case, we will use different modes to set up (a) a low-magnification crystal search by whole-grid atlas, (b) a Record beam for data collection, and (c) a low-dose diffraction-defocus beam (offset from the Record beam) [6, 7] for manually centering the crystal as necessary.

3 Methods

3.1 Microscope and Software Setup

A well-aligned TEM electron-optical system for imaging in the SA magnification range is the ideal starting point for microscope alignment for the diffraction experiment [5, 8, 9]. Once switched to the diffraction mode, the user selects the appropriate camera length, adjusts dose (C2 intensity), and focuses the direct beam to a sharp, condensed point via diffraction focus. Astigmatism of the diffraction beam should be checked at this point. The result is the diffraction beam used for data collection and is ultimately saved in the SerialEM's Low Dose Mode as the Record beam.

1. Start SerialEM. Make sure the phosphor screen is down and the beam is visible.

2. Enable the Low Dose Mode by clicking its box under "Low Dose Control".

3. To set the diffraction beam for data collection, go to "Rec." mode by clicking its button: (a) Enable "Continuous update". (b) Set the Low Dose Record beam as mentioned above in Subheading 3.1. (c) Disable "Continuous update" to save the setting.

4. Setting Low Dose View mode for the defocused-diffraction beam used for crystal centering (*see* **Note 1**): (a) Make sure you are in Low Dose Record Mode. (b) Copy the settings to Low Dose View Mode by pressing the "V" button in the Low Dose Control. (c) Go to Low Dose View Mode by clicking "Vie." (d) Enable "Continuous update". (e) Change the defocus offset in the View mode by clicking on the arrows for it until the image of the crystal shows on the phosphor screen. (f) When ready to save, disable "Continuous update".

5. Centering the selected-area (SA) aperture. Ideally, the SA aperture is chosen based on crystal size and is used to select for diffracting electrons from the crystal itself and not the surrounding background. At this point, the defocused-diffraction beam is the easiest way to accurately center the SA aperture with the diffraction beam: (a) While the beam is in defocused-diffraction (set above in **step 4**, Subheading 3.1), use the stage to move the sample in view and center it on a unique feature. If there is no sample on the stage, use the defocused-diffraction beam edge as the feature. (b) Select the appropriate SA aperture size and insert the aperture into the column. (c) Center the SA aperture with the feature mentioned in **step 5a** (Subheading 3.1). (d) Retract the SA aperture if continuing with low-magnification TEM imaging.

6. Setting up imaging for the whole-grid atlas. Commonly called the "full montage" or "LMM" for "Low Magnification Montage", the whole-grid atlas is a collection of images taken of the grid at low magnification, while the stage rasterizes across the sample (*see also* **Note 1**): (a) Exit Low Dose Mode by unchecking its box. (b) Enable "Continuous update". (c) Set the microscope to the lowest magnification where the beam edges are not seen when taking an image using the camera. (d) Disable "Continuous update" to save the setting. (e) Adjust camera parameters as necessary for the Record setting. (f) Use SerialEM's "Setup full montage" function to collect the grid map.

3.2 Crystal Screening

1. Use SerialEM's "Setup full montage" function to collect the grid map.

2. Use the grid map to find crystals.

3. Add crystal targets with the Navigator using the "Add Points" function.

3.3 Testing Crystals for Diffraction in Batch

1. Go to a saved crystal coordinate in the Navigator using "Go to XY".

2. Center the crystal in the SA aperture.

3. Remove the current crystal coordinate.

4. Click "Add Stage Pos" to update the (x,y) position of the current crystal.

5. Repeat **steps 1** through **4** (Subheading 3.3), above for the remaining crystal targets.

6. Set crystal targets to "Acquire" in the Navigator.

7. Adjust camera parameters as necessary for the Record beam.

8. Open a new file to save images.

9. Use the "Acquire at Items" function to collect the Record image by choosing the Primary Task as "Just acquire and save image or montage". The microscope collects diffraction images at the saved crystal locations and saves them into the opened file for later review. Crystal targets with good diffraction are kept in the Navigator for automated data collection.

3.4 Preparing Crystals for Data Collection (See Note 2)

1. Ensure crystals are at eucentric height for each target. Similar to Subheading 3.3. above, this can be automated in SerialEM using the "Acquire at Items" function.

   (a) Exit the Low Dose Mode.

   (b) Turn off diffraction mode if activated, and select an imaging magnification that clearly shows the crystal target in view.

   (c) In the Navigator, set crystal targets to "Acquire".

   (d) Adjust camera parameters as necessary for the Record beam.

   (e) Open a new file to save images if desired.

   (f) In the "Acquire at Items" dialog, choose the Primary Task to run the Eucentricity function for each target. Click "GO" to start. The microscope goes to each crystal location, runs the automated Eucentricity routine, and updates the Z-height for each target.

2. Check and decide on the tilt range for data collection.

   (a) Go to a crystal target on the Navigator.

   (b) Go to the defocused-diffraction beam set in **step 4** (Subheading 3.1), set as the Low Dose View Mode.

   (c) While using the phosphor screen, or live imaging, change the stage alpha tilt as desired to check the maximum tilt range for the crystal and note the imaging limits of the crystal.

(d) Repeat above **steps 2a** through **2c** (Subheading 3.4) for each crystal and determine a tilt range for a set of crystals to acquire.

3. In the Navigator, set crystal targets to Acquire for a particular tilt range.

3.5 Data Acquisition (See Note 3)

1. Adjust camera parameters as necessary for the Record setting.

2. Use the CRmov script [5] to collect data.

3. The script must be edited with parameters appropriate to your data collection.

4. Acquired datasets, in the form of TIF files, must be converted to a crystallographic format. Software published by the Gonen laboratory [9] can read these TIF files generated by SerialEM and output the corresponding IMG files in SMV format.

4 Notes

1. In **step 4** (Subheading 3.1), during the setup of defocused-diffraction, note that recent versions of SerialEM have the Low Dose View Mode defocus offset setting disabled for diffraction mode. In this case, simply set the defocus offset using the focus knob while in diffraction mode, then use the "Continuous update" checkbox to save it to the Low Dose View mode. Additionally, the Low Dose Focus mode can be used for this purpose (this feature available in SerialEM 3.8 beta); for convenience, this would free the Low Dose View mode to be set up at the low magnification for the whole-grid atlas, as mentioned in **step 6** (Subheading 3.1).

2. In Subheading 3.1, the microscope must be in imaging mode in order for the stage to properly move to targets and for the Eucentricity routine to work.

3. Note that the CRmov script collects each batch of targets at the same tilt range. Crystals that need to be collected at different tilt ranges must be run separately.

4. Initially, only microscopes from Thermo Fisher Scientific had been tested to work with this method because the SerialEM BackgroundTilt command (used in the CRmov script) was originally written for that manufacturer's microscope scripting interface. However, as of SerialEM version 3.7.5, recent JEOL TEMs controlled by the TemExt server now have access to advanced microscope scripting via SerialEM, including the BackgroundTilt command.

5. Refer to the protocol mentioned previously [5] for more details on this procedure.

Acknowledgments

I would like to thank Tamir Gonen for access to a Thermo Fisher Talos Arctica TEM for testing and data collection, and Shian Liu and M. Rhyan Puno for their critical reading of this manuscript. This work is supported by the NIH/NCI Cancer Center Support Grant P30 CA008748 and the Sloan Kettering Institute.

References

1. Tan YZ, Cheng A, Potter CS et al (2016) Automated data collection in single particle electron microscopy. Microscopy (Oxf) 65:43–56

2. Mastronarde DN (2005) Automated electron microscope tomography using robust prediction of specimen movements. J Struct Biol 152:36–51

3. Yonekura K, Ishikawa T, Maki-Yonekura S (2019) A new cryo-EM system for electron 3D crystallography by eEFD. J Struct Biol 206:243–253

4. Suloway C, Pulokas J, Fellmann D et al (2005) Automated molecular microscopy: the new Leginon system. J Struct Biol 151:41–60

5. de la Cruz MJ, Martynowycz MW, Hattne J et al (2019) MicroED data collection with SerialEM. Ultramicroscopy 201:77–80

6. Cheng A (2013) Automation of data acquisition in electron crystallography. In: Schmidt-Krey I, Cheng Y (eds) Electron crystallography of soluble and membrane proteins, Methods in Molecular Biology, vol 355. Humana Press, Totowa NJ, pp 307–312

7. Nakamura N, Shimizu Y, Shinkawa T et al (2010) Automated specimen search in cryo-TEM observation with DIFF-defocus imaging. J Electron Microsc 59:299–310

8. Shi D, Nannenga BL, de la Cruz MJ et al (2016) The collection of MicroED data for macromolecular crystallography. Nat Protoc 11:895–904

9. Hattne J, Reyes FE, Nannenga BL et al (2015) MicroED data collection and processing. Acta Crystallogr Sect A Found Adv 71:353–360

Chapter 17

Ab Initio Determination of Peptide Structures by MicroED

Chih-Te Zee, Ambarneil Saha, Michael R. Sawaya, and Jose A. Rodriguez

Abstract

Structural elucidation of small macromolecules such as peptides has recently been facilitated by a growing number of technological advances to existing crystallographic methods. The emergence of electron micro-diffraction (MicroED) of protein nanocrystals under cryogenic conditions has enabled the interrogation of crystalline peptide assemblies only hundreds of nanometers thick. Collection of atomic or near-atomic resolution data by these methods has permitted the ab initio determination of structures of various amyloid-forming peptides, including segments derived from prions and ice-nucleating proteins. This chapter focuses on the process of ab initio structural determination from nano-scale peptide assemblies and other similar molecules.

Key words Peptide, Nanocrystal, Amyloid, Ab initio, Direct methods

1 Introduction

Electron micro-diffraction (MicroED) is a cryoEM technique well-suited to the determination of peptide and small macromolecule structures [1–4]. The first novel macromolecular structure solved by MicroED [5], a peptide segment of the Parkinson's-associated protein alpha synuclein, established the method's utility as a means of rapid structural analysis of crystals as small as a hundred nanometers thick. The subsequent determination of four ab initio structures of prion peptides demonstrated the approach's expediency in obtaining atomic-resolution structures of peptides [6].

Since these initial breakthroughs, the investigation of peptides by MicroED has grown and now includes over 20 structures; 14 structures at better than 1.2 Å resolution [3, 7–11]. Because of the success of ab initio methods for structural determination of peptide assemblies by MicroED [3, 6], we focus this chapter on procedures for the collection of high-resolution MicroED data from peptide nanocrystals and atomic-resolution determination of ab initio structures from said crystals. This approach hinges on the availability of well-ordered crystals hundreds of nanometers thick as

Tamir Gonen and Brent L. Nannenga (eds.), *CryoEM: Methods and Protocols*, Methods in Molecular Biology, vol. 2215, https://doi.org/10.1007/978-1-0716-0966-8_17, © Springer Science+Business Media, LLC, part of Springer Nature 2021

well as the concurrent use of cryogenic techniques to minimize radiation damage [1]. We demonstrate the implementation of this technique on GSNQNNF, a routinely studied heptapeptide segment of the prion protein [12], which forms needle-like crystals several microns in length but no more than a few hundred nanometers thick and wide.

As with the interrogation of larger macromolecular structures by MicroED, data collection from frozen hydrated crystals proceeds by measuring diffraction from a selected area of an illuminated crystal while that crystal is rotated unidirectionally in the electron beam [1, 13]. Diffraction movies recorded during this process are then converted into formats accessible to conventional crystallographic processing programs, which reduce MicroED data to intensities [13] used for ab initio phasing by direct methods. Solutions obtained by this approach contain collections of atoms which display correlations with the measured data. These in turn can generate electrostatic potential maps that inform the placement of residues along a polypeptide chain. Fourier difference density in these structures may reveal the positions of riding hydrogens, a feat difficult to accomplish via conventional X-ray crystallographic methods [5, 6, 14].

The procedures outlined in this chapter assume a certain degree of practical and theoretical knowledge related to operation of transmission electron microscopes, cryoEM, crystallography, and diffraction theory. For a primer on these subjects the reader is encouraged to read additional chapters in this book, or one of several recent reviews or books [1, 15–18].

2 Materials

2.1 Sample-Preparation Tools and Consumables

1. 200–400 mesh grids, covered with continuous or perforated carbon.

2. Ultra-fine grade reverse self-closing tweezers.

3. Vitrobot fine-tip tweezers.

4. Whatman No. 1 filter paper for specimen blotting and standard Vitrobot filter paper.

5. Glass petri dish with lid for grid storage and transport.

6. 2, 10, 20 μL micropipettors for sample manipulation and application.

7. Crystal slurries prepared in batch, or in hanging or sitting drops via vapor diffusion or other crystal growth and storage system.

8. Cryogen-resistant grid storage boxes with screw-top tool.

9. Personal protective equipment: several procedures require use of safety goggles, gloves, and cryogen protection including cryogen gloves and face shields.

2.2 Grid Surface Treatment

1. Several surface treatment systems are adequate [19]. We describe use of a PELCO easiGlow, glow discharge, and plasma generation system capable of 30 mA plasma currents, with vacuum control and containing a platform with a recess suited for 25 × 75 mm slide. This equipment also requires a rotary vacuum pump and gas and vent inlets.

2. Clean a standard glass micro-slide, 75 × 25 mm, 1 mm thick onto which grids can be placed for glow discharging. Other grid holders can be used instead.

2.3 Cryo Storage and Cryo-Transfer Systems

1. Several sample vitrification tools are available to perform rapid sample cooling following the Dubochet method [20, 21]. Here, we describe use of a Vitrobot Mark IV cryo-sample plunger capable of automated sample vitrification into liquid ethane.

2. While a number of sample transfer holders are available, we describe use of the Gatan 626 cryo-transfer holder with a high tilt range and low sample drift.

3. The Gatan cryo-transfer holder pairs with Gatan dry pumping station model 655 and Gatan cryo-transfer station.

2.4 Microscope Components

1. Microscope types include Tecnai series FEI microscopes such as the F20 TEM (200 keV) [13], F30 TEM (300 keV) [22], FEI Talos Arctica S/TEM (200 keV) [4] or Titan (300 keV), or JEOL JEM-3200, JEM-2200/2100 or cryo ARM series [23]. Here, we describe use of a Tecnai F30 TEM (300 keV) in microprobe mode.

2. Electron source: Field-emission gun operating with accelerating voltages in the 100–300 keV range (*see* **Note 1**).

3. Selected area aperture: Aperture diameters in the 50–150 μm range are recommended. Depending on lens system, aperture sizes can correspond to ~0.5–4.0 μm in diameter at the image plane.

4. Sample stage: Stage accuracy tilt range and tilt speed should be carefully calibrated to maintain eucentricity at the sample. Typical stage parameters include a tilt range of ±70° and tilt speed of 0.1–10°/s.

2.5 Cameras and Computing

1. Several detectors are suitable for data collection including some that have been developed specifically for MicroED. We focus on use of the TVIPS TemCam-XF416: a 16-megapixel CMOS sensor with integrated electronics delivering 16-Bit images at up to 48 frames/s with correlated double sampling [22] (*see* **Note 2**).

2. The camera is controlled by EMMenu4 with two pre-set camera recording settings. One may be a full-chip readout with subsampling by 2, the second a full-chip readout with subsampling by 4; both in rolling shutter mode.

3. Computing infrastructure is essential for both data collection and processing. Here, we describe software operating in a Linux environment on a server system powered by 40 processing threads (Intel® Xeon ® CPU E5-2680 v2 @ 2.80 GHz) with 256 GB of RAM, connected to a storage server with 400 TB of redundant (RAID6) NFS storage.

2.6 Data Reduction and Structure Determination Software

1. Data parsing and visualization software: Software for conversion of diffraction movies to crystallographic format diffraction images is limited. We describe conversion of Tietz Video Image Processing Systems (TVIPS) files to SMV format using a custom program: tvips2smv (https://cryoem.ucla.edu/pages/MicroED) [13]. Diffraction image visualization can be achieved using the Adxv program (www.scripps.edu/tainer/arvai/adxv.html) [24].

2. Data processing programs (see **Note 3**): Indexing, integration, merging, and scaling can be achieved through a variety of software programs. We describe use of the XDS package (http://xds.mpimf-heidelberg.mpg.de/) [25].

3. Ab initio structure determination is achieved using various direct methods programs. We describe use of the SHELX-97 suite, programs SHELXD and SHELXT (http://shelx.uni-ac.gwdg.de/) [26].

4. X-ray crystallographic refinement programs suitable for processing electron diffraction data must have the capacity to use electron scattering factors. Among those described here are the SHELX-97 suite [26]; Phenix suite (https://www.phenix-online.org/) [27]; CCP4 suite: REFMAC (http://www.ccp4.ac.uk/) [28].

5. Programs for dataset visualization, model building, and structural analysis include the ViewHKL program in the CCP4 suite for dataset visualization [29], the Coot model building program (https://www2.mrc-lmb.cam.ac.uk/personal/pemsley/coot/) [30] and PyMOL (https://pymol.org/2/) for model building and structure analysis [31].

3   Methods

This section provides guidance on the experimental workflow of electron diffraction of peptide nano- and microcrystals, from sample preparation to structure determination (Figs. 1 and 2).

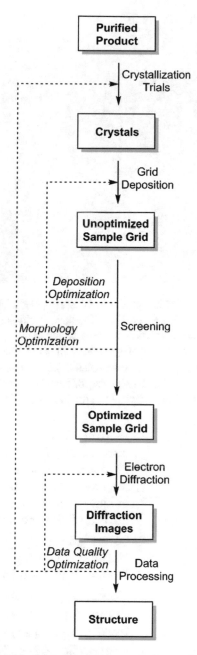

Fig. 1 General workflow of experiments to prepare samples, collect diffraction patterns, and process data for structure determination

3.1 Grid Preparation and Freezing (See Note 4)

1. Transfer EM grid from box to a clean glass slide using ultra-fine self-closing tweezers.

2. Glow discharge grid for 20–60 s per side at ~30 mA, following manufacturer's instructions.

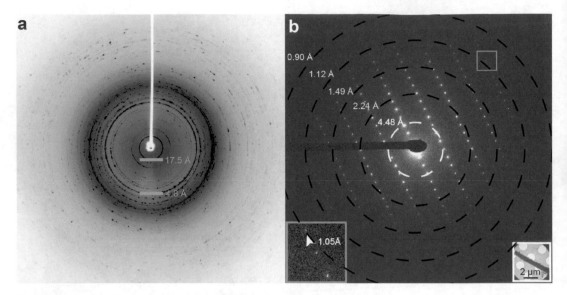

Fig. 2 X-ray powder diffraction (**a**) and single crystal MicroED pattern (**b**) of L-GSNQNNF (unit cell dimensions: $a = 4.86$ Å, $b = 14.12$ Å, $c = 17.71$ Å). The powder diffraction in **a** integrates a 5° scan of a loop filled with small crystals. Powder diffraction patterns which show gradual radial decay of signal in high-resolution regions suggest the presence of crystalline material, even if certain unit cell vectors lack an intense, corroborating ring. The pattern in **b** samples a 0.6° wedge of reciprocal space. Blue squares correspond to magnified regions of the pattern showing diffraction between 1.0 and 1.2 Å. White inset shows the over-focused image of the crystal irradiated to produce the diffraction pattern

3. Prepare Vitrobot for freezing following manufacturer's instructions, filling the sample storage reservoir with liquid nitrogen, placing a grid storage box in the liquid nitrogen reservoir, placing the copper ethane cup at its center, and linking them via a heat transfer apparatus. Condense ethane into cup following safety protocols (*see* **Note 5**).

4. Secure grid using Vitrobot tweezers and place into Vitrobot assembly following manufacturer's instructions, as per the Dubochet method [21].

5. Collect 5 μL or more of a slurry of crystal fragments of sub-micron thickness suspended in mother liquor.

6. Place 1–3 μL of crystal slurry per side, on either one or both sides of the grid (*see* **Note 6**).

7. Immediately follow Vitrobot procedures for blotting and plunge freezing of grid into liquid ethane [32] (*see* **Note 7**), then transfer the grid to a liquid nitrogen storage box.

8. Transfer grid boxes to permanent storage or to cryo holder transfer station for immediate use.

**3.2 Cryo-Transfer
(See Note 4)**

1. Cool cryo-transfer station with liquid nitrogen and transfer sample grid box containing sample grids into the transfer station sample platform.

2. Having followed procedures for preparation of the 626 cryo-transfer holder—in particular regeneration of the zeolite in the holder dewar—remove the holder from the 655 pumping station following manufacturer's instructions and place into the cryo-transfer station.

3. Fill sample holder dewar with liquid nitrogen and refill transfer holder station with liquid nitrogen to a level above the holder rod. Pull back protective copper blades to allow sample placement onto holder.

4. Remove grid from sample box and place onto sample holder with pre-cooled fine tip tweezers.

5. Place copper clip ring on top of grid and secure onto sample holder. Cover sample with protective blades.

6. Remove transfer holder from station and place into microscope CompuStage at the correct orientation, making sure to minimize the amount of time the holder spends in air.

**3.3 Grid Screening
and Preliminary
Diffraction**

1. Assuming operation of a fully calibrated instrument in microprobe mode, perform eucentric height alignment by wobbling the sample and adjusting the z-height.

2. Focus the system on the back focal plane of the objective lens by adjusting the focus knob at a magnification greater than $25,000 \times$ while viewing the edge of a hole or a distinctive feature on the carbon support.

3. Expand the beam using the intensity knob to a near parallel state, and set spot size to 11 (*see* **Note 8**).

4. Press diffraction button on console to operate the microscope in diffraction mode. Ensure the beam is free of astigmatism (*see* **Note 9**).

5. Position the beam at the center of the detector and calibrate the location of the selected area aperture (SAA) as needed.

6. For sample detection, expand the beam by adjusting the power to the condenser lens used to focus the beam. This can be achieved, for example, by turning the intensity knob to enter overfocused diffraction mode. One could pre-set this configuration using low-dose settings. Scan the grid for crystals by moving the stage x, y location using the trackball or joystick. If available, low-dose mode can automate the focus adjustments to minimize electron exposure on samples while the grid is examined. Where low dose is available, visual inspection of the grid can be performed in overfocused diffraction under

the "focus" or "search" mode, while diffraction patterns can be recorded with the beam at crossover in "exposure" mode on an FEI microscope.

7. Find an appropriate object on the grid to use as a reference and fine-tune the specimen's eucentric height by adjusting the z-height while wobbling.

8. Diffract from a target crystal by (a) inserting the SAA; (b) inserting and adjusting the beamstop; (c) returning the beam to crossover; (d) capturing an initial diffraction pattern with an exposure ranging from 0.1 to 10 s (*see* **Note 10**); (e) repeating **steps a** to **d** as necessary to (1) assess the resolution to which the sample appears to diffract and (2) diagnose whether or not this particular crystal diffracts well enough to warrant collection of a full dataset.

9. Verify if the detector distance is appropriate for sample and adjust if necessary, using the magnification knob (*see* **Note 11**).

10. Fine-tune the z-axis height in overfocused diffraction mode while tilting the sample between the upper and lower limits planned for data collection. The crystal should always remain within the SAA throughout the desired tilt range (*see* **Note 12**).

11. To collect diffraction movies as the stage rotates from one high angle tilt to the other, initiate data recording on the camera while independently setting a stage trajectory at a desired rotation rate (*see* **Note 13**).

12. Repeat **step 11** for subsequent data collection from different samples (*see* **Note 14**).

Subheadings 3.4–3.6 provide detailed instructions to execute data processing with select software programs (Fig. 3).

3.4 Data Conversion and Processing

1. Use a script such as tvips2smv to convert raw movie files from native ".tvips" format (Tietz Video and Image Processing Systems, generic output from TemCam detectors) or other proprietary format into a series of images in standard crystallographic format (e.g., SMV, with a ".img" suffix) readable by XDS. Before executing conversion, specify the oscillation range and detector distance used for the dataset in the command line [13].

2. Visualize the images using Adxv to cycle through the converted frames. Select an appropriate image to use for identification of the beam center (*see* **Note 15**) Repeat this process on several different images to obtain a more precise value.

3. Generate or download an appropriate XDS input template (http://xds.mpimf-heidelberg.mpg.de/html_doc/xds_prepare.html).

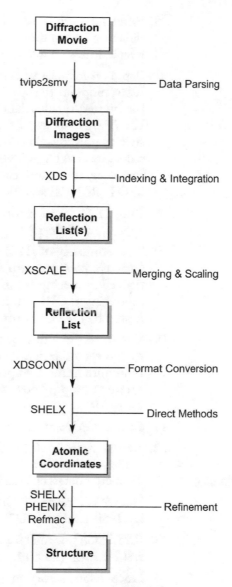

Fig. 3 Sample workflow for structure determination *via* direct methods. Subheading 3.3 covers diffraction; Subheading 3.4 covers indexing and integration; Subheading 3.5 covers merging, scaling, and format conversion; Subheading 3.6 covers direct methods and refinement

4. To specify the location of the beam center in a dataset, use the values calculated in **step 2** as the inputs for origin parameters in the XDS.INP file.

5. Enter the correct relativistic wavelength of the incident electron beam as the X-RAY_WAVELENGTH. This parameter is dependent on the specific accelerating voltage employed by the instrument (*see* **Note 16**).

6. Enter the correct ROTATION_AXIS values according to the direction of rotation and predetermined detector geometry, which differs by instrument.

7. Based on a visual inspection of the stack of diffraction images in Adxv, pinpoint the specific range of frames which incorporates the highest quality data to be used for indexing. Set SPOT_-RANGE as this set of frames. Avoid frames with visible defects such as polycrystalline reflections or overly elongated, ellipsoidal spots. DATA_RANGE specifies the range of images used for integration and can include all the frames specified in SPOT_RANGE and more.

8. Run XDS. If no errors occur, proceed to **step 9**. If errors appear, *see* **Note 17**.

9. The stringency of IDXREF may result in the frequent appearance of errors, prematurely halting further processing. If confidence in indexing is high, these halts can be circumvented by modifying the JOB settings in the XDS.INP input file as suggested by the error output and rerunning XDS (*see* **Note 18**).

10. Optimize the indexing solution by adjusting parameters such as the set of frames used for indexing and integration, the minimum number of pixels in a peak, the thresholds for background and peak pixel selection, and the mosaicity model (*see* **Note 19**).

11. Run XSCALE on the optimized indexing solution.

12. Repeat **steps 1–10** for each movie collected.

3.5 Dataset Merging

1. To merge datasets, create a new XSCALE.INP input file.

2. Specify the respective file paths of each desired XDS_ASCII. HKL file using INPUT_FILE (*see* **Note 20**).

3. Run XSCALE. Analyze the statistics of the merged solution in XSCALE.LP (*see* **Note 21**) (Fig. 4a).

4. Once a satisfactory merge is successfully implemented, open XDSCONV.INP and set the output type and file extension to SHELX and ".hkl," respectively.

5. Run XDSCONV to convert the scaled reflection file generated by XSCALE (with an ".ahkl" extension) to ".hkl" format. This renders it readable by SHELX.

6. Alternatively, open XDSCONV.INP and set the output type and file extension to CCP4_I+F and ".ascii," respectively. Run XDSCONV. XDSCONV may prompt the user for additional CCP4 commands to be executed, generating a binary list of reflections with the extension ".mtz".

7. Visualize the ".mtz" file (Fig. 4b) with viewHKL and search for systematic absences to aid in space group determination.

A

Resolution Limit	Number of Reflections			Completeness	R-Factor	I/Sigma	CC(1/2)
	Observed	Unique	Possible				
5.00	15	6	15	40.0%	11.7%	8.72	97.6
3.00	313	67	79	84.8%	11.5%	11.66	97.3
2.00	713	177	220	80.5%	12.6%	9.89	98.0
1.50	1411	341	421	81.0%	13.4%	9.09	96.9
1.40	590	141	173	81.5%	15.5%	7.72	96.7
1.30	714	175	210	83.3%	16.9%	7.15	95.6
1.20	903	226	287	78.7%	15.7%	6.80	97.2
1.10	1602	378	458	82.5%	18.9%	6.35	94.9
1.05	928	227	276	82.2%	21.3%	5.39	95.4
1.00	928	234	325	72.0%	25.1%	4.34	92.4
Total	**8117**	**1972**	**2464**	**80.0%**	**14.6%**	**7.20**	**98.3**

B

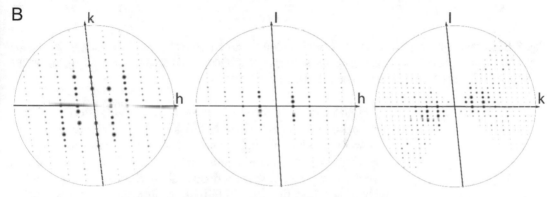

Fig. 4 Data reduction statistics for the merging and scaling of partial datasets from six peptide crystals (**a**). This example demonstrates reduction of data in the lowest symmetry space group, P1. The lack of symmetry is confirmed by a lack of systematic absences along *hkl* zones of the merged and scaled data (**b**; $l = 0$ left, $k = 0$ middle, $h = 0$ right) in ViewHKL

3.6 Structure Determination

1. Generate a SHELX instruction file (extension ".ins") by using SHELXPRO [26] (*see* **Note 22**).

2. Execute SHELXD/T with the newly made instruction file and the list of reflections generated in **step 5** in the previous section (".hkl"). During the SHELX trials, keep track of the best CFOM values; these form a distribution whose high end may be indicative of correct solutions (Fig. 5) (*see* **Note 23**).

3. Open the direct methods solution in Coot (".pdb" or ".res" files tolerated) or PyMOL (accepts ".pdb" files) to visualize the atoms. Generally, a solution is considered valid when each atom in the polypeptide backbone is placed in accordance with well-established restraints on bond geometries (Fig. 6a).

4. While structure determination programs do not routinely assign the absolute configuration of individual stereocenters from electron diffraction data, special routines can be implemented to exploit potential differences in Friedel mates [33]. In the event that absolute structure is known but SHELXD/T outputs its inverse, the PDBSET program [29] can rectify the handedness of the coordinate file by inverting the input coordinates.

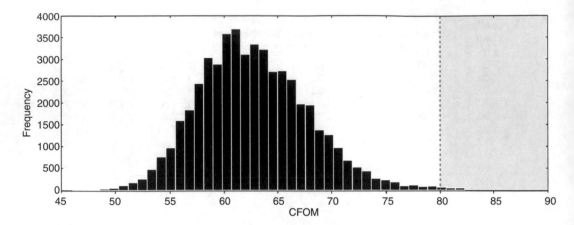

Fig. 5 Histogram of the combined figure of merit (CFOM) scores for 50,000 trials by SHELXD, indicating the approximate frequency of correct solutions. The shaded region, where CFOM scores exceed 80 [6], represents an area in which solutions have a high probability of being correct

5. SHELXL can use the original list of reflections and the atomic coordinates outputted by SHELXD/T to generate a density map for model building. Make a copy of the ".res" output from SHELXD/T and rename it as an ".ins" file. Now rename the ".hkl" file used for direct methods calculations so that it matches the new ".ins" file. Run SHELXL, calling the new ".ins" and ".hkl" files as inputs. This will produce a calculated density map as an ".fcf" file (also readable in Coot).

6. Build a polypeptide model in Coot or PyMOL by sequentially assigning specific residues according to the data (Fig. 6b).

7. Refine the structure using the newly built model in tandem with the original list of reflections supplied by XDSCONV in **step 5** of Subheading 3.5. Several standard programs can execute refinement; Phenix and REFMAC natively support electron scattering factors (*see* **Note 24**).

4   Notes

1. Recommended instruments employ field-emission guns (FEGs) as electron sources. These contain elemental tungsten (W) tips roughly 15 nm in diameter coated with a thin layer of zirconium oxide (ZrO). The apex of the emitter approaches 1800 K during excitation, augmenting ZrO conductivity and facilitating thermionic emission by diminishing the work function of the cathode. Schottky-type FEGs generate a highly coherent beam, with an energy spread ΔE of 0.3–0.7 eV. Alternative sources such as W hairpin filaments or lanthanum hexaboride crystals exhibit ΔE values of 1.5–3 eV and 1–2 eV, respectively. Systems operating at a higher accelerating voltage

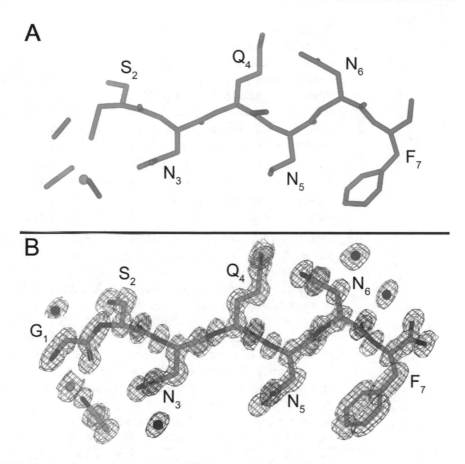

Fig. 6 Ab initio atomic coordinates of L-GSNQNNF outputted by SHELXD (**a**) and the corresponding atomic coordinates refined by Phenix (**b**). The $2F_o-F_c$ map, represented by the black mesh, is contoured at 1.2 σ. Modelled water molecules are present as red spheres, with hydrogens omitted for clarity. The final structure has an R_{work} of 0.172 and an R_{free} of 0.205

are preferred since these can also accommodate lower accelerating voltages (i.e., a 300 keV instrument can be driven at 200 keV if necessary). Although cold FEGs (in instruments such as the JEOL JEM-2100) can achieve even better spatial coherence, Schottky-type FEGs may offer greater durability, lower noise, and steadier beam current [34].

2. Other detectors include the CMOS-based CetaD 16 M (Thermo Fisher Scientific, [35]), a previous generation TVIPS TemCam-F416 (Tietz Video and Image Processing, [13]), and two direct electron detectors, Falcon in integrating mode [35] and K2 in counting mode (Gatan Inc., [36]). Hybrid pixel detectors such as the Dectris EIGER detector [37] and the Medipix2 [38] allow for fast and high dynamic range data collection. In principle, collection of MicroED data using a slow-scan CCD is possible, but inherent deficits in readout speed may severely limit data quality.

3. Because continuous rotation of the sample stage during MicroED is analogous to rotation of a mounted crystal on an X-ray diffractometer, software originally developed for X-ray crystallography can also process MicroED data. However, numerous parameters and settings (e.g., the wavelength of the incident radiation and the values of the relevant atomic scattering factors) must be adjusted accordingly to account for the discrepancies between X-rays and electrons.

4. To avoid redundancy, we minimize our description of procedures that may be more extensively covered by other chapters. Please refer to other chapters in this volume for a more thorough explanation of these topics and/or procedures.

5. Due to liquid nitrogen's relatively low heat capacity, samples plunge-frozen in liquid nitrogen often suffer from co-crystallization of adventitious ice. This is because interaction with the room-temperature sample causes an insulating layer of nitrogen to evaporate immediately upon contact as described by the Leidenfrost effect, prolonging the flash-freezing process and allowing water sufficient time to crystallize. Conversely, liquid ethane exhibits much higher heat capacity and therefore functions as a ruthlessly efficient cryogen that captures samples in vitreous ice. However, this property also renders it far more hazardous. Liquid ethane causes severe cryogenic burns immediately upon contact with exposed skin. When condensing ethane, exercise extreme caution and wear appropriate personal protective equipment, such as cryogenic gloves and eye protection.

6. Certain types of EM grids exhibit incompatibility with particular solvents. For example, formvar, a common plastic support film, is soluble in chloroform, dichloroethane, and 1,4-dioxane. Also, copper and some other metal meshes are highly susceptible to degradation by acids and bases.

7. Some cryo-protectants can increase the difficulty of an experiment by obstructing blotting efficiency. In general, an evaluation of solvent choice and sample compatibility with grids and freezing conditions prior to sample deposition is critical. For example, certain materials routinely used in X-ray diffraction can be incompatible with conventional sample preparation for MicroED. Common, non-volatile cryoprotectants (such as paratone oil, sucrose, trimethylamine N-oxide (TMAO), lipidic cubic phases, and perfluoropolyether oils) can cause particular challenges for both freezing and sample interrogation by an electron beam.

8. The size of the beam will determine the flux on the sample in electrons per Ångstrom squared. An image of the electron beam propagating through vacuum can help estimate the dose at low electron counts.

9. To calibrate the diffraction stigmator, adjust the C2 lens intensity. A fully condensed beam sits at a crossover point: adjusting the C2 intensity in either direction should result in an expansion. The beam should expand and collapse concentrically as a nearly circular ellipse. Shape distortions should be corrected by adjusting the console multifunction knobs while the stigmator adjustment icon is active.

10. While the XF416 detector is capable of full-chip readout at nearly 50 frames/s, not all applications warrant this readout speed. In particular, certain applications such as shutter-free data collection may benefit from slower readouts that limit the accumulation of readout noise. Likewise, spatial subsampling or binning of the collected frames can enhance signal over noise. The collection strategies described here rely on 2×2 binning of full-frame diffraction images, but larger binning can be employed if desired, with a caveat. Binning reduces sampling of the reciprocal lattice and therefore constrains measurements, as lattice spacings shrink and reflections risk being overlapped. This is also influenced by the selected camera length as described in the following note.

11. The ideal detector distance places the finest resolution reflections at or near the edge of the detector, maximizing the collection of reflections at both coarse- and fine-resolution shells. Crystals with larger unit cell dimensions will exhibit a more compact reciprocal lattice and require longer detector distances to prevent overlapping spots from affecting integrated intensities, which is influenced by detector sampling of the reciprocal lattice.

12. Visual obstruction of the sample by grid bars, other crystals, or debris is commonplace at high tilts, effectively limiting the upper and lower bounds of rotation. Keep in mind that as the x- and y-axes approach the stage's outer boundaries, high angle tilts become more likely to cause physical collisions between the microscope and the sample holder.

13. Coordination of stage rotation with diffraction data collection requires intervention. The software module controlling stage rotation can be accessed through the Temspy dialog box, under dialogs, CompuStage. The parameters for saving recorded movies can be set by accessing the Recorder Settings dialog box from the toolbar menu. Video capture occurs by pressing the camcorder icon on the EMMenu4 user interface after having set a desired integration time per frame. A second icon (red record button on the EMMenu4 user interface) allows for storage of the captured video frames when pressed as video is being captured.

14. To achieve optimal data quality, parameters such as camera length, continuous rotation rate, exposure time, and crystal orientation must be judiciously selected. A rotation rate of 0.1°–0.5°/s is recommended. Faster oscillations may result in overlapping spots, while slower oscillations may result in an insufficient number of spots due to sample degradation over time. Additionally, exposure times between 0.1 and 10 s are recommended. Longer exposures can increase the signal-to-noise ratio, but may place reflections from large unit cells in danger of overlapping, even at slow oscillations. Data from crystals that demonstrate orientation bias when deposited on a grid is limited in its sampling of reciprocal space (i.e., via collection of diffraction patterns at high tilt). Sufficient sampling is critical for obtaining high-completeness data.

15. Several procedures exist for identifying the location of the beam center in a diffraction image. We describe a procedure based on the relationship between the approximate locations of Friedel mates. Here, one would draw lines connecting pairs of Friedel mates to estimate the spatial coordinates of the beam center. The position of the beam center may change or drift between or within datasets. If this occurs, recalculate its value accordingly.

16. XDS requires the user to supply the wavelength of the incident radiation. While this is denoted X-ray for historical reasons, its name has no bearing on the calculations performed by the program. Calculation of this key parameter is straightforward for X-rays (characteristic values include 0.7107 Å for Mo Kα and 1.5418 Å for Cu Kα). In contrast, typical TEM accelerating voltages (i.e., >100 keV) suitable for MicroED produce a high-energy beam in which each constituent electron is forcibly propagated through a potential drop, ultimately generating a beam of quanta traveling at a velocity greater than half the speed of light. Therefore, an accurate calculation of the electron wavelength must incorporate relativistic contraction, as follows:

$$\lambda = \frac{h}{\sqrt{2m_0 eV \left(1 + \frac{eV}{2m_0 c^2}\right)}}$$

In the above equation, h is Planck's constant, m_0 is the rest mass of the electron, eV is the kinetic energy imparted by the accelerating voltage, and c is the speed of light in a vacuum. *See* Table 1 for characteristic values for the relativistic wavelength of the incident electron beam, which include 0.0251 Å at 200 keV and 0.0197 Å at 300 keV [34]. Discrepancies between the relativistic and non-relativistic calculations widen significantly as the accelerating voltage rises. Furthermore, recall that

Table 1
Relativistic and non-relativistic electron wavelengths at energies usable in MicroED experiments [41]

Accelerating voltage (keV)	Non-relativistic wavelength (Å)	Relativistic wavelength (Å)	Percent error (%)
100	0.0386	0.0370	4.15
120	0.0352	0.0335	4.83
200	0.0273	0.0251	8.06
300	0.0223	0.0197	11.66
400	0.0193	0.0164	15.03

the radius $1/\lambda$ of the corresponding Ewald sphere scales inversely with the wavelength of the incident quanta. As a result, given the markedly shorter wavelengths inherent to electron diffraction, the curvature of the Ewald sphere is flattened, and its gently sloping arc is reliably approximated as a plane at >100 keV. (This stands in stark contrast to X-ray diffraction, in which the geometry of the Ewald sphere stays true to its name.) Each observed reflection represents a cross-section between the surface of the Ewald sphere and a reciprocal lattice vector. Therefore, a wide, flattened Ewald sphere can in principle accommodate many reciprocal lattice points per angle. In practice, this advantage is attenuated by the restricted tilt range available to TEM instruments.

17. If IDXREF produces an error beginning with "INSUFFICIENT PERCENTAGE (<50%) OF INDEXED REFLECTIONS," proceed to **step 9**. If XDS fails to find an adequate indexing solution ("IER = 0"), inspect the values for the parameters set in **steps 3–5**. Specifically, consider supplying a different range of frames for SPOT_RANGE. If XDS cannot determine an adequate indexing solution after multiple iterations of adjustments, it may indicate poor data quality.

18. The IDXREF error described in **Note 17** may be caused by many factors, including a limited tilt range. Unlike a multi-axis goniometer, TEM instruments exhibit a restricted range of motion. Thus, MicroED datasets may not always include a large number of reflections within a given wedge of reciprocal space, leading to indexing ambiguities. Other potentially causative factors include non-eucentric sample rotation and lens aberrations [18].

19. Statistical analysis of each processing task implemented by XDS (such as IDXREF) is stored in its respective log file (with an ".LP" extension, such as IDXREF.LP). These log files contain a preponderance of potentially useful information which could enhance or facilitate an indexing solution. As an illustrative

example, CORRECT.LP supplies detailed statistics describing several alternative solutions generated by indexing smaller subsets of frames within the previously specified SPOT_RANGE. This provides insight into which specific set of frames is truly ideal for SPOT_RANGE.

20. Due to hardware restrictions imposed by the instrument, individual MicroED datasets rarely yield sufficiently high values of overall completeness to generate tractable direct methods solutions on their own. Merging several datasets can compensate for this. For n independent datasets, there exist 2^n-1 unique ways to exhaustively merge them together. Note that XSCALE can also truncate each individual dataset at a given resolution range using INCLUDE_RESOLUTION_RANGE. Only select individual datasets that generate highly convergent indexing solutions (i.e., datasets which exhibit strong agreement in parameters such as unit cell dimensions and Laue symmetry). Forcibly merging inherently discordant datasets will undeniably prove fruitless.

21. Obtaining an optimal dataset may require systematic screening of parameters (i.e., adjusting resolution range, re-indexing individual movies, and merging disparate combinations of movies) to improve statistics. Attempt to minimize R-factors and maximize completeness and I/SIGMA. For an example of the summary statistics produced by XSCALE, *see* Fig. 4.

22. Many methods exist for generating instruction files for SHELXD. For example, a custom program (referred to here as shelxdpro) can facilitate direct methods solutions of protein/peptide structures. To achieve this, shelxdpro prompts users for information needed to generate instruction files for running SHELXD or SHELXT. First, shelxdpro prompts the user to enter the protein sequence, the space group number, and the unit cell parameters. Shelxdpro uses this information to calculate Matthews coefficients and suggest feasible choices for the number of molecules in the asymmetric unit. The user selects a value, and then shelxdpro calculates the number of carbons, nitrogens, oxygens, and other atoms required to write the SFAC, UNIT, FIND, and PLOP statements. Lastly, shelxdpro asks for the name of the SHELX formatted diffraction data file (.hkl) for this project and an indication of whether the file contains amplitudes or intensities. Shelxdpro writes this information, including space group operators in a format that SHELXD/T accepts. The user can use this instruction file to run SHELXD/T immediately. The benefit of such a program is time savings in omitting a search inputs such as space group operators and number of atoms of each element in the unit cell. Such a program also relieves the user of remembering the cryptic lattice code specific to the space group and eliminates the possibility for formatting errors.

23. SHELXD requires the user to supply an exact space group for direct methods and is therefore unlikely to yield a reasonable solution if the user's prediction is incorrect. Conversely, SHELXT extracts the Laue symmetry of the indexing solution and determines the space group through multiple considerations [39]. Depending on the number of atoms, solutions with a combined figure of merit (CFOM) greater than 80 will have a high probability of being correct (Fig. 5).

24. Programs such as SHELXLE require manual input of electron scattering factors as SFAC commands [40]. Due to artefacts unique to electron diffraction, the R-factors of refined MicroED structures generally exceed those of their X-ray counterparts. An acceptable solution could exhibit R_{work} and R_{free} values near or less than 0.25, with a split below 0.05 (Fig. 6b).

Acknowledgments

We thank Drs. Duilio Cascio, Marcus Gallagher-Jones, and Ms. Lee Joon Kim (UCLA) for careful reading of this chapter and for their insightful comments and discussion. This work is supported by the STROBE National Science Foundation Science & Technology Center, Grant No. DMR-1548924, DOE Grant DE-FC02-02ER63421 and NIH-NIGMS Grant R35 GM128867. A.S. is supported as an NSF Graduate Research Fellow under Grant No. DGE-1650604. C.Z. is supported by a UCLA dissertation year fellowship. J.A.R. is supported as a Searle Scholar, a Pew Scholar, a Packard Fellow and a Beckman Young Investigator.

References

1. Shi D, Nannenga BL, de la Cruz MJ et al (2016) The collection of MicroED data for macromolecular crystallography. Nat Protoc 11:895–904

2. Shi D, Nannenga BL, Iadanza MG et al (2013) Three-dimensional electron crystallography of protein microcrystals. elife 2:e01345

3. Rodriguez JA, Eisenberg DS, Gonen T (2017) Taking the measure of MicroED. Curr Opin Struct Biol 46:79–86

4. Jones CG, Martynowycz MW, Hattne J et al (2018) The cryoEM method MicroED as a powerful tool for small molecule structure determination. ACS Cent Sci 4:1587–1592

5. Rodriguez JA, Ivanova MI, Sawaya MR et al (2015) Structure of the toxic core of α-synuclein from invisible crystals. Nature 525:486–490

6. Sawaya MR, Rodriguez J, Cascio D et al (2016) Ab initio structure determination from prion nanocrystals at atomic resolution by MicroED. PNAS 113(40):11232–11236

7. Zee C, Glynn C, Gallagher-Jones M et al (2019) Homochiral and racemic MicroED structures of a peptide repeat from the ice-nucleation protein InaZ. IUCrJ 6:1502238

8. Gallagher-Jones M, Glynn C, Boyer DR et al (2018) Sub-ångström cryo-EM structure of a prion protofibril reveals a polar clasp. Nat Struct Mol Biol 25:131–134

9. de la Cruz MJ, Hattne J, Shi D et al (2017) Atomic-resolution structures from fragmented protein crystals with the cryoEM method MicroED. Nat Methods 14:399–402

10. Guenther EL, Cao Q, Trinh H et al (2018) Atomic structures of TDP-43 LCD segments

and insights into reversible or pathogenic aggregation. Nat Struct Mol Biol 25:463–471

11. Hughes MP, Sawaya MR, Boyer DR et al (2018) Atomic structures of low-complexity protein segments reveal kinked β sheets that assemble networks. Science 359:698–701

12. Hattne J, Shi D, Glynn C et al (2018) Analysis of global and site-specific radiation damage in cryo-EM. Structure 26:759–766.e4

13. Hattne J, Reyes FE, Nannenga BL et al (2015) MicroED data collection and processing. Nat Struct Mol Biol 71:353–360

14. Palatinus L, Brázda P, Boullay P et al (2017) Hydrogen positions in single nanocrystals revealed by electron diffraction. Science 355:166–169

15. Zou X, Hovmöller S, Oleynikov P (2011) Electron crystallography: electron microscopy and electron diffraction. Oxford University Press, Oxford

16. Zuo JM, Spence JCH (2017) Advanced transmission electron microscopy: imaging and diffraction in nanoscience. Springer-Verlag, New York

17. Frank J (2006) Three-dimensional electron microscopy of macromolecular assemblies: visualization of biological molecules in their native state. Oxford University Press, Oxford

18. Heidler J, Pantelic R, Wennmacher JTC et al (2019) Design guidelines for an electron diffractometer for structural chemistry and structural biology. Acta Cryst D 75:458–466

19. Passmore LA, Russo CJ (2016) Specimen preparation for high-resolution cryo-EM. Methods Enzymol 579:51–86

20. Knapek E, Dubochet J (1980) Beam damage to organic material is considerably reduced in cryo-electron microscopy. J Mol Biol 141:147–161

21. Dubochet J, Adrian M, Chang J-J et al (1988) Cryo-electron microscopy of vitrified specimens. Q Rev Biophys 21:129–228

22. Jones C, Asay M, Kim LJ et al (2019) Characterization of reactive organometallic species via MicroED. ACS Cent Sci 5(9):1507–1513

23. Yonekura K, Ishikawa T, Maki-Yonekura S (2019) A new cryo-EM system for electron 3D crystallography by eEFD. J Struct Biol 206:243–253

24. Arvai A (2015) Adxv—A program to display X-ray diffraction images

25. Kabsch W (2010) XDS. Acta Crystallogr D Biol Crystallogr 66:125–132

26. Sheldrick GM (2008) A short history of SHELX. Acta Cryst 64:112–122

27. Adams PD, Afonine PV, Bunkóczi G et al (2010) *PHENIX*: a comprehensive Python-based system for macromolecular structure solution. Acta Cryst 66:213–221

28. Murshudov GN, Skubák P, Lebedev AA et al (2011) *REFMAC* 5 for the refinement of macromolecular crystal structures. Acta Cryst 67:355–367

29. Winn MD, Ballard CC, Cowtan KD et al (2011) Overview of the CCP4 suite and current developments. Acta Crystallogr D Biol Crystallogr 67:235–242

30. Emsley P, Lohkamp B, Scott WG et al (2010) Features and development of Coot. Acta Crystallogr D Biol Crystallogr 66:486–501

31. Delano W (2016) The PyMOL molecular graphics system. LLC, Schrödinger

32. Grassucci RA, Taylor DJ, Frank J (2007) Preparation of macromolecular complexes for cryo-electron microscopy. Nat Protoc 2:3239–3246

33. Brázda P, Palatinus L, Babor M (2019) Electron diffraction determines molecular absolute configuration in a pharmaceutical nanocrystal. Science 364:667–669

34. Reimer L, Kohl H (2008) Transmission Electron Microscopy: Physics of Image Formation. Springer-Verlag, New York

35. Hattne J, Martynowycz MW, Gonen T (2019) MicroED with the Falcon III direct electron detector. IUCrJ 6:921–926

36. Gallagher-Jones M, Ophus C, Bustillo KC et al (2019) Nanoscale mosaicity revealed in peptide microcrystals by scanning electron nanodiffraction. Commun Biol 2:26

37. Tinti G, Fröjdh E, van Genderen E et al (2018) Electron crystallography with the EIGER detector. IUCrJ 5:190–199

38. Nederlof I, van Genderen E, Li Y-W et al (2013) A Medipix quantum area detector allows rotation electron diffraction data collection from submicrometre three-dimensional protein crystals. Biol Crystallogr 69:1223–1230

39. Sheldrick GM (2015) SHELXT—integrated space-group and crystal-structure determination. Acta Crystallogr A Found Adv 71:3–8

40. Peng L-M (1999) Electron atomic scattering factors and scattering potentials of crystals. Micron 30:625–648

41. Williams DB (1996) Transmission electron microscopy—a textbook for materials science. Springer, New York. https://www.springer.com/us/book/9780387765006

INDEX

Tamir Gonen and Brent L. Nannenga (eds.), *CryoEM: Methods and Protocols*, Methods in Molecular Biology, vol. 2215,
https://doi.org/10.1007/978-1-0716-0966-8, © Springer Science+Business Media, LLC, part of Springer Nature 2021